Industrial Power Systems

Computational Intelligence in Engineering Problem Solving

Series Editor:
Nilanjan Dey

Computational Intelligence (CI) can be framed as a heterogeneous domain that harmonizes and coordinates several technologies, such as probabilistic reasoning, artificial life, multi-agent systems, neuro-computing, fuzzy systems, and evolutionary algorithms. Integrating several disciplines, such as Machine Learning (ML), Artificial Intelligence (AI), Decision Support Systems (DSS), and Database Management Systems (DBMS), increases the CI power and impact in several engineering applications. This book series provides a well-standing forum to discuss the characteristics of CI systems in engineering. It emphasizes the development of CI techniques and their role as well as the state-of-the-art solutions in different real-world engineering applications. The book series is proposed for researchers, academics, scientists, engineers, and professionals who are involved in the new techniques of CI. CI techniques including artificial fuzzy logic and neural networks are presented for biomedical image processing, power systems, and reactor applications.

Applied Machine Learning for Smart Data Analysis
Nilanjan Dey, Sanjeev Wagh, Parikshit N. Mahalle, Mohd. Shafi Pathan

IoT Security Paradigms and Applications
Research and Practices
Sudhir Kumar Sharma, Bharat Bhushan, Narayan C. Debnath

Applied Intelligent Decision Making in Machine Learning
Himansu Das, Jitendra Kumar Rout, Suresh Chandra Moharana, Nilanjan Dey

Machine Learning and IoT for Intelligent Systems and Smart Applications
Madhumathy P, M Vinoth Kumar and R. Umamaheswari

Industrial Power Systems: Evolutionary Aspects
Amitava Sil and Saikat Maity

For more information about this series, please visit: https://www.crcpress.com/Computational-Intelligence-in-Engineering-Problem-Solving/book-series/CIEPS

Industrial Power Systems
Evolutionary Aspects

Amitava Sil
Saikat Maity

CRC Press
Taylor & Francis Group
Boca Raton London New York

CRC Press is an imprint of the
Taylor & Francis Group, an **informa** business

First edition published 2022
by CRC Press
6000 Broken Sound Parkway NW, Suite 300, Boca Raton, FL 33487-2742

and by CRC Press
2 Park Square, Milton Park, Abingdon, Oxon, OX14 4RN

CRC Press is an imprint of Taylor & Francis Group, LLC

ISBN: 9781032138770 (hbk)
ISBN: 9781032138794 (pbk)
ISBN: 9781003231240 (ebk)

DOI: 10.1201/9781003231240

Typeset in Times
by codeMantra

Dedicated to all well-wishers and family members

Dr. Amitava Sil

and

Dr. Saikat Maity

Contents

Preface

The power sector has witnessed a number of transformational changes since 1882 when the first coal-based thermal power generation started its journey. During the same time, the world has witnessed another mode of power generation called hydroelectricity on the Fox River in Appleton, Wisconsin, USA. In fact, this was the first transition in power generation. In 1907, in Switzerland the first use of pumped storage power generation began. In 1949, mankind discovered new kind of fuel, i.e. gas, and using this as fuel, gas-based power generation commenced. In 1954, in Obninsk, Soviet Union, the first nuclear plant was built.

This was the era of large-scale power generation and transmission of electricity to drive the economy, which was made possible by a series of inventions and discoveries. Initially, the need for power was modest. With the passage of time, the power requirement escalated, almost doubling every 10 years with declining production cost and thereby increasing the generation facilities. Such widespread generation has ignited a realization into the mind that dependence of electricity generation largely from fossil fuels like coal to be shifted to have the sustainability, although a mix of power generation technologies was used like hydropower and then nuclear power. Attention turned to the environmental aspects also as the global warming phenomenon pinched to most of mankind. The technology for an alternate mode of power generation was available by then.

In 1888, the first known wind turbine was created for electricity production by Charles Brush to provide electricity for his mansion in Ohio, USA. Between 1888 and 1941, the wind power generation was mostly concentrated to stand-alone operation, and in 1941 the first 1.25-MW megawatt wind turbine was connected to a local electrical distribution grid. During this time, solar power generation technology was available as another option. Additionally, with these modes of power generation, termed renewable, environmental degradation can be mitigated. This has led to a dramatic transformation in the energy sector worldwide. Thus, the conventional mode of centralized electricity generation experienced a threat, although renewable electricity has a degree of dispersion, which is an intrinsic characteristic offset only by its unlimited and sustainable character. There are various modes of power generation besides wind and solar, such as biomass, tidal, bio-fuels, etc. With these modes of power generation, a terminology in power generation came, which is called distributed generation. Even after distributed generation, we have got dispersed generation.

The transmission and distribution of electricity was inevitable so that mankind can use it. In earlier years, the distribution systems were low voltage and short ranged located in the area surrounding where the electricity was generated using direct current. In fact, in the year 1889, the first long-distance transmission of electricity took place, linking a powerhouse at Willamette Falls to a string of lights in Portland, Oregon, at a distance 14 miles apart. The transition from direct current (DC) to alternating current (AC) was possible after a war of current; whether AC or DC, by then a transformer was invented (1882). In 1886, high-voltage long-range AC transmission systems of Ganz & Co. of Budapest began. Over the century, there was a growing need for electricity to travel greater distances around the world to meet the growing requirement. The transmission/distribution system of electricity was built, since 1882, as vertically integrated utilities, where incharge of electricity generation was responsible to create transmission facilities to transport their electricity to customers. The transmission system grew from local to regional to national to international. The system was called network or grid. Over the centuries, there have been consistent technological innovation and advancement for bulk power transmission through grids. What the world has experienced for decades is that most countries rely on large grids to fulfill electricity needs. However, with the rapid development of power generation through renewable/distributed generation, which is located near the demand centers, there are significant requirements for smaller grids all over the world in recent years, especially to fulfill the demands in remote areas. Smaller grids comprise mini-grids, micro-grids and nano-grids. In the era of internet, the new development in the power system is to develop smart grids having robust faster self-healing fail-safe protection. Besides this, at the beginning of the electricity distribution, there was a great battle between AC and DC, and we have got AC from DC. Nowadays, as the size of electricity supply systems increased, for very large transmission requirements, high-voltage DC is preferred.

Electric power is transmitted and distributed either by overhead transmission systems using overhead conductors or by underground cables. Over the years, there have been significant technological developments both for overhead conductors and underground cables. The major areas to be concentrated for these technological advancements are materials with their requisite electrical properties, voltage grade and insulation. Many of the natural materials were extensively modified to improve their insulation characteristics, besides the conducting materials for overhead conductors and underground cables. The opportunity and technological improvement are still available for all these materials.

So, in all the three pillars of the power system, i.e. generation, transmission and distribution of electricity that adds light and a means of communication to the world, with control engineering as the fourth pillar, the phenomenal developments that have taken place fascinated the authors to write this book following in the footsteps of books in electrical engineering and seminal works by the eminent technologists and researchers in this field. As an engineer associated with the academic and industrial arena for a long time, I feel it is necessary to disseminate the monumental works undertaken by the predecessors and the present-day stalwarts to the young engineers in a concise form for their education.

The authors are indebted to a number of individuals who, directly or indirectly, assisted in the work and wish to express thanks for their helpful criticism and suggestions. Thanks are also due to all the past students who over several years enriched the idea by expressing critical concepts in a simple way.

Dr. Amitava Sil and Dr. Saikat Maity
Kolkata

Authors

Dr Amitava Sil, Chartered Engineer, earned his bachelor of engineering degree in electrical engineering from Bengal Engineering College (presently IIEST, Shibpore), Shibpore, in 1984. He pursued integrated PhD in electrical engineering [power systems] from Jadavpur University, Kolkata, and earned the degree in 2012. His area of work includes power system dynamics, electrical machine and modern electrical control systems. He completed training courses approved by the Directorate General of Shipping, Ministry of Surface Transport, Govt. of India, and meets the requirements laid down under STC'95 Convention and Meta Manual Volume – II for Trainers and Assessors at Sensea Maritime Academy in 2011.

Dr. Sil is associated with the Department of Electrical and Electronics Engineering, at the National Institute of Technology Management and Science (formerly ITME) as Head of the Department and is associated with the Department of Electrical Engineering at the Academy of Technology as Associate Professor and later as Principal. He attended the workshop MISSION 10X of Wipro that empowers teachers from engineering colleges to teach various subjects in an inspiring way and emphasizes learning for a better career and fulfilling life experience as a team along with improvement of interpersonal communication and listening skills. He spent over one and a half decades in high-profile medium-voltage electrical equipment manufacturing industry, used predominantly by the power sector, held key portfolios and traveled extensively and adopted necessary Techno-Managerial skills. He is associated with Industrial and Management Consultancy House and mastered macro-studies of industrial sectoral activities including Snap Technology Status Assessment Study of wide range of industries. He has attended various workshops and seminars for career building and self-development with improvement in interpersonal skill and communication.

Dr Saikat Maity, Professor and HOD (Computer Science & Engineering, JIS University), CCNA, PhD (Computer Science & Engineering), is a Professor and Head of Computer Science and Engineering Department at JIS University from June 2019 with a total of 20 years of work experience. He earned a BTech in computer science and engineering from Kalyani University and has completed MTech in computer science and engineering from Calcutta University. He pursued PhD in computer science and engineering from the Indian Institute of Engineering Science and Technology (IIEST), Howrah, 2018. His main area of experience has been best practices in software engineering and artificial intelligence. He is working primarily in the domain of networking and multimedia soft computing and image processing technologies. His managerial forte is to act as a Coordinator of AICTE and UGC and NBA accreditation and Brand Ambassador of IIC and Incubation Centre establishment for self-financed universities along with Industry Academia program initiation. He is also designated as Tech Head for maintaining PARAM SHAVAK, a supercomputer from CDAC to work with Onama, CHReME and Ganglia.

He is also the board member of Academic Council, JIS University. Dr Maity is in charge of digital learning through Myperfectice and Zoom along with Cisco Webex and AudioVisual Head. He is also a CCNA professional and also Fellow IE; Senior Member of IEEE; Senior Member of IEEE Systems, Man, and Cybernetics Society; Life Member of ISTE, CSI, and ICS and Member of ACM.

1 Introduction

1.1 EVOLUTION OF ELECTRICAL POWER SYSTEM

Power systems have been operating for the last 100 years. All power systems have one or more sources of power. The history of power generation is long and convoluted, marked by myriad technological milestones, conceptual and technical, from hundreds of contributors. Coal has been used for power generation, and the first coal-fired steam generators provided low-pressure saturated or slightly superheated steam for steam engines driving direct current (DC) dynamos. Sir Charles Parsons, who built the first steam turbine generator (with a thermal efficiency of just 1.6%) in 1884, improved its efficiency 2 years later by introducing the first condensing turbine, which drove an Alternating Current (AC) generator. By the early 1900s, coal-fired power units featured outputs in the 1–10 MW range. By the 1910s, the coal-fired power plant cycle was improved which boosted net efficiency to about 15%. The demonstration of pulverized coal steam generators at the Oneida Street Station in Wisconsin in 1919 vastly improved coal combustion. Reheat steam turbines became the norm in the 1930s, when unit ratings soared to 300-MW output level. Main steam temperatures consistently increased through the 1940s, and the decade also ushered in the first attempts to clean flue gas with dust removal. The 1950s and 1960s were characterized by more technical achievements to improve efficiency with a supercritical main stream pressure. In 1878, the world's first hydroelectric power project was created in the Cragside country house in Northumberland, England. Most large hydroelectric power plants generate electricity by water stored in vast reservoirs behind dams. Water from the reservoirs flows through turbines to generate electricity. Hydropower generations create difficult trade-offs when considering the impact on wildlife, climate change and other issues. In 1951, electricity generation using nuclear energy was started experimentally in Idaho, USA. The first commercial electricity-generating plant powered by nuclear energy was operated in Shippingport, Pennsylvania, in 1957 with a capacity of 60 MW. Nuclear plants are different from coal-based power generation as they do not burn anything to get the heat to generate steam. Instead, they split radioactive atoms like uranium by a process called fission to generate the necessary heat. As a result, unlike other coal-based energy sources, nuclear power plants do not release carbon or pollutants like nitrogen and sulfur oxides into the air.

The growing attention to the environmental impact and the consequent rise in new policies that support the usage of renewable energy sources, especially non-hydro sources such as wind and solar power, decentralized in nature, are real game changers. Three points can immediately make the difference clear: (i) renewable energy is derived from natural processes that are replenished constantly; (ii) generation from renewable energy sources is not perfectly predictable, and furthermore only in a limited sense controllable; and (iii) generation from renewable energy sources pushes to a more decentralized approach. The energy from renewable sources is generally located where the primary source (e.g., wind or solar irradiation) is maximum, which hardly ever matches the location of maximum load, resulting in a challenge for the transmission system, which is not developed to operate in this way. The volatility adds on the challenge, as it is not a good match for a system that was developed to operate "as planned" rather than "as it comes", the latter requiring real-time monitoring and control. Due to co-existent traditional (nonrenewable) and renewable generating technologies, system operators have to coordinate the operation of the generation plants and ensure the stable and secure operation of the system. This has necessitated Wide-Area Measurement System enabled by communication technologies to control the operation of the generating stations and has created smart grids that enable bidirectional flows of energy and use two-way communication and control capabilities.

DOI: 10.1201/9781003231240-1

Smart grid is a large "System of Systems", where each functional domain consists of three layers: (i) the power and energy layer, (ii) the communication layer and (iii) the IT/computer layer. Layers (ii) and (iii) are the enabling infrastructure that makes the existing power and energy infrastructure "smarter". In smart grid domain energy-efficient transmission network will carry the power from the generation sites to the power distribution systems, which have communication interface between the transmission network and the generating stations, system operator, power market and the distribution system. The transmission network is monitored in real-time and protected against any potential disturbance. Further, substation automation and distribution automation are the key enablers for the smart distribution systems. There exists communication infrastructure to exchange information between the substations and a central distribution management system. Information exchange between the distribution system operator and the customers for better operation of the distribution system is a prime feature of the smart distribution systems. In smart grid domain building or home automation system monitors and controls the power consumption at the consumer premises in an intelligent way. In smart grids, customers play a supporting and pivotal role through demand response by peak-load shaving, valley-filling and emergency response for better operation of the distribution system.

Smart grid components include: (i) intelligent appliances capable of deciding when to consume power based on preset customer preferences, (ii) smart power meters featuring two-way communications between consumers and power providers to automate billing data collection, (iii) smart distribution that is self-healing, self-balancing and self-optimizing, (iv) smart generation capable of "learning" the unique behavior of power generation resources to optimize energy production and to automatically maintain voltage, frequency and power factor standards based on feedback from multiple points in the grid and (v) universal access to affordable, low-carbon electrical power generation (e.g., wind turbines, concentrating solar power systems, photovoltaic (PV) panels) and storage.

1.2   THERMAL POWER PLANT

The location of thermal power plants depends on the following: (i) availability of cooling water, (ii) availability of fossil fuel in the command area, (iii) transport facilities, (iv) availability of land and its character, (v) ash disposal and (vi) availability of manpower and security considerations. Thermal power plants are of two types: (i) subcritical, where plants operate at a pressure around 170 bar and temperature around 550°C, and (ii) supercritical, where plants operate at very high temperature around 650°C and pressure around 300 bar which increase the efficiency (around 45%) in comparison to subcritical operation, where efficiency is around 38%. The increase in efficiency directly leads to reductions in unit cost of power and CO_2 emission. In the supercritical operation there is a significant reduction in NOx, SOx and particulate emissions. "Supercritical" is a thermodynamic expression describing the state of a substance where there is no clear distinction between the liquid and the gaseous phase (i.e., they are a homogeneous fluid). There are Ultra Supercritical plants where temperature is 593°C and efficiency is 42% and Advanced Supercritical thermal power plants where temperature is 700°C and efficiency is 49%.

Advantages of coal-based thermal power plants are as follows: they can respond to rapidly changing loads without difficulty, a portion of the steam generated can be used as a process steam in different industries (cogeneration plant), steam engines and turbines can work under 25% of overload continuously, the fuel used is cheaper, reliable – both as supply of power during peak demand as base power or as off peak power, high load factor and low capital cost. Disadvantages include the following: maintenance and operating costs are high, long time is required for erection and putting into action, a large quantity of water is required, great difficulty is experienced in coal handling, the presence of troubles due to smoke and heat in the plant, unavailability of good quality coal, maximum of heat energy is lost and the problems of ash removal.

Typical components of thermal power plants are water preparation system, coal preparation system, boiler and auxiliaries, air pre-heater, re-heater, turbine, generator, condenser, cooling tower,

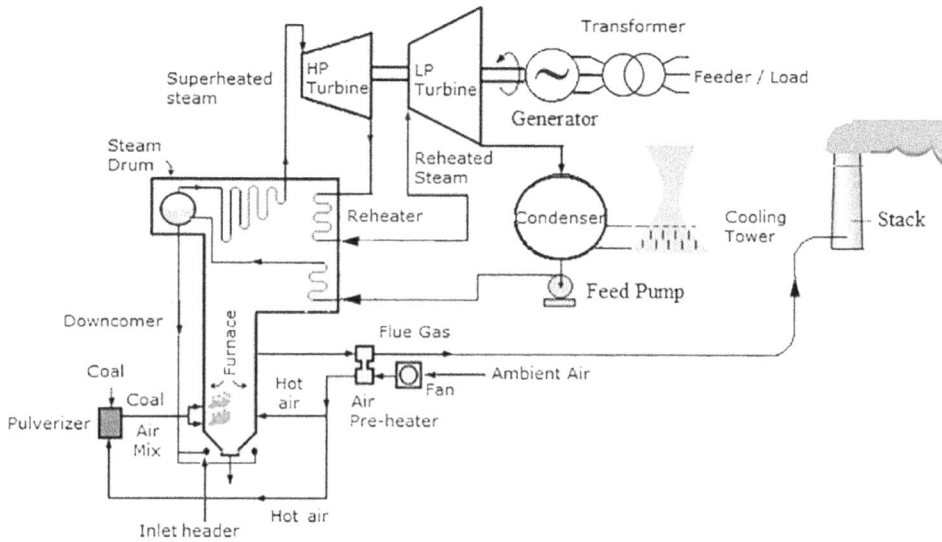

FIGURE 1.1 Thermal power plant.

fan or draught system and ash handling system. A line diagram of thermal power plant is given in Figure 1.1.

A thermal power plant has the following sections:

Water Preparation – The total feed water in a thermal power plant consists of re-circulated condensate water and purified makeup water to replace water lost through sampling systems, steam losses, evaporation from cooling and blow-down. Water softeners and ion exchange demineralizers using membrane technology to remove dissolved impurities and produce ultra-high-pure makeup water that it coincidentally becomes an electrical insulator.

Coal Preparation – Coal preparation is the removal of undesirable material from the Run-of-Mine coal by employing separation processes. Coal from the coal storage area is first crushed into small pieces and then conveyed to the coal feed hoppers at the boilers. Before entering the boiler, the coal is pulverized into a very fine powder, so that coal will undergo complete combustion during combustion process. Dryers are used in order to remove the excess moisture from coal that is mainly wetted during transport. The presence of moisture will result in fall of efficiency due to incomplete combustion and also result in CO emission. Magnetic separators are used to remove tramp iron pieces or separate iron particles from coal. Crushers are used for breaking coal into pieces of required feed size, which in pulverizing mill is 30 mm or below. The crushing is done in ring crusher and the hammer mill. Pulverizing of coal is done in either Ball and Tube Mill Pulverizer or Ring and Ball Pulverizer. Breaking a given mass of coal into smaller pieces in a pulverizer exposes more surface area for combustion, which allows faster combustion as more coal surface is exposed to heat and oxygen. This reduces the excess air required to ensure complete combustion and the required fan power also. A wide variety of low-grade coal can be burnt more easily when the coal is pulverized. Pulverized coal gives faster response to load changes as the rate of combustion can be controlled easily and immediately. Further, pulverized fuel systems are nowadays universally used for large capacity plants and use low-cost (low grade) fuel as it gives high thermal efficiency and better control as per the load demand.

Boiler and Auxiliaries – Boilers may be of two types. (i) Fire Tube type where hot gases pass through the tubes and boiler feed water in the shell side is converted into steam and is generally used for relatively small steam capacities up to 12,000 kg/hour and low to medium steam pressures up to 18 kg/cm^2. Fire tube boilers are available for operation with oil, gas or solid fuels. (ii) Water Tube Boiler where boiler feed water flows through the tubes and enters the boiler drum. The circulated water is heated by the combustion gases (capacity range 4,500–120,000 kg/hour of steam) and converted into steam at the vapor space in the drum. These boilers are selected when the steam demand and steam pressure requirements are high as 125 kg/cm^2. Further boilers may be classified as: (i) Fluidized Bed Combustion type that has the advantages of high efficiency. Reduction in boiler size can be achieved due to high heat transfer rate over a small heat transfer area immersed in the bed with reduction in pollution control as SO_2 formation is greatly minimized by the addition of limestone or dolomite. Fluidized bed boilers have a wide capacity range of – 0.5 to over 100 T/hour. (ii) Atmospheric Fluidized Bed Combustion type, where atmospheric air, which acts as both the fluidization air and combustion air, is delivered at a pressure and flows through the bed after being preheated by the exhaust flue gases. (iii) Combined Fluidized Bed Combustion type, where coal is crushed to a size of 6–12 mm depending on the rank of coal and type of fuel feed fed into the combustion chamber. A Combined Fluidized Bed Combustion could be a good choice if the capacity of boiler is large to medium.

In the steam-generating process, the furnace or burner systems provide controlled, efficient conversion of the chemical energy to heat energy, which in turn is transferred to the heat-absorbing surfaces of the steam generator. To do this, the firing system introduces fuel and air for combustion, mixes these reactants, ignites the combustible mixture and distributes the flame envelope and the products of combustion. Furnaces are of two types: (i) Grate-fired furnace is suitable for burning solid fuels like coal. Grate is provided for supporting the solid fuel and is so designed that it can also allow air to admit in the solid fuel for combustion. (ii) Fuel bed furnace is suitable for burning pulverized coal. Fluidization is a method of mixing fuel and air in a specific proportion, for obtaining combustion. A fluidized bed may be defined as the bed of solid particles behaving as a fluid. It operates on the principle that when an evenly distributed air is passed upward through a finely divided bed of solid particles at low velocity, the particles remain undisturbed, but if the velocity of air flow is steadily increased, a stage is reached when the individual particles are suspended in the air stream. If the air velocity is further increased, the bed becomes highly turbulent and rapid mixing of particles occurs which appears like formation of bubbles in a boiling liquid and thus the process of combustion as a result is known as fluidized bed combustion.

Stokers is a mechanical device which is used for supplying solid coal to furnace to maintain uniform operating condition, higher burning rate and greater efficiency. They may be of overfeed or underfeed type. It is determined by the feeding rate of coal below or above the level at which primary air is admitted.

The Rankine cycle is the fundamental operating cycle of all thermal power plants for steam generation. The operation of the cycle includes: (i) water from the condenser at low pressure is pumped into the boiler at high pressure, and this process is reversible adiabatic; (ii) water is converted into steam at constant pressure to the final (saturation) temperature by the addition of heat in the boiler; (iii) reversible adiabatic expansion of steam in the steam turbine; and (iv) constant pressure heat rejection in the condenser to convert condensate into water. By lowering the condenser pressure, superheating the steam to high temperatures and increasing the boiler pressure, the efficiency of the Rankine cycle can be increased.

Economizer is located in the boiler and above pre-heater to improve the efficiency of boiler by extracting heat from flue gases to heat the boiler feed water. In air pre-heater, the heat carried out with the flue gases that come out of economizer is further utilized for preheating the air before supplying to the combustion chamber. It is necessary equipment for supply of hot air for drying the coal in pulverized fuel systems to facilitate grinding and satisfactory combustion of fuel in the furnace.

From the boiler, steam goes to steam turbine. Turbines may be classified as non-condensing or backpressure type and condensing type. Backpressure turbines operate with an exhaust equal to or in excess of atmospheric pressure. The exhaust steam is used for lower pressure steam process applications. Condensing-type turbines operate with an exhaust pressure less than atmospheric pressure. It is costlier than the non-condensing type. The steam turbines are mainly divided into two groups: impulse turbine and impulse-reaction turbine. The turbine generator consists of a series of steam turbines interconnected to each other and a generator on a common shaft. There is a high-pressure turbine at one end, followed by an intermediate pressure turbine, two low-pressure turbines and the generator. The steam at high temperature (536°C–540°C) and pressure (140–170 kg/cm^2) is expanded in the turbine.

The condenser condenses the steam from the exhaust of the turbine into liquid to allow it to be pumped. If the condenser can be made cooler, the pressure of the exhaust steam is reduced and efficiency of the cycle increases. The functions of a condenser are: (i) to provide lowest economic heat rejection temperature for steam, (ii) to convert exhaust steam to water for reserve, thus saving on feed water requirement and (iii) to introduce makeup water. Boiler feed pump is a multistage pump provided for pumping feed water to economizer. Boiler feed pump is the biggest auxiliary equipment after boiler and turbine. It consumes about 4%–5% of total electricity generated. The cooling tower is a semi-enclosed device for evaporative cooling of water by contact with air. The hot water coming out from the condenser is fed to the tower on the top and allowed to tickle in the form of thin sheets or drops. The air flows from the bottom of the tower or perpendicular to the direction of water flow and then exhausts to the atmosphere after effective cooling.

Electrostatic Precipitator, or ESP, is a particulate collection filtration device that removes particles or fly ash from the exhaust called flue gas using the force of an induced electrostatic charge. It is a filtration device. ESP collects fly ash and other solid suspended particles from exhaust gases of a coal-fired boiler furnace. The dust-laden flue gas is passed between the oppositely charged conductors and becomes ionized as the voltage applied between the conductors is sufficiently large (30–60 kV depending upon the electrode spacing). As the dust-laden gas is passed through the highly charged electrodes, both negative and positive ions are formed. The ionized gas is further passed through the collecting unit which consists of a set of metal plates. Alternate plates are charged and earthed. As the alternate plates are grounded, high intensity electrostatic field exerts a force on the positive charged dust particles and drives them toward the ground plate. The deposited dust particles are removed from the plates by rapping hammer (dry ESP), scraping brush (dry ESP) or flushing water (wet ESP).

1.3 HYDROPOWER PLANT

The location of hydropower plants is usually predetermined by: (i) the availability of water, (ii) the water must be available at a usable head. The head is the vertical height from the top of the penstock to the bottom of the penstock. (iii) The water must be available in sufficient quantity and (iv) if the flow is not regular enough for continuous supply, there must be accommodation for a reservoir at a reasonable cost; the lower the head, the larger the reservoir must be. The three types of hydropower facilities are as follows: (i) impoundment – which is typically a large hydropower system where water is released from the dam that stores the water and flows through a turbine, spinning it, which in turn activates a generator to produce electricity. (ii) Diversion – which is called run-of-river type that channels a portion of a river through a canal or penstock dispensing the requirement of a dam and the natural flow rate is used to generate the power. Run-of-river type can be classified as micro – generation limited to less than 100 KW, mini – generation limited to less than 1 MW, and small – generation limited to less than 50 MW; (iii) pumped storage – where water is pumped to a higher elevation reservoir when there is a surplus of electricity and the stored water is then released into lower elevation reservoirs to generate electricity when needed.

The flow is the volume of water which flows in 1 second. Energy $= mgH$, where $m =$ mass of water in kg $= \rho Q$, $g = 9.81$ m/s² (the acceleration due to gravity), ρ is the water density, 1,000 kg/m³, Q is the flow or discharge, m³, and H is the height water fall. Hence, Energy $= \rho QgH$. The hydraulic power in watts $P = g\rho(dQ/dt)H$, where $dQ/dt =$ rate of discharge $=$ m³/s. The electrical energy produced in kWh can then be $W = 9.81 \times 1,000 \times Q \times H \times \eta \times t$, where t is the operating time in hours and η is the efficiency of the turbine-generator assembly, which generally varies between 0.5 and 0.9.

Advantages of hydropower are as follows: (i) water source is perennially available, (ii) running cost is low as compared to thermal and nuclear power plant, (iii) pollution free and no waste disposal problem, (iv) hydro turbines can be switched on and off in a very short time, (v) modern hydropower equipment has a greater life expectancy (can be more than 50 years), (vi) hydropower can be used as the ideal spinning reserve in a system mix of thermal, hydro and nuclear power stations and (vii) manpower requirement is low. A typical diagram of hydropower plant is given in Figure 1.2.

Typical components of hydropower plants are: (i) Reservoir/Dam – to store water during excess flow period and supply during lean period. It creates an artificial head. (ii) Surge Tank – a small storage tank or reservoir required in the hydropower plants for regulating the water flow during load reduction and sudden increase in the load on the hydro generator (water flow transients in penstock) and thus reducing the pressure on the penstock. (iii) Penstocks –pipes that carry water from the reservoir to the turbines inside power station. (iv) Turbine/Generator, (v) Spillway – a way for spilling of water from dams that provide the controlled release of flows from a dam into a downstream area and (vi) Trail race – the channel into which the turbine discharges the water.

Hydraulic turbines are of two types: (i) Impulse turbine – generally suitable for high head and low flow applications, normally used for more than 250 m of water head. This turbine generally uses the velocity (kinetic energy) of the water to move the runner and discharges to atmospheric pressure. The water stream hits each bucket on the runner through well-positioned nozzle. Impulse turbines are of three types: (a) Pelton wheel developed by American engineer Laster A. Pelton that consists of a wheel with a series of split buckets placed around its rim uniformly which is driven by jets of water being discharged at atmospheric pressure from pressure nozzles. They are horizontal or vertical shaft. Pelton wheel is only considered for heads above 150 m, but for micro-hydro applications Pelton turbines can be used effectively at heads down to about 20 m. (b) Cross-flow turbine is drum-shaped and uses an elongated, rectangular section nozzle directed against curved vanes on a cylindrically shaped runner. They are only horizontal shaft. The cross-flow was developed to accommodate larger water flows and lower heads than the Pelton. They work on net heads from just 1.75–200 m. (c) Turgo turbine is designed to have a higher specific speed. (ii) Reaction turbine develops power from the combined action of pressure and moving water. The runner is placed directly in the water stream flowing over the blades rather than striking each individually. Reaction turbines are generally used for sites with lower head and higher flows than compared with the impulse turbines. Reaction turbines are of four types: (a) Francis turbine was developed by an American Engineer, James B. Francis, which is used for head varying between 2.5 and 450 m.

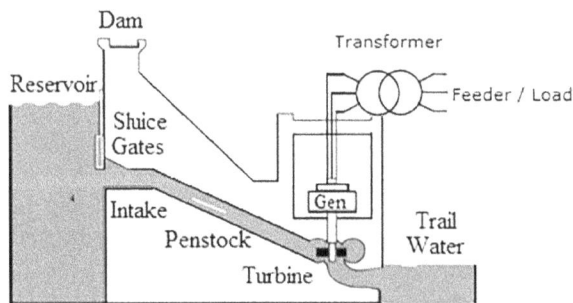

FIGURE 1.2 Typical diagram of hydropower plant.

It has guide vanes also known as wicket gates or stay vanes which convert a part of pressure energy of the water at its entrance to the kinetic energy and then to direct water on to the runner blades at the angle appropriate to the design. They are vertical shafts. (b) The Kaplan turbine designed by Australian engineer Viktor Kaplan is a special type of a propeller turbine which is used for higher specific speed and lower heads varying between 1.5 and 70 m. In this turbine, the individual runner blades are fixed on the hub so that their inclination may be adjusted during operation responding to changes of load. (c) Bulb turbines have higher full-load efficiency and higher flow capacity as compared to Kaplan turbine. It has a relatively lower construction cost. (d) Deriaz turbine developed by Paul Deriaz is a type of water turbine similar to a Kaplan turbine but it has inclined blades to make it more suitable for higher heads. It is particularly suitable for the head range between 20 and 100 m.

1.4 NUCLEAR POWER PLANT

Location of nuclear power plant is generally determined by: (i) geology and seismology, (ii) atmospheric extremes and dispersion, (iii) population consideration, (iv) water quality and availability and (v) transportation facility.

Typical line diagram of nuclear power plant is shown in Figure 1.3.

Typical components of nuclear power plants are: (i) nuclear fuel – reactor grade uranium usually enriched to about 3.5% U235 is fed in the form UO_2 pellets in zircalloy tubes, and (ii) nuclear reactor – a reactor primarily utilizing thermal neutrons for fission is called a thermal reactor. A Breeder Reactor is a nuclear reactor that "breeds" fuel. A Breeder consumes fissile and fertile material at the same time as it creates new fissile material. A breeder reactor must be specifically designed to create more fissile material than it consumes. Thus a reactor is a device to initiate and control a sustained nuclear chain reaction. Heat from nuclear fission is passed to a working fluid (water or gas), which runs through turbines. A reactor primarily utilizing thermal neutrons for fission is called a thermal reactor. Reactor has Reactor Vessel which is a robust steel vessel containing the reactor core and moderator/coolant. Neutrons produced by fission have high energies and move extremely quickly. These so-called fast neutrons do not cause fission as efficiently as slower-moving ones so they are slowed down in most reactors by the process of moderation. A liquid or gas moderator commonly uses water or helium that cools the neutrons to optimum energies for causing fission. Often Graphite or Beryllium is used in a reactor to slow down fast neutrons, (iii) Control rods/Neutron poison is made with neutron-absorbing material, such as cadmium, hafnium or boron, and is inserted or withdrawn from the core to control the rate of reaction, or to halt it, (iv) Secondary control systems made of other neutron absorbers, usually boron in the coolant – its concentration can be adjusted over time as the fuel burns up, (v) Neutron Howitzer provides steady source of neutron to reinitiate reaction after shutdown. Californium is generally used. (vi) Boiler feed water pump and (vii) Steam generator and Heat exchanger.

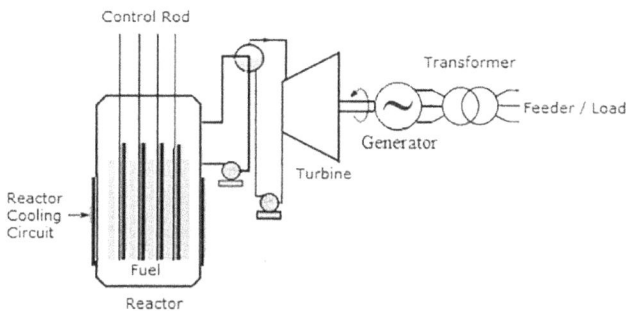

FIGURE 1.3 Typical diagram of nuclear power plant.

1.5   GAS TURBINE

A gas turbine is an internal combustion engine that can convert natural gas or other liquid fuels to mechanical energy. This energy then drives a generator that produces electrical energy. The gas turbine has a second turbine that acts as an air compressor, which can be either axial flow or centrifugal flow, mounted on the same shaft. Axial flow compressors are more common in power generation because they have higher flow rates and efficiencies. The air turbine (compressor) draws in air, compresses it and feeds it at high pressure into the combustion chamber with the fuel increasing the intensity of the burning flame. Gas turbines can utilize a variety of fuels, including natural gas, fuel oils and synthetic fuels.

Typical diagram of gas turbine based power generation is shown in Figure 1.4.

Essential components of gas turbine are: (i) Low pressure (LP) air compressor where atmospheric air is drawn in and passed through the air filter which then flows into the low pressure compressor. Major percentage of power developed (66%) by the turbine is used to run the compressor. The power required to run the compressor can be reduced by compressing the air in two stages, i.e., in low pressure and high pressure compressor, and also by incorporating an intercooler between the two. (ii) Intercooler is used to reduce work of the compressor and increase the efficiency. The energy required to compress air is proportional to the air temperature at inlet. Therefore if intercooling is carried out between the stages of compression, total work can be reduced. Normally *Brayton Cycle* Gas turbine is used for power generation. (iii) From the intercooler, the compressed air enters the high-pressure compressor, where it is further compressed to high pressure, which is then passed into the regenerator. (iv) Regenerator where the heat of the turbine exhaust gases that goes as waste is utilized. (v) Combustion chamber where hot air from regenerator flows and fuel (natural gas or coal gas or kerosene or gasoline) is injected into the combustion chamber and burns in the steam of hot air. The products of combustion comprise a mixture of gases as high temperature and pressure are passed to the turbine, which turns the alternator to produce electricity. When the heat is given to the air by mixing and burning the fuel in the air and the gases that come out of the turbine are exhausted to the atmosphere, the cycle is known as "open cycle system". If the heat to the working medium (air or any other suitable gas) is given without directly burning the fuel in the air and the same working medium is used again and again, the cycle is known as "closed cycle system".

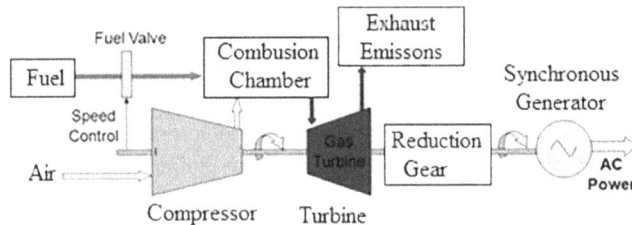

FIGURE 1.4   Typical diagram of gas-based power generation.

1.6   WIND ENERGY

Wind turbines harness the power of the wind and use it to generate electricity. The advantages are as follows: (i) wind is free, and newer technologies make the extraction of wind energy much more efficient; (ii) the energy it produces does not cause green house gases or other pollutants; (iii) it takes up only a small plot of land; (iv) remote areas that are not connected to the electricity power grid can use wind turbines to produce their own supply; and (v) the cost of producing wind energy has come down steadily over the last few years.

Normally wind turbines are of two types: (i) Horizontal axis – the most commonly used type in which each turbine possesses two or three blades and the rotating axis of which is horizontal, or parallel to the ground. The turbine can harness wind energy when the wind blows parallel to the ground.

(ii) Vertical axis – the rotating axis of which is vertical, or perpendicular to the ground. The turbine can harness from wind blowing in any direction and are usually made with blades that rotate around a vertical pole. The components of a wind power turbine of horizontal axis are shown in Figure 1.5.

The components of wind power generations are: (i) anemometer – measures the wind speed and transmits wind speed data to the controller. (ii) Turbine and rotor blades – can be either horizontal axis, where the blades rotate at an axis parallel to the ground, or vertical axis, where the blades rotate at an axis perpendicular to the ground. (iii) Nacelle – contains the key components of the wind turbine, including the gearbox and the electrical generator. (iv) Low-speed shaft – the low-speed shaft of the wind turbine connects the rotor hub to the gearbox. On a modern 1,000 kW wind turbine the rotor rotates relatively slowly, about 19–30 revolutions per minute (rpm). (v) Gearbox – connects the low-speed shaft to the high-speed shaft and increases the rotational speeds from about 30–60 rotations per minute (rpm) to about 1,000–1,800 rpm, approximately 50 times faster than the low-speed shaft. (vi) High-speed shaft – the high-speed shaft rotates with approximately 1,500 revolutions per minute and drives the electrical generator. (vii) Generator – usually of induction type. On a modern wind turbine the maximum electric power is usually between 600 and 3,000 kW. (viii) Yaw mechanism – to turn the nacelle with the rotor against the wind. The yaw mechanism is operated by the electronic controller which senses the wind direction using the wind vane. (ix) Brake system – stops the rotor mechanically, electrically or hydraulically, during emergencies. (x) Controller. (xi) Tower. (xii) Wind vane – measures wind direction and communicates with the yaw drive to orient the turbine properly with respect to the wind. The turbine blade itself can be as long as 75 m, while the entire rotor assembly measures 154 m in diameter. As it spins, the blades cover an area of 18,600 m^2 – that's roughly two and a half soccer fields – at a brisk 80 m/s, or 180 MPH at the tips. Rotor blade tips rotating at a speed larger than 80 m/s will be subject to erosion of the leading edges from their impact with dust or sand particles in the air, and will require the use of special erosion-resistant coatings much like in the design of helicopter blades. Further, it will produce noise and will be subject to vibration.

Types of Wind Turbines Generators (WTG) are as follows: (i) Type 1 WTG is equipped with a squirrel-cage induction generator and is connected to the step-up transformer directly. The turbine speed is fixed (or nearly fixed) to the electrical grid's frequency, and generates real power (P) when the turbine shaft rotates faster than the electrical grid frequency creating a negative slip (positive slip and power are motoring convention). Torque is controlled by adjusting pitch. Type 1 turbines typically

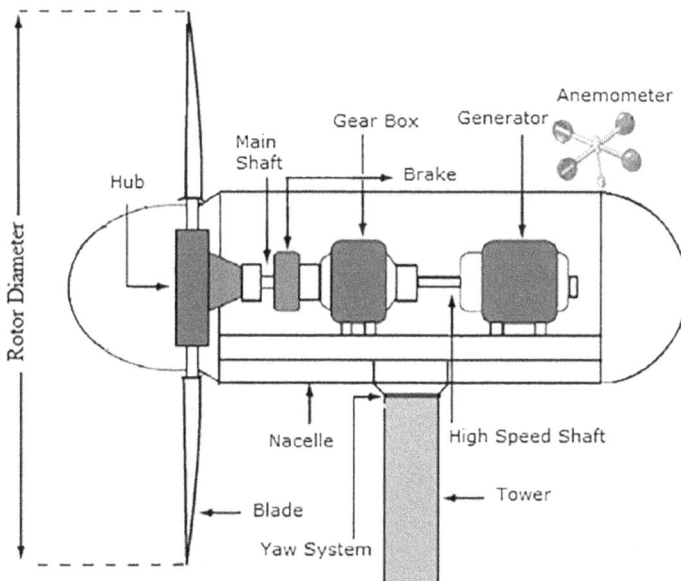

FIGURE 1.5 Typical diagram of horizontal axis wind turbine.

operate at or very close to a rated speed, thereby having a limitation. (ii) Type 2 WTG is equipped with a wound rotor induction generator and is connected directly to the step-up transformer similar to Type 1 with regard to the machine's stator circuit, and also includes a variable resistor in the rotor circuit. The variable resistors control the rotor currents so as to keep constant power even during gusting conditions, and can influence the machine's dynamic response during grid disturbances. Torque is controlled by adjusting this resistance; (iii) Type 3 WTG is equipped with Doubly Fed Induction Generator or Doubly Fed Asynchronous Generator that takes the Type 2 design to the next level, by adding variable frequency ac excitation (instead of simply resistance) to the rotor circuit. The additional rotor excitation is supplied via slip rings by a current regulated, voltage source converter, which can adjust the rotor currents' magnitude and phase nearly instantaneously. This rotor-side converter is connected back-to-back with a grid side converter, which exchanges power directly with the grid. Between the two converters a dc-link capacitor is placed, as energy storage, in order to keep the voltage variations (or ripple) in the dc-link. With the machine-side converter it is possible to control the torque or the speed of the Doubly Fed Induction Generator and also the power factor at the stator terminals, while the main objective for the grid-side converter is to keep the dc-link voltage constant.

The relationship between the wind speed and the rate of rotation of the rotor is characterized by a non-dimensional factor, known as the Tip Speed Ratio or λ = Speed of blade tip/wind speed = $\omega r/v$, where ω is the s the angular velocity [radian/s], r is rotor radius [m] and v is the wind speed [m/s]. Power coefficient C_p is defined as the power extracted by the turbine relative to that available in the wind stream and is the ratio of the extracted power P_t to the power generated P. The maximum achievable power factor is 59.26%, and is designated as the Betz limit.

1.7   SOLAR ENERGY

Two major technologies have been developed to harness solar energy: (i) Solar PV and (ii) Solar thermal. Solar PV directly converts sunlight into electricity through solar PV cells. Components of solar PV–based power generation are shown in Figure 1.6. Components of Solar PV power generation are as follows:

Solar Panels – These consist of a series of solar cells. There are several types of semiconductor technologies currently in use for solar panels such as the following: (i) mono-crystalline, which is efficient and yet expensive having conversion efficiency of 15%–20%; (ii) polycrystalline solar cells which are less energy efficient than mono-crystalline cells; and (iii) thin film technology having energy conversion efficiency of 8%–12%. Three main types of thin film solar cells are used: (a) amorphous silicon, (b) cadmium telluride and (c)

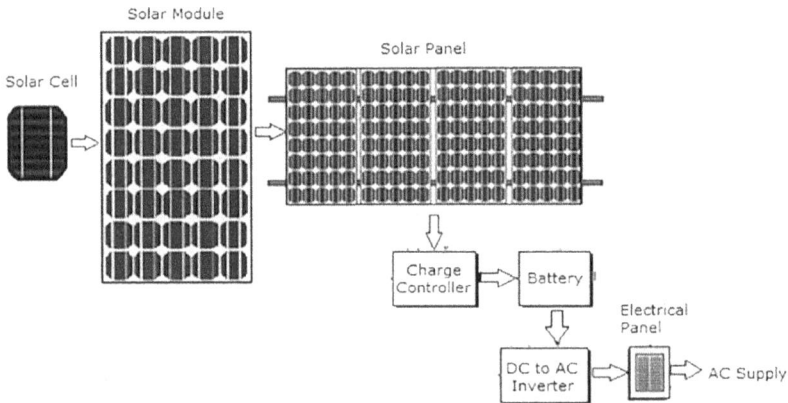

FIGURE 1.6   Solar PV–based power generation.

copper, indium and selenide (CIGS). Solar cells are typically combined into modules that hold about 40 cells; a number of these modules are mounted in PV arrays. PV arrays are mounted on a stable, durable structure.

Array Junction Box – This is used to connect solar module to inverter. Array junction box is having blocking diode to protect the panel from reverse current. Array junction box provides protection of the system against overcurrent and overvoltages.

Inverter – This converts the produced DC (Direct current) output of the solar module into AC (Alternating current) supply which is used for transmission of electricity, as well as most appliances in homes. PV systems either have one inverter that converts the electricity generated by all of the modules, or micro-inverters that are attached to each individual module. A single inverter is generally less expensive and can be more easily cooled and serviced when needed.

Solar Racking and Mounting (Module Support) – These are of two types. Fixed mounts are stationary and less expensive. This type of mounts cannot move with the change in angle of the sun; hence they are less efficient. Track mounts are flexible and rotate as per angle of the sun. This type of mounting is costly but more efficient as compared to a fixed mount system. Track mounts require more maintenance and are suitable for larger ground mount plants. One-axis trackers are typically designed to track the sun from east to west. Two-axis trackers allow for modules to remain pointed directly at the sun throughout the day.

Cabling Systems – These are of two types: (i) DC cable that is used to connect the solar panels to each other and to the inverter, and delivers the DC power generated by solar panels to the inverter; and (ii) AC cable is used for connecting solar power inverters to the grid through other protection components, and delivers AC power from the inverter to the grid/load.

Distribution Box

Storage Battery – This is used for an off-grid system or any stand-alone solar system where energy is stored in the battery during the daytime when the solar panel generates electricity and supplies power during night time or as needed. Batteries used for solar systems undergo frequent charging and discharging, hence rechargeable batteries are used.

Charge Controller – This is responsible for determining the adequate amount of charge to be withdrawn and supplied to the battery improving the efficiency of the battery bank by protecting it from overcharging.

Metering and Protection Device

Solar thermal technology captures the sun's heat; hence solar thermal systems differ from solar PV systems, which generate electricity rather than heat. This energy is used to heat water or other fluid. The main source of heat generation is through solar thermal collectors (panels) mounted on a roof, shade structure or other location that absorb solar energy. Water or fluid to be heated is circulated through the collectors by a low-energy pump and delivers heat to a water or fluid which is kept in a storage tank.

A solar pond is a body of water that collects and stores solar energy. Solar energy will warm a body of water (that is exposed to the sun), but the water loses its heat unless some method is used to trap it. Water warmed by the sun expands and rises as it becomes less dense. Once it reaches the surface, the water loses its heat to the air through convection, or evaporates taking heat with it. The colder water, which is heavier, moves down to replace the warm water, creating a natural convective circulation that mixes the water and dissipates the heat. The design of solar ponds reduces either convection or evaporation in order to store the heat collected by the pond. They can operate in almost any climate. To store solar heat much more efficiently solar pond water is mixed with salt which collects and stores solar thermal energy.

1.8   DISTRIBUTED POWER GENERATION

In a conventional power model, energy is generated in massive quantities at a large-scale power plant and then distributed across a transmission grid to end users. By contrast distributed power generation [1,2], commonly known as DG for short, refers to power that is generated on-site at, or very close to, the location where it will be used that can reduce the peak demand mitigating the loss in distribution. The modular nature of distributed generation system coupled with low gestation period enables the easy capacity additions when required. These generations are based on the technologies, mainly renewable, including, but not limited to, wind turbines, PV cells, geothermal energy and micro-hydropower plants, and eliminate the costs associated with the transmission and distribution of power over long distances. Distributed generation systems can reduce the peak demand and offer an effective solution to the problem of high peak load shortages. It is to provide power to remote and inaccessible areas. With independence from utility grid systems, distributed generation systems offer easy maintenance of power, voltage and frequency. It also offers the possibility of combining energy storage and management systems, with reduced congestion.

Currently available DG technologies in the 5 kW–5 MW size range include: (i) Reciprocating engines – The engines range in size from less than 5 to over 5,000 kW. Development efforts remain focused on improving efficiency and on reducing emission levels. Reciprocating engines are used primarily for backup power, peaking power and in cogeneration applications. (ii) Micro-turbines – A new and emerging technology, micro-turbines are currently available ranging from 30 to 200 kW. Micro-turbines consist of a compressor, combustor, turbine and a generator. (iii) Industrial combustion turbines – A mature technology, combustion turbines range from 1 to over 5 MW. They have low capital cost, low emission levels and also usually low electric efficiency ratings. (iv) Fuel cells – Although the first fuel cell was developed more than 150 years ago, currently fuel cells are commercially available in the 5–1,000+ kW size range. Fuel cell emission levels are quite low. The few fuel cells currently used provide premium power or are in applications subsidized by the government or gas utilities.

1.9   ENERGY STORAGE

Storage provides electricity at short notice, which overcomes intermittent nature of renewable source besides peak shaving. The process involves converting and storing electrical energy from an available source into another form of energy, which can be converted back into electrical energy when needed. Energy storage systems are classified into four types: mechanical, chemical, electrical and electrochemical. The benefits of energy storage are: (i) improved power quality and the reliable delivery of electricity to customers, (ii) improved stability and reliability of transmission and distribution systems and (iii) increased use of existing equipment, thereby deferring or eliminating costly upgrades. Energy storage can be classified as:

Pumped Hydropower – Pumped hydroelectric storage facilities store energy in the form of water in an upper reservoir, pumped from another reservoir at a lower elevation. During periods of high electricity demand, power is generated by releasing the stored water through turbines in the same manner as a conventional hydropower station. During periods of low demand (usually nights or weekends when electricity is also lower cost), the upper reservoir is recharged by using lower-cost electricity from the grid to pump the water back to the upper reservoir. Reversible pump-turbine/motor-generator assemblies can act as both pumps and turbines.

Compressed Air Energy Storage – It is equivalent to pumped-hydropower plants in terms of their applications. But, instead of pumping water from a lower to an upper pond during periods of excess power, in a compressed air energy storage plant, ambient air or another gas is compressed and stored under pressure of about 70 bar in an underground hermetically sealed cavern or container. When needed the compressed air is mixed with natural gas and burned and expanded in a modified gas turbine that drives the generator for power production.

Thermal Energy Storage – It can be either Pumped Heat Energy Storage, which is analogous to pumped hydro storage, but rather than pumping water to a reservoir placed up, heat is pumped from one thermal storage tank (–160°C) to another (+500°C) using a reversible heat pump/heat engine.

Fuel Cells – A fuel cell converts chemical energy to electrical energy by electrochemical reactions. Fuel is continuously supplied to one electrode and an oxidant (usually oxygen) to the other electrode. A simple hydrogen-oxygen fuel cell diffuses hydrogen gas through a porous metal electrode (nickel). In practical cells, conversion efficiencies of 80% have been attained. A major use of the fuel cell could be in conjunction with a future hydrogen energy system.

Solid State Batteries – They use solid electrodes and solid electrolytes that have high ionic conductivity with sufficient mechanical strength and offer high performance and safety at low cost. These batteries also have low flammability, higher electrochemical stability and higher energy density as compared to liquid electrolyte or polymer gel batteries.

Flow Batteries – They are fully rechargeable liquid electrolyte electrical energy storage batteries where the energy is stored directly in the electrolyte solution for longer cycle life, and quick response times. In flow batteries, electrolytes flow through one or more electrochemical cells from one or more tanks. The redox reactions during charge and discharge take place at the electrodes of the half cells.

Flywheels – They are mechanical devices that harness rotational energy to deliver instantaneous electricity.

Super Capacitor – A super capacitor is a double-layer capacitor or ultra capacitor that has very high capacitance but low voltage limits. It has a value up to 10,000 Farad at 1.2 V. This bridges the gap between electrolytic capacitors and rechargeable batteries. A super capacitor typically stores 10–100 times more energy per unit volume than electrolytic capacitor and can be cycled hundreds of thousands times with minimal change in performance. A Li-ion battery has a specific power of 1–3 kW/kg, whereas the specific power of a typical super capacitor is around 10 kW/kg. Super capacitors do not heat as much due to their low internal resistance.

1.10 TARIFF

The rate at which energy is sold to the consumers is called tariff with a view to ensure to recover the total cost of producing power and also to have some profits. The tariff development process shall be carried out in two phases: (i) calculation of annual revenue requirement of the companies and (ii) cost analysis, classification, selection of tariff structure and calculation of tariff rates. Electricity prices are usually highest for residential and commercial consumers because it costs more to distribute electricity to them. Industrial consumers use more electricity and can receive it at higher voltages, so supplying electricity to these customers is more efficient and less expensive. Factors affecting the tariff are:

Connected Load – The sum of the continuous ratings of all the equipments connected to the power system is called *connected load*. The load is classified into three types: domestic, commercial and industrial. If a consumer has connections of 5 # 60 W lamps and 4 # 75 W fans, then connected load of the consumer is $5 \times 60 + 4 \times 75 = 600$ W.

The greatest of all the demands (loads) which occur during a given period is called *maximum demand*. It can be expressed in KW, Kvar.

Demand Factor = Maximum demand/Connected load. It is always less than 1.0. If a residence consumer has 10 # lamp of 40 W each, it is possible that only 9 # lamps are used at the same time. Here Total Connected load is 10×40 = 400 W. Consumer maximum demand is 9×40 = 360 W. Demand Factor of this Load = 360/400 = 0.9% or 90%.

The total electrical energy (in WH or KWH) delivered in a given period divided by the time (in hours) in that period is called *average load*.

Load Factor = Average load/Maximum load. There may be *daily load factor* if the period of time is a day, *monthly load factor* if the period of time is a month and *yearly load factor* if the period of time is a year. If a plant operated for 24 hours, then load factor = (Average load×24)/(Maximum Demand×24 hours) = Unit generated in 24 hours/(Maximum Demand×24 hours). Higher the load factor of the power station, lesser will be the cost per unit generated.

Diversity Factor = Sum of individual maximum demands/maximum demand on the system. This factor gives the time diversification of the load and is used to decide the installation of sufficient generating and transmission plants. If all the demands came at the same time, i.e., unity diversity factor, the total installed capacity required would be much more. The diversity factor will always be greater than 1.

Plant Capacity Factor – The capacity factor of a power plant is the ratio of its average output power over a period of time, to its maximum possible power that could be produced. It can be determined as average demand (KW)/installed capacity (KW). The plant capacity factor is an indication of the reserve capacity of the plant.

Load Curve – A load curve or load profile is a chart illustrating the variation in demand/ electrical load over a specific time. A typical load curve is shown in Figure 1.7.

The load on a power station is never constant; it varies from time to time. These load variations during the whole day (i.e. 24 hours) are recorded half-hourly or hourly and are plotted against time on the graph. The curve thus obtained is known as daily load curve as it shows the variations of load against the time during the day. The area under the daily load curve gives the number of units generated in the day. The highest point on the daily load curve represents the maximum demand on the station on that day. The area under the daily load curve divided by the total number of hours gives the average load on the station in the day.

Utilization Factor – The utilization factor is the ratio of the maximum demand of a system to the rated capacity of the system, UF indicates the degree to which the system is being loaded during peak load periods with respect to its capacity.

UF = Maximum Demand of the system/Rated capacity of the system.

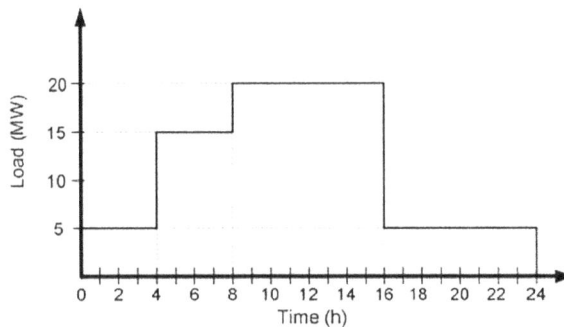

FIGURE 1.7   Typical load curve.

Various factors involving the tariff are as follows: (i) Tariff should ensure the recovery of the total cost of generation, transmission, distribution, etc.; (ii) tariff should be simple, cheap and capable of easy explanation to consumers; (iii) tariff should be attractive so that consumers are encouraged to make more extended use of electrical energy; (iv) tariff must send economic signals that promote efficient operation and investment; (v) tariff should promote efficiency in consumption of the good in the short and long term; and (vi) tariff should have Equity or Non-Discrimination in the allocation of costs to consumers: Same charge should apply to the same provision of a service, regardless of the end use of the electricity. One of the objectives of the Tariff Policy is to promote competition, efficiency in operations and improvement in quality of supply. The different types of tariffs are as follows:

a. **Simple Tariff** – The tariff, in which the rate per unit of energy is fixed, is called *simple tariff*. The rate per unit of energy consumed by the consumer is fixed irrespective of the quantity of energy consumed by a consumer. Energy consumed is measured by installing an energy meter.

   Advantages of such tariff are as follows: (i) It is in simplest form and easily understood by the consumers and (ii) the consumer is to pay as per his consumption. The disadvantages are as follows: (i) The consumer is to pay the same rate per unit of energy consumed irrespective of the number of units consumed by him. Hence, consumers are not encouraged to consume more energy. (ii) The cost of energy per unit delivered is high; and (iii) the supplier does not get any return for the connection given to the consumer if consumer does not consume any energy in a particular month.

b. **Flat Rate Tariff** – The tariff in which different types of consumers are charged at different per unit rates is called *flat rate tariff*. Here consumers are grouped into different classes and each class of consumer is charged at a different per unit rate.

   Advantages of such tariff are as follows: (i) Tariff is similar to simple tariff and is more fair to different types of consumers, since consumers are grouped; and (ii) it is quite simple in calculations. The disadvantages are as follows: (i) Consumers are not encouraged to consume more energy because same rate per unit of energy consumed is charged irrespective of the quantity of energy consumed; (ii) separate meters are required to measure energy consumed for light loads and power loads; and (iii) suppliers do not get any return for the connection given to the consumer if he does not consume any energy in a particular period or month.

c. **Block Rate Tariff** – In this type of tariff, the energy units are divided into numbers of blocks and the rate per unit of energy is fixed for each block. The rate per unit of energy for the first block is the highest and reduces progressively with the succeeding blocks.

   Advantage of such tariff is that by giving an incentive, the consumers are encouraged to consume more energy. This increases the load factor of the power system and hence reduces per unit cost of generation. But the disadvantage is that the supplier does not get any return for the connection given to the consumer if the consumer does not consume any energy in a particular period.

d. **Two-Part Tariff** – The tariff in which electrical energy is charged on the basis of both maximum demand of the consumer and the units consumed by consumers. In this tariff, the total charges to be made from the consumer are split into two components, namely, fixed charges and running charges. The fixed charges are independent of energy consumed by the consumer but depend upon the maximum demand, whereas the running charges depend upon the energy consumed by the consumer. The maximum demand of the consumer is assessed on the basis of the kW capacity of all the electrical devices owned by a particular consumer or on the connected load. The charges made on maximum demand recover the fixed charges of generation such as interest and depreciation on the capital cost of building and equipment, taxes and a part of operating cost which is independent of energy generated. Whereas the charges made on energy consumed recover operating cost which varies with variation in generated (or supplied) energy.

The advantages are as follows: (i) It is easily understood by the consumers; and (ii) the supplier gets the return in the form of fixed charges for the connection given to the consumer even if he does not consume any energy in a particular period. The disadvantages are as follows: (i) If a consumer does not consume any energy in a month, even then he has to pay the fixed charges; and (ii) since the maximum demand of consumer is not measured, there is always conflict between consumer and the supplier to assess the maximum demand.

e. **Maximum Demand Tariff** – The tariff in which electrical energy is charged on the basis of maximum demand of the consumer and the units consumed by him is called maximum demand. This tariff is actually similar to two-part tariff with the only difference that the maximum demand is actually measured by installing a maximum demand indicator meter. Thus the drawback of two-part tariff is removed.

f. **Power Factor Tariff** – The tariff in which the power factor of the consumer's load is also taken into consideration while fixing it is called power factor tariff. Consumers are advised to operate their loads at higher power factor since low power factor increases the rating of power plant equipment and gives higher losses.

g. **KVA Maximum Demand Tariff** – In this case the fixed charges are made on the basis of maximum demand in KVA instead of KW. Therefore, a consumer having low power factor has to pay more fixed charges. Thus the consumers are encouraged to operate their loads at higher power factor.

h. **KWH and RKVAH Tariff** – In this tariff, the consumers are charged for KWH and KVARH separately. Therefore, a consumer having low power factor shall have to pay more charges.

i. **Availability Based Tariff** – This depends on the availability of power rather than on MWh/MVAh output or peak MW/MVA as in conventional two-part tariff [3]. The tariff is based on frequency which tends to make the power system more stable and reliable. When the total generation of a system is less than the demand at a particular time, the frequency drops and vice versa; the frequency rises when there is a generation surplus exceeding the demand during off-peak hour.

This tariff mechanism has three parts: energy charge is the charge for the amount of energy delivered to the system and usually covers the variable cost+capacity charge, which is paid for the declared MW output capability of the station for a particular time block and is meant to cover the total fixed cost of a station+UI (Unscheduled interchange) that is payable for deviations of power injections to the system, and the drawals from the respective schedule and the deviations are linked to average frequency in particular time block, which is maximum when the grid frequency falls. The UI charges are payable/receivable if the utility overdraws power that causes reduction of frequency, and if the utility under-draws power that causes increase of frequency, the generating stations generate more than the schedule, thereby increasing the frequency, and the generating stations generate less than the schedule, thereby decreasing the frequency.

Generally the fixed cost and the variable cost of a generating station are charged to the consumers in proportion to the actual energy drawn by them during that period. In the Availability Based Tariff system [4,5], the fixed charge for a period is to be pro-rated among the consumers in the ratio of their entitlement for power from supplier. The logic is that the power station is created for catering to these beneficiaries. Hence its fixed cost has to be borne by them according to their share in the capacity so created. As with regard to energy charges, they are proposed to be charged only to the extent of the scheduled drawal by the consumers. By bifurcating the method of charging Capacity Charges (fixed) and Energy Charges (variable), the incentive for trading in power is enhanced.

If everything goes well, power demand is equal to power supplied and the system is stable and frequency is 50 Hz. But practically this rarely happens. One or more state overdraws or one or more Generating Stations (GS) under supplies. This led to deviation in frequency and system stability. If demand is more than supply, frequency dips from normal and vice versa. Before Availability Based Tariff (ABT) was introduced, the tariff mechanism did not provide any incentive to reduce generation under high frequency or to maximize generation under low frequency. In other words, the tariff mechanism encouraged grid indiscipline. But introduction of ABT has streamlined the operation of regional grids. ABT provides any incentive for either backing down generation during off-peak hours or for reducing consumer load/enhancing generation during peak-load hours.

UI charges are incentives provided or penalties imposed on the generating stations. If the frequency is less than 50 Hz, it implies demand is more than supply, and then the GS which supplies more power to the system than committed is given incentives. On the other hand, if frequency is above 50 Hz, it implies supply is more than demand, and incentives are provided to GS for backing up the generating power. Hence it tries to maintain the system stability.

1.11 LEARNING OUTCOME

The power sector has come a long way since over the last century and has evolved from being DC systems to highly interconnected three-phase AC systems. The need of the hour is to develop high intensity transmission corridor (MW per meter right of way) suitable for bulk power transfer over long distances in an environmental-friendly manner. This is possible through upgrading/uprating of existing transmission system using technology suitable for high capacity EHV/UHV AC system and HVDC system. Bulk power systems consist of a backbone of a extra high-voltage transmission grid with generators pooling their generation into it, and the pooled power is withdrawn by loads at lower voltage levels. Considering the operational regime of the various Regional Grids, asynchronous connection between the Regional Grids was necessary to enable them to exchange large regulated quantum of power. Power generation consisted primarily of thermal power that used fossil fuels for steam generation and hydropower. With the passage of time, nuclear power generation was adopted for steam generation to meet the base load demand. The focus of the present-day generation of electricity has shifted use of efficient, sustainable and environment-friendly renewable generating sources. The renewable/nonrenewable sources of power is being generated in a decentralized way closed to the load centers called distributed generation. Since the power generated from the renewable source is intermittent and of unpredictable nature, it has resulted in the creation of energy storage devices in order to maintain the energy balance within the renewable energy system. All these are being done only to attract new generation capacity and enhance per capita availability and sustainability of electricity as the tariff policy is under change.

In this chapter, the readers have been able to learn all the above aspects.

2 Transmission and Distribution Systems

2.1 LINE DIAGRAM

The electrical energy is produced at the generating stations, and through the transmission network, it is transmitted to the consumers. The transmission lines should transmit power over the required distance economically and efficiently. In transmitting power, it would be of prime importance to maintain the limit of given regulation, efficiency and losses. Between the generating stations and the distribution stations, three different levels of voltage (transmission, sub-transmission and distribution level of voltage) are used.

The high voltage is required for long distance transmission, and the low voltage is required for utility purposes. The voltage level is going on decreasing from the transmission system to the distribution system. The generation voltage is usually 11 and 33 KV. This voltage is too low for transmission over long distance. It is, therefore, stepped up to 132, 220, 400 KV, or more by step-up transformers. At that voltage, the electrical energy is transmitted to the bulk power substation where energy is supplied from several power substations. The transmission voltage is, to a very large extent, determined by economic considerations. High-voltage transmission requires conductors of smaller cross section which results in economy of copper or aluminum. But at the same time cost of insulating the line and other expenses are increased. At the distribution level, the transmitted power is received at the distribution substation where the voltage is stepped down to generally 66 KV and fed to the sub-transmission system for onward transmission to the different distribution substations. These substations are located in the region of the load centers. The voltage is further stepped down to 33 and 11 KV. The large industrial consumers are supplied at the primary distribution level of 33 KV while the smaller industrial consumers are supplied at 11 KV. Voltage is stepped down further by a distribution transformer located in the residential and commercial area, where it is supplied to these consumers at the secondary distribution level of 400 V for three phase and 230 V for single phase. A complete diagram of power system representing all the components through their accepted symbols is called line diagram.

2.1.1 SINGLE-LINE DIAGRAM

Electric power systems are supplied by three-phase generators. Ideally, the generators are supplying balanced three-phase loads. A balanced three-phase system is always solved as a single-phase circuit composed of one of the three lines and the neutral return. Often the diagram is simplified further by omitting the neutral and by indicating the component parts by standard symbols rather than by their equivalent circuits. Such a simplified diagram of electric system is called a one-line diagram. It is also called as single-line diagram (SLD). An SLD is thus the concise form of representing a given power system. It is to be noted that a given SLD will contain only such data that are relevant to the system analysis/study under consideration.

2.1.2 IMPEDANCE DIAGRAM

The impedance diagram on single-phase basis for use under balanced conditions is drawn from the SLD. In drawing the impedance diagram, the assumptions are as follows: (i) The single-phase transformer equivalents are shown as ideals with impedances on appropriate side (LV/HV); (ii) the magnetizing reactances of transformers are negligible; (iii) the generators are represented as constant

DOI: 10.1201/9781003231240-2

voltage sources with series resistance or reactance; (iv) the transmission lines are approximated by their equivalent π or T Models; and (v) the loads are assumed to be passive and are represented by a series branch of resistance or reactance, and since the balanced conditions are assumed, the neutral grounding impedances do not appear in the impedance diagram.

2.1.3 REACTANCE DIAGRAM

By having more simplifications on the impedance diagram, reactance diagram is drawn. The assumptions are as follows: (i) the resistance is small and is omitted; (ii) the loads are omitted; and (iii) transmission line capacitances are ineffective and magnetizing currents in a transformer are small and are omitted.

2.2 PER UNIT REPRESENTATION

Per unit value of any quantity is defined as the ratio of actual value in any unit and the base or reference value in the same unit.

A well-chosen per unit system [6] can minimize computational effort, simplify evaluation and facilitate understanding of system characteristics. The product of two quantities expressed in per unit is expressed in per unit itself. The base values are chosen independently and quite arbitrarily, while others follow automatically. It is usual to assume the base values. Some base quantities are chosen independently and quite arbitrarily, while others follow automatically. Normally, the base values are chosen on that the principal variables will be equal to per unit under rated condition. In case of synchronous machine, per unit system may be used to remove arbitrary constants and simplify mathematical equations, so that it may be expressed in terms of equivalent circuit. It is to be noted that out of the four quantities – voltage, current, impedance and volt-ampere – if we specify two quantities, the other two quantities can be calculated from:

Single-Phase System	Three-Phase System
Base volt-amperes $=(VA)_B$ or Base megavolt-amperes $=(MVA)_B$ or $(KVA)_B$	Base megavolt-amperes $=(MVA)_B$
Base voltage $= V_B$ V or $(KV)_B$ V	Line to Line base KV $= KV_B$
Base current $I_B = (VA)_B/V_B$ A or $[1,000 \times (MVA)_B]/(KV)_B$ A	Base Current $I_B = [1,000 \times (MVA)_B]/[\sqrt{3} \times KV_B]$
Base impedance $Z_B = V_B/I_B = (V_B)^2/(VA)_B$ Ω or $[1,000 \times (KV)_B]/I_B = (KV_B)^2/(MVA)_B$ Ω	Base impedance $Z_B = [(1,000 \times KV_B)/(\sqrt{3}I_B)] =$ $(KV_B)^2/(MVA)_B = [1,000 \times (KV_B)^2]/(KVA_B)\Omega$
If the actual impedance is Z Ω, its per unit value will be $Z_{pu} = Z/Z_B = [Z \times (VA)_B]/(V_B)^2 = [Z \times (MVA)_B]/(KV_B)^2$	If the actual impedance is Z Ω, its per unit value will be $Z_{pu} = Z/Z_B = [Z \times (MVA)_B]/(KV_B)^2$

Change of Base in Per Unit Quantities: Sometimes to know the per unit impedance of a component based on a particular base values, we need to find the per unit value of that component based on some other base values. It is to be noted that the per unit impedance is directly proportional to base MVA and inversely proportional to $(base\ KV)^2$. Therefore, to change from per unit impedance on a given base to per unit impedance on a new base, the following equation applies:

$$Z(pu)_{new} = Z(pu)_{old} \frac{(MVA)_{B,new}}{(MVA)_{B,old}} \frac{(KV)^2_{B,old}}{(KV)^2_{B,new}} \qquad (2.1)$$

Advantages of per unit are (i) dealing with numeric near unity rather than over a wide range (ii) provides more meaningful comparison of parameters of machines with different ratings (iii) the avoidance of $\sqrt{3}$ factor in the calculation for a three-phase system (iv) enable the use of symmetrical components for electrical fault analysis (v) ideal transformers are eliminated as circuit elements. This results in a large saving in component representation and reduces computational burden.

2.3   CALCULATION OF FAULT LEVEL

In a power system, the Fault Level is maximum fault current (or fault MVA) that can flow into a zero impedance. The fault level is usually expressed in MVA (or corresponding per-unit value), or in short circuit KA. The $(\mathrm{MVA})_{\mathrm{Base}}$ and the $(\mathrm{MVA})_{\mathrm{Fault}}$ can be expressed as $\sqrt{3}$ × Nominal Voltage (KV) × Base Current $(\mathrm{I_B})$ (KA)   and $\sqrt{3}$ × Nominal Voltage (KV) × Short Circuit Current × Short Circuit Current Current $(\mathrm{I_{SC}})$ (KA), respectively. Thus, the fault level is

$$\text{Fault Level} = \frac{(\mathrm{MVA})_{\mathrm{Fault}}}{(\mathrm{MVA})_{\mathrm{Base}}} = \frac{\mathrm{I_{SC}}}{\mathrm{I_B}} = \mathrm{I_{SC\ (PU)}} = \frac{\text{Voltage}_{\text{nominal, PU}}}{Z_{\mathrm{PU}}} \tag{2.2}$$

Per unit voltage for nominal value is unity, so that Fault Level $(\mathrm{PU}) = 1/Z_{\mathrm{PU}}$ and is independent of complex power base.

2.4   CHOICE OF VOLTAGE

The choice of voltage is linked with the conductor size and performance of the line as expected within permissible percentage losses and the regulation of the lines. The regulation would give the drop of volts between the sending end and the receiving end. Electricity is transmitted at high voltages because (i) with the increase in the transmission voltage, the size of the conductors is reduced (cross section of the conductors reduces as current required to carry reduces). The volume of copper required is equal to $(3P^2\rho l^2)/(\mathrm{W}V^2\cos^2\theta)$, which is inversely proportional to square of transmission line voltage (V), square of the power factor $(\cos\theta)$ and the total copper loss (W). P is the power transmitted, ρ is the resistivity and l is the length of the conductor. (ii) With the reduction in current-carrying requirement losses reduces results in better efficiency $= \eta = 1 - \left[\left(\sqrt{3}\rho lJ\right)/(\mathrm{V}\cos\theta)\right]$, where J is the current density of the conductor. (iii) Due to low current, line drop will be less so voltage regulation improves. Percent line drop $= (\rho lJ/\mathrm{V}) \times 100$. (iv) By using high voltages, the resulting lower currents make it possible to use conductors with reduced cross-sectional areas that allow the conductors to be light enough to be suspended from towers or poles.

Thus selection criteria for voltage depends upon the following: (i) Quantum of power to be transmitted, (ii) Length of the line, (iii) Voltage regulation, (iv) Power loss in transmission, (v) Initial and operating cost and (vi) Present and future voltage in the neighborhood of transmission of power. The economic voltage is given by $5.5\sqrt{[(\mathrm{L}/1.6) + (\mathrm{KVA}/150)]}$.

However, with the increase in transmission line voltage (i) the insulation required between the conductors and the earthed tower increases, thereby increasing the cost of line support and (ii) more clearance is required between conductors and ground, thereby increasing the tower height; and (iii) with increase in the voltage transmission, more distance is required between the conductors. Therefore cross arms should be long. (iv) Cost of electrical equipment like transformers, switchgears and other associated equipment is higher at higher voltages and (v) Corona effect is higher.

2.5   CHOICE OF FREQUENCY

In the early days of electrification, so many frequencies were used that no one value prevailed (London in 1918 had ten different frequencies). As the 20th century continued, more power was produced at 60 Hz (North America) or 50 Hz (Europe and most of Asia). Standardization allowed international trade in electrical equipment. Much later, the use of standard frequencies allowed interconnection of power grids. Generators can only be interconnected to operate in parallel if they are of the same frequency and wave-shape. By standardizing the frequency used, generators in a geographic area can be interconnected in a grid, providing reliability and cost savings.

2.6   CHOICE OF CONDUCTOR

Choice of conductor depends upon the following: (i) Mechanical Requirement – Tensile Strength (for tension)/Strain Strength (for vibration) and (ii) Electrical Requirement – continuous current rating/short time current-carrying rating/voltage drop/power loss/minimum diameter to avoid corona/length of line/charging current.

Conductors for overhead transmission lines are usually stranded unless the section is small. Solid conductors are susceptible to mechanical fatigue and finally fracture due to continuous swinging and vibration at the point of connection of a large solid wire to an insulator. In case of stranded conductors, adjacent layers are spiraled with the result that the layers are bounded to one another. Aluminum is now the most commonly employed conductor material. It has the advantages of being cheaper and lighter than copper though with less conductivity and tensile strength. Low density and low conductivity result in larger overall conductor diameter, which offers another advantage in high voltage lines. Increased diameter results in reduced electrical stress at conductor surface for a given voltage so that the line is corona free. Typical stranded conductor used for overhead transmission is shown in Figure 2.1.

Types of Conductors

There is no unique process by which all transmission and/or distribution lines are designed. It is clear, however, that all major cost components of line design depend upon the conductor electrical and mechanical parameters. There are five major types of overhead conductors used for electrical transmission and distribution.

- **AAC** – All Aluminum Conductor, sometimes referred to as Aluminum Stranded Conductor, is made up of one or more strands of 1350 Alloy Aluminum in the hard drawn H19 temper. One thousand three hundred and fifty Aluminum Alloy, previously known as electrical conductor (EC) grade or electrical conductor grade aluminum, has a minimum conductivity of 61.2% International Annealed Copper Standard (IACS). Because of its relatively poor strength-to-weight ratio, AAC has had limited use in transmission lines and rural

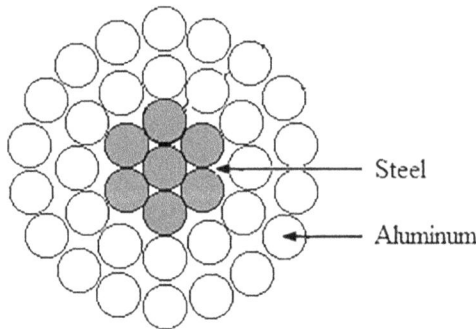

Steel

Aluminum

FIGURE 2.1   Typical stranded conductor.

distribution because of the long spans utilized. However, AAC has seen extensive use in urban areas where spans are usually short but high conductivity is required. The excellent corrosion resistance of aluminum has made AAC a conductor of choice in coastal areas.

- **AAAC** – All Aluminum Alloy Conductor, a high strength Aluminum-Magnesium-Silicon Alloy Cable was developed to replace the high strength 6/1 ACSR (Aluminum Conductor Steel Reinforced) conductors. Originally called AAAC, this alloy conductor offers excellent electrical characteristics with a conductivity of 52.5% IACS, excellent sag-tension characteristics and superior corrosion resistance to that of ACSR.
- **ACSR** – It consists of a solid or stranded steel core surrounded by one or more layers of strands of 1350 aluminum. Historically, the amount of steel used to obtain higher strength soon increased to a substantial portion of the cross section of the ACSR, but more recently, as conductors have become larger, the trend has been to less steel content. The total number of strands (N) in concentrically stranded cables with total annular space filled with strands of uniform diameter (d) is given by $N = 3x^2 - 3x + 1$, where x is the number of layers wherein the single central strand is counted as the first layer. The overall diameter (D) of a stranded conductor is $D = (2x - 1)d$.
- **ACAR** – Aluminum Conductor Aluminum-Alloy Reinforced.
- **ACCC** – Aluminum Conductor Composite Core belongs to high temperature low sag category, where central core is replaced by Carbon Fiber Thermoset Polymer Composite material and trapezoidal strands made of annealed aluminum. It is light in weight and almost 2–2.5 times current-carrying capacity of the same size ACSR conductor and can operate at temperatures as high as 250°C against the maximum operating temperature of 85°C of ACSR conductor.

In recent years, AAAC conductor has been a popular choice for transmission lines due to its high electrical carrying capacity and high mechanical tension to mass ratio. The high tension to mass ratio allows AAAC conductors to be strung at a higher tension and longer spans than traditional ACSR (Aluminum Conductor Steel Reinforced) conductors. Unfortunately, the self-damping of conductor decreases as tension increases. The wind power into the conductor increases with span length. Hence AAAC conductors are likely to experience more severe vibration than ACSR. The "Stockbridge" type vibration damper is commonly used to control vibration of overhead conductors. The vibration damper has a length of steel messenger cable. Two metallic weights are attached to the ends of the messenger cable. The center clamp, which is attached to the messenger cable, is used to install the vibration damper onto the overhead conductor.

Bundled Conductors – A bundled conductor arrangement with two or more sub-conductors in parallel per phase (which reduces effective resistance of the conductor), spaced a short distance apart is frequently used for HV and EHV transmission lines. The advantages of bundled conductors are as follows: (i) reduced corona loss due to larger cross-sectional area, (ii) reduced interference with the communication circuits, (iii) reduced inductance per phase due to increased Geometric Mean Radius per phase which in turn reduces the net series reactance, (iv) improved voltage regulation and improved stability margin and (v) increased power transmission capacity

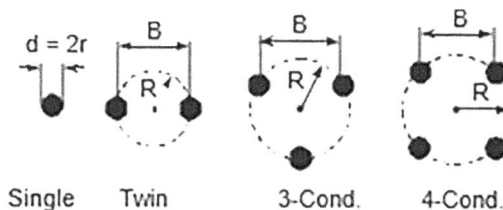

FIGURE 2.2 Bundle conductor configuration.

with reduced power loss leading to increased to efficiency. Conductor configuration of bundle is shown in Figure 2.2. In almost all cases, the sub-conductors of a bundle are uniformly distributed on a circle of radius R. The spacing between adjacent sub-conductors is termed "Bundle Spacing" and denoted by B. The radius of each sub-conductor is r with diameter d.

2.7 TRANSMISSION AND DISTRIBUTION TOPOLOGIES

Transmission line helps to transmit electricity from generating station to the substations. The part of system by which electric power is distributed among various consumers for local use from the substation is known as distribution system. The amount of power that can be passed through a transmission network from one place to another refers to "transfer capability". The main parts are as follows: (i) feeders which are electric lines that connect generating station (power station) or substation to distributors. These are conductors that are never tapped. The feeder current always remains constant. (ii) Distributors are electric lines that connect distribution substations or feeding points. These are conductors which are tapped to supply various loads of the consumers by service mains. (iii) Service Mains, which is a line, connect the consumer to the distributor.

Radial Distribution System – only one/single path exists between each distribution and substation which is called radial Distribution system. If fault occurs either on feeder or a distributor, all the consumers connected to that distributor will get affected. This is a tree shape topology where no close loops exist. The system is shown in Figure 2.3.

The advantages of such system are as follows: (i) simple and low cost, (ii) simple in planning, design and operation, (iii) useful when the generation is at low voltage, (iv) station is located at the center of the load, (v) requirement of small land area and (vi) easily expandable. The disadvantages are as follows: (i) distributor nearer to the feeding end is heavily loaded and (ii) the consumers at the far end of the feeder would be subjected to voltage fluctuations with the variations in load.

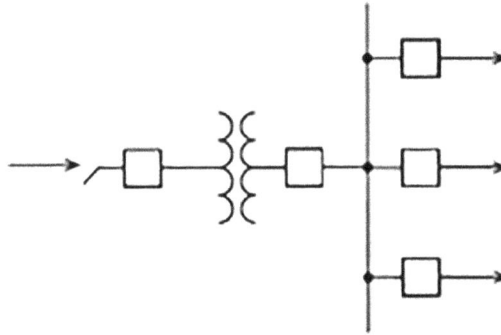

FIGURE 2.3   Radial system.

Ring Main System – Feeder covers the whole area of supply in the ring fashion and finally terminates at the substation from where it is started. It is like a closed loop form and looks like a ring. The system is shown in Figure 2.4.

Advantages of such system are as follows: (i) flexible operation; (ii) double feed to each circuit, no main buses; (iii) less conductor material is required as each part of the ring carries less current than in the radial system; and (iv) less voltage fluctuations. Disadvantages are the following: (i) t is difficult to design when compared to the designing of a radial system and (ii) the cost is more.

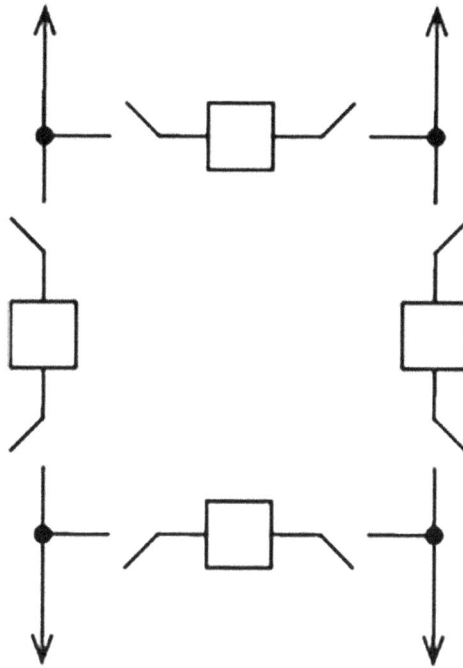

FIGURE 2.4   Ring main system.

Mesh – Mesh networks are the most complicated and most reliable method of distributing electric power. The network involves multiple paths between all the points in the network. Power flow between any two points is split along several paths. If a failure occurs, power instantly re-routes itself. Meshed distribution networks are usually employed in high density urban areas where the system must be placed underground and where repairs and maintenance are difficult because of traffic and other considerations.

2.8   ELECTRICAL GRID

An electrical grid [7,8] is an interconnected network for delivering electricity from producers to consumers. The interconnection provides the best use of power resource and ensures great security to supply. A grid is a transmission corridor, although besides transmitting electricity, transmission system does system integration and interconnection. Development of grid system emphasizes to enable scheduled/unscheduled exchange of power as well as for providing open access to encourage competition in power market. The interconnected grid system offers reliability, flexibility and economic competition. The electrical grid is mainly classified into two types: (i) Regional Grid which is formed by interconnecting the different transmission system of a particular area through the transmission line and (ii) National Grid which is formed by interconnecting the different regional grid.

Besides developing large centralized robust grids that have provided most of our electricity for the last century, to cater local electricity demand, micro-grid system is being developed which is a self-sufficient energy system. Micro-grids are small and localized versions of a power grid for regions with no or poor central grid connection. A micro-grid can disconnect from the central grid and operate independently. It has the potential to boost the economy by bringing electricity to remote areas. The choice of microgrids highly depends upon the criteria like stability, connectivity, total capacity, available sources and available infrastructure.

The interconnection between networks is mainly classified into two types, i.e., the HVAC link and HVDC link.

- **HVAC (High-Voltage Alternating Current) Interconnection** – In HVAC link the two AC systems are interconnected by an AC link. For interconnecting the AC system, it is necessary that there should be sufficiently close frequency control on each of the two systems. For the 50 Hz system, the frequency should lie between 48.5 and 51.5 Hz. Such an interconnection is known as synchronous interconnection or synchronous tie. The AC link provides a rigid connection between two AC system to be interconnected. But the AC interconnection has certain limitations. Currently, the maximum voltage for AC transmission is 765 KV beyond which power dissipation through dielectric loss becomes significant. At high voltage, non-resistive power dissipation via dielectric losses and/or through corona discharge becomes severe. The interconnection of an AC system has suffered from the following problems:
  - The interconnection of the two AC networks is the synchronous tie. The frequency disturbances in one system are transferred to the other system.
  - The power swings in one system affects the other system. Large power swing in one system may result in frequent tripping due to which major fault occurs in the system. This fault causes complete failure of the whole interconnected system.
  - There is an increase in the fault level if an existing AC system is connected with the other AC system with an AC tie line. This is because the additional parallel line reduces the equivalent reactance of the interconnected system. If the two AC systems are connected to the fault line, then the fault level of each AC system remains unchanged.
- **HVDC (High-Voltage Direct Current) Interconnection** – The DC interconnection or DC tie provides a loose coupling between the two AC system to be interconnected. The DC tie between two AC systems is non-synchronous (Asynchronous). The DC interconnection has the certain advantages. They are as follows:
  - The DC interconnection system is asynchronous thus the system which is to be interconnected is either of the same frequency or at the difference frequency. The DC link thus provides the advantages of interconnection of two AC network at different frequencies. It also enables the system to operate independently and to maintain their frequency standards.
  - The HVDC links provide fast and reliable control of magnitude and direction of power flow by controlling the firing angle of converters. The rapid control of power flow increases the limit of transient stability.
  - The power swings in the interconnected AC networks can be damped rapidly by modulating the power flow through the DC tie. Thus, the stability of the system is increased.

2.9   LEARNING OUTCOME

Once the complicated network of power system is formed, the performance of the system has to be analyzed under load condition and also upon the occurrence of the fault. To analyze the system a simplified diagram is to be prepared using the various system components connected to the system. This is called SLD that provides significant information. From the SLD, the reactance diagram is formed. Since an electrical system/network operates at different voltage level taking the advantage of transformer as per requirement, to calculate various operating parameters per unit representation is required. Continuous changes to meet the requirement of the increased demand and the uninterrupted power supply necessitated continuous search of suitable cost-effective conductive material for bare overhead conductors, which is oldest and exists till today. The high temperature low sag conductor for transmission and distribution that is a truly interconnected network is gradually entering in place of conventional ACSR conductor.

In the early days of the transmission of electric power conductors were usually copper, but aluminum conductors have completely replaced copper for overhead lines because of the much lower cost and lighter weight of aluminum conductors compared with copper conductors of the same resistance. Options are being explored to increase the transmission capacity with adequacy to increase the voltage which is going up and up from 132 to 220 to 400 KV and more. Nowadays modern micro-grids have been part of the grid that can be regarded as a controlled entity within the power system and can be operated as a single aggregated load.

In this chapter, readers have been able to learn all the above aspects.

3 Overhead Transmission Line Constants

The primary constants of a segment of transmission line having constant cross-section area along its length has three circuit constants: resistance, inductance and capacitance [9,10]. In any circuit analysis these constants are expressed in per unit length. In the metric system resistance is expressed in ohms per meter (Ω/m), inductance in henries per meter (H/m) and capacitance in farads per meter (F/m), respectively. In most transmission lines the effects due to inductance and capacitance tend to dominate because of the relatively low series resistance of the transmission line.

3.1 LINE RESISTANCE

The dc resistance of a wire is given by $R_{dc} = \rho l / A\,\Omega$, where ρ is the resistivity of the wire in Ω meter, l is the length in meter and A is the cross-sectional area in m^2. The resistance of an overhead conductor is not the same as that given by the said expression. When alternating current flows through a conductor, the current density is not uniform over the entire cross section but is somewhat higher at the surface. This is called the *skin effect* and this makes the ac resistance a little more than the dc resistance forcing more current flow near the outer surface of the conductor. The higher the frequency of current, the more noticeable skin effect would be. At frequencies of our interest (50–60 Hz), however, skin effect is not very strong. Moreover, in a stranded conductor, the length of each strand is more that the length of the composite conductor. This also increases the value of the resistance from that calculated as above.

The resistivity of a conductor is a fundamental property of the material that the conductor is made from. It varies with both type and temperature of the material. At the same temperature, the resistivity of aluminum is higher than the resistivity of copper.

Material	Conductivity	Resistivity ρ in Ω m at 20°C
Annealed copper conductor	100%	1.724×10^{-8}
Hard drawn copper conductor	97.3%	1.78×10^{-8}
Aluminum conductor	61%	2.86×10^{-8}
Iron or steel conductor	17.4%	12.2×10^{-8}

The resistance of the transmission line increases with temperature. The lines are subjected to temperature variation corresponding to the location and the variation of temperature due to seasons and climatic changes. The rise in resistance will depend on the temperature coefficient of the conductor material of the lines. The resistance of a conductor R_t at temperature t is given by the relation $R_t = R_0 (1 + \alpha_0 t)$, where R_0 is the resistance of the conductor at 0°C and α_0 is the temperature coefficient of the conductor at 0°C.

3.2 LINE INDUCTANCE

The series inductance of a transmission line consists of two components: internal and external inductances, which are due the magnetic flux inside and outside the conductor, respectively. The

DOI: 10.1201/9781003231240-3

inductance of a transmission line is defined as the number of flux linkages [Wb-turns] produced per ampere of current flowing through the line: $L = \psi/I$. The inductance of the transmission line depends on the arrangement of conductors and their size. The conductors can be solid or stranded; and the single-phase or three-phase lines have uniform size of conductors. The arrangement of conductor spacing can be symmetrical and equidistant or the spacing may be at unequal distance.

3.2.1 INTERNAL INDUCTANCE

Consider a conductor of radius r carrying a current I amps as shown in Figure 3.1. Assume next that the fraction of the current I_x enclosed in the circle at a distance x meter assuming the current is distributed uniformly in the conductor: $I_x = \left(\pi x^2/\pi r^2\right)I$ amp.

Ampere's law determines the magnetic field intensity $H_x = Ix/2\pi x = Ix/2\pi r^2$ ampere turns per m. The flux density at a distance x from the center of the conductor is $B_x = \mu_0 H_x = \left[\mu_0 xI/2\pi r^2\right]$ Wb/m², where μ_0 is the permeability of the free space and is given by $4\pi \times 10^{-7}$ H/m. Consider now an infinitesimal tubular element of thickness dx and length 1 m. Let the flux along the circular strip be denoted by $d\phi_x = B_x dx \cdot 1 = \left[\mu_0 xI/2\pi r^2\right] dx$ Wb.

The entire conductor cross section does not enclose the above flux. The ratio of the cross-sectional area inside the circle of radius x to the total cross section of the conductor can be thought about as fractional turn that links the flux $d\phi_x$. Therefore, the flux linkage is $\psi_x = \left(\pi x^2/\pi r^2\right) d\phi_x = \left(\mu_0 I/2\pi r^4\right) x^3 dx$. Integrating over the range of x, i.e., from 0 to r, we get the internal flux linkage as $\psi_{int} = \mu_0 I/8\pi = \left(I/2\right)10^{-7}$ Wb/m. Thus, the internal inductance per unit length is $L_{int} = \left(1/2\right)10^{-7}$ H/m.

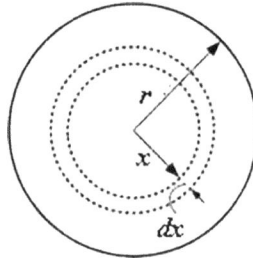

FIGURE 3.1 Flux linkage due to internal flux.

3.2.2 EXTERNAL INDUCTANCE

Let us consider an isolated straight conductor carries a current I amp. Assume that the tubular element at a distance x meter from the center of the conductor has a field intensity H_x. Since the circle with a radius of x encloses the entire current, the mmf around the element is given by $2\pi x H_x = I$ and hence the flux density at a radius x becomes $B_x = \mu_0 I/2\pi x$. This is shown in Figure 3.2.

The entire current I is linked by the flux at any point outside the conductor. Since the distance x is greater than the radius of the conductor, the flux linkage $d\psi_x$ is equal to the flux $d\phi_x$. Therefore for 1 m length of the conductor we get $d\psi_x = d\phi_x = B_x dx \cdot 1 = \left[\mu_0 I/2\pi x\right] dx$. The external flux linkage between any two points D_1 and D_2 that is external to the conductor is

$$\psi_{ext} = \left(\mu_0 I/2\pi\right)\int_{D_1}^{D_2}\left(1/x\right)dx = 2 \times 10^{-7} I \ln\left(D_2/D_1\right) \text{Wb/m} \tag{3.1}$$

The inductance between any two points outside the conductor is $L_{ext} = 2 \times 10^{-7} \ln\left(D_2/D_1\right)$ H/m, and the total inductance due to internal and external flux is $L_{ext} = (1/2)10^{-7} + 2 \times 10^{-7} \ln\left(D/r\right) = 2 \times 10^{-7}\left[(1/4) + \ln\left(D/r\right)\right] = 2 \times 10^{-7}\left(\ln\left[D/r_1'\right]\right)$ H/m; where $r_1' = re^{-1/4}$.

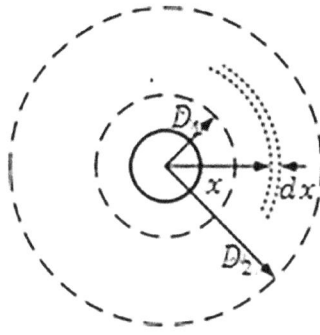

FIGURE 3.2   Flux linkage due to external flux.

3.2.3 INDUCTANCE OF A SINGLE-PHASE TWO-WIRE LINE

Consider two solid round conductors 1 and 2 as shown in Figure 3.3 with radii r_1 and r_2. One conductor is the return circuit for the other, like single-phase two-wire conductor. This implies that if the current in conductor 1 is I_1, then the current in conductor 2 is I_2 and $I_1 + I_2 = 0$.

First let us consider conductor 1. The current flowing in the conductor will set up flux lines. External flux from r_1 to $(D - r_2)$ links the current I_1 in conductor 1. External flux from $(D - r_2)$ to $(D + r_2)$ links a current whose magnitude progressively reduces from I_1 to zero. Moreover, since $D \gg r_1$ and r_2, it can be assumed that the flux from $(D - r_2)$ to the center of conductor 2 links all the current I_1 and the flux from the center of conductor 2 to $(D + r_2)$ links zero current. The inductance of conductor 1 due to internal and external flux is $L_1 = 2 \times \left(\ln \left[D/r_1' \right] \right) = 4 \times 10^{-7} \left[\ln \left(1/r_1' \right) - \ln \left(1/D \right) \right]$ H/m. Similarly, $L_2 = 2 \times 10^{-7} \left(\ln \left[D / r_2' \right] \right)$ H/m.

Therefore, the inductance of the complete circuit is $L_1 + L_2 = 2 \times 10^{-7} \left(\ln \left[D / r_1' \right] \right) + 2 \times 10^{-7} \left(\ln \left[D / r_2' \right] \right) = 2 \times 10^{-7} \left(\ln \left[D^2 / \sqrt{r_1' r_1' r_2' r_2'} \right] \right) = 4 \times 10^{-7} \left(\ln \left[D / \sqrt{r_1' r_2'} \right] \right)$ H/m. If $r_1' = r_2' = r'$, then total inductance becomes $= 4 \times 10^{-7} \left[\ln \left(1 / r' \right) - \ln \left(1 / D \right) \right]$ H/m. The greater the spacing between the phases of a transmission line, the greater the inductance of the line. The greater the radius of the conductors in a transmission line, the lower the inductance of the line. Since the phases of a high-voltage overhead transmission line must be spaced further apart to ensure proper insulation, a high-voltage line will have a higher inductance than a low-voltage line.

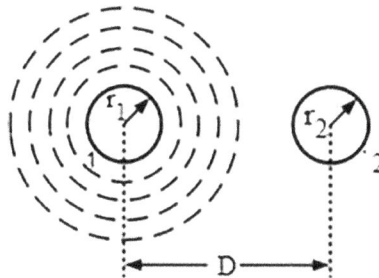

FIGURE 3.3   Single phase line.

3.2.4 INDUCTANCE OF ONE CONDUCTOR IN A GROUP

Consider a group of n parallel conductors carrying phasor currents $I_1, I_2, \ldots, I_n$, as shown in Figure 3.4 such that $I_1 + I_2 + I_3 \ldots + I_n = 0$ as the conductors form a closed circuit. Distances of these conductors from a point P are $D_1, D_2, \ldots, D_n$.

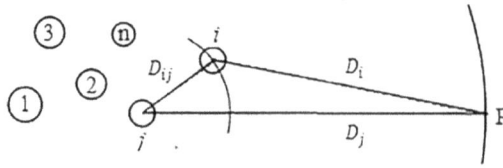

FIGURE 3.4 Group of conductor.

The flux linkage of ith conductor of the group due to its own current I_1 is $\psi_{ii} = 2 \times 10^{-7} I_i \ln(D_i / r_i')$. The flux linkage of conductor i due to current in conductor j is $\psi_{ij} = 2 \times 10^{-7} I_j \ln(D_j / D_{ij})$ Wb/m. Total flux linkages of conductor i due to flux up to the point P are

$$\psi_i = \psi_{i1} + \psi_{i2} + \psi_{i3} + \cdots + \psi_{ii} + \cdots \psi_{in}) = 2 \times 10^{-7} \left(I_1 \ln \frac{D_1}{D_{i1}} + I_2 \ln \frac{D_2}{D_{i2}} + \cdots + I_i \ln \frac{D_i}{D_{ii}} + \cdots + I_n \ln \frac{D_n}{D_{in}} \right)$$

$$= 2 \times 10^{-7} \left[\left(I_1 \ln \frac{1}{D_{i1}} + I_2 \ln \frac{1}{D_{i2}} + \cdots + I_i \ln \frac{1}{D_{ii}} + \cdots + I_n \ln \frac{1}{D_{in}} \right) \right.$$

$$\left. + \left[(I_1 \ln D_1 + I_2 \ln D_2 + \cdots + I_i \ln D_{1i} + \cdots + I_n \ln D_n) \right] \right]$$

$$= 2 \times 10^{-7} \left[\left(I_1 \ln \frac{1}{D_{i1}} + I_2 \ln \frac{1}{D_{i2}} + \cdots + I_i \ln \frac{1}{D_{ii}} + \cdots + I_n \ln \frac{1}{D_{in}} \right) \right.$$

$$\left. + \left[\left(I_1 \ln \frac{D_1}{D_n} + I_2 \ln \frac{D_2}{D_n} + \cdots + I_i \ln \frac{D_i}{D_n} + \cdots + I_{n-1} \ln \frac{D_{n-1}}{D_n} \right) \right] \right] \tag{3.2}$$

since $I_1 + I_2 + I_3 \cdots + I_n = 0$. If P is at infinity, then $D_1/D_n = D_2/D_n = \ldots = D_{n-1}/D_n = 1$ and $\ln 1 = 0$. The resultant linkage

$$\psi_i = 2 \times 10^{-7} \left[\left(I_1 \ln \frac{1}{D_{i1}} + I_2 \ln \frac{1}{D_{i2}} + \cdots + I_i \ln \frac{1}{D_{ii}} + \cdots + I_n \ln \frac{1}{D_{in}} \right) \right] \tag{3.3}$$

The inductance of conductor 1 will be the ratio of flux linkage about the conductor 1 due to the currents in all the conductors to the current of conductor 1.

3.2.5 INDUCTANCE OF A COMPOSITE CONDUCTOR

Future 3.5 shows a single-phase two-wire line comprising two sets of composite conductors A and B with A having n parallel strands and B having m' parallel strands as shown in Figure 3.5. Neglecting the minor differences of inductance of each strand, it will be sufficiently accurate to assume that the current is equally divided among the strands of each composite conductor. It is further assumed that the spacing between conductors is large compared to the diameter of each conductor. Thus, current in each strand of A can be taken as I/n, while current in each strand of conductor B will be current of I/m'.

$$\psi_i = 2 \times 10^{-7} \frac{I}{n} \left(\ln \frac{1}{D_{i1}} + \ln \frac{1}{D_{i2}} + \cdots + \ln \frac{1}{D_{ii}} + \cdots + \ln \frac{1}{D_{in}} \right) - 2 \times 10^{-7} \frac{I}{m'} \left(\ln \frac{D_1}{D_{i1'}} + \ln \frac{D_2}{D_{i2'}} + \cdots + \ln \frac{D_{n-1}}{D_{im'}} \right)$$

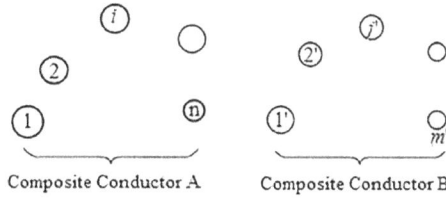

FIGURE 3.5 Group of conductors of single-phase line.

$$= 2 \times 10^{-7} I \ln \left(D_{i1'} \cdot D_{i2'} \ldots D_{im'} \right)^{1/m'} \Big/ \left(D_{i1} \cdot D_{i2} \cdot D_{i3} \ldots D_{in} \right)^{1/n} \text{ Wb/m} \tag{3.4}$$

The inductance of strand i is then $L_i = \psi_i / (I/n) = 2n \times 10^{-7} \ln \left[\left(D_{i1} \cdot D_{i2} \ldots D_{im'} \right)^{1/m'} \right] / \left[\left(D_{i1} \cdot D_{i2} \cdot D_{i3} \ldots D_{in} \right)^{1/n} \right]$ H/m. The average inductance of filaments of conductor A is $L_{\text{avg}} = (L_1 + L_2 + \cdots + L_n)/n$ H/m. The conductor A is composed of n parallel strands, and its inductance is

$$2 \times 10^{-7} \ln \left[\left(D_{i1'} \cdot D_{i2'} \ldots D_{im'} \right) \left(D_{i1'} \cdot D_{i2'} \ldots D_{im'} \right) \ldots \left(D_{i1'} \cdot D_{i2'} \ldots D_{im'} \right) \right]^{1/m'n}$$

$$\Big/ \left[\left(D_{i1} \cdot D_{i2} \cdot D_{i3} \ldots D_{in} \right) \left(D_{i1} \cdot D_{i2} \cdot D_{i3} \ldots D_{in} \right) \ldots \left(D_{i1} \cdot D_{i2} \cdot D_{i3} \ldots D_{in} \right) \right]^{1/n^2} = 2 \times 10^{-7} \ln \left(D_m / D_S \right) \text{ H/m}$$

The numerator of the argument of the aforesaid equation is the $m'n$th root of $m'n$ terms, which are the products of all possible mutual distances from the n strands of conductor A to m' strands of conductor B. It is mutual geometric mean distance (GMD) between conductors A and B and is denoted as D_m. The denominator of the argument of the aforesaid equation is the n^2 root of n^2th terms, which are the products of all possible self-distances from the n strands of conductor A. It is self-geometric mean distance or self-GMD of conductor A and is denoted as D_S. For a single-phase two-wire solid conductor, $D_m = 0.778 \ r$, where r is the radius of the conductor, and $D_S = d$, where d is the spacing between two conductors.

3.2.6 Inductance of a Three-Phase Line with Asymmetrical Spacing

It is rather difficult to maintain symmetrical spacing as shown earlier while constructing a transmission line. With asymmetrical spacing between the phases, the voltage drop due to line inductance will be unbalanced even when the line currents are balanced. Consider the three-phase asymmetrically spaced line shown in Figure 3.6, where the radius of each conductor is assumed to be r. The distances between the phases are denoted by D_{ab}, D_{bc} and D_{ca}.

The flux linkages for the three phases are $\psi_a = 2 \times 10^{-7} \left[I_a \ln \left(1/r' \right) + I_b \ln \left(1/D_{ab} \right) + I_c \ln \left(1/D_{ca} \right) \right]$; $\psi_b = 2 \times 10^{-7} \left[I_b \ln \left(1/r' \right) + I_a \ln \left(1/D_{ab} \right) + I_c \ln \left(1/D_{bc} \right) \right]$; and $\psi_c = 2 \times 10^{-7} [I_c \ln \left(1/r' \right) + I_a \ln \left(1/D_{ca} \right) + I_b \ln \left(1/D_{bc} \right)]$ Wbt/m. Let us define the operators $a = e^{j120^0} = 1 \angle 120^0$, $a^2 = e^{j240^0} = e^{-j120^0} = a^*, 1 + a + a^2 = 0$. If the current is balanced, then $I_b = a^2 I_a$ and $I_c = a I_a$.

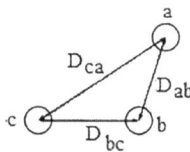

FIGURE 3.6 Three-phase line (asymmetrically spaced).

Substituting I_b and I_c, the inductance of the three phases are as follows: $L_a = 2 \times 10^{-7} \left[\ln(1/r') + a^2 \ln(1/D_{ab}) + a \ln(1/D_{ca}) \right]$, $L_b = 2 \times 10^{-7} \left[\ln(1/r') + a \ln(1/D_{ab}) +, a^2 \ln(1/D_{bc}) \right] L_c = 2 \times 10^{-7} \left[\ln(1/r') + a^2 \ln(1/D_{ca}) + a \ln(1/D_{bc}) \right]$.

For symmetrical spacing $D_{ab} = D_{bc} = D_{ca} = D$ (say). Hence the inductance of phase a is given as $L_a = 2 \times 10^{-7} \ln(D/r')$ H / m. Due to symmetry, the inductances of phases b and c will be the same as that of phase a given above, i.e., $L_b = L_c = L_a$.

3.2.7 INDUCTANCE OF A THREE-PHASE TRANSPOSED LINE

The inductances that are given just above are undesirable as they result in an unbalanced circuit configuration. One way of restoring the balanced nature of the circuit is to exchange the positions of the conductors at regular intervals. This is called transposition of line and is shown in Figure 3.7. In this each segment of the line is divided into three equal sub-segments.

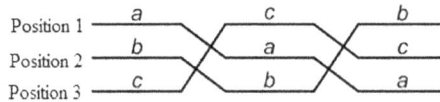

FIGURE 3.7 Transposed line.

The conductors of each of the phases a, b and c are exchanged after every sub-segment such that each of them is placed in each of the three positions once in the entire segment. For example, the conductor of the phase a occupies positions in the sequence 1, 2 and 3 in the three sub-segments while that of the phase b occupies 2, 3 and 1. The transmission line consists of several such segments. When transposed and let the conductor radius of each phase is equal to r, then for three consecutive positions as shown in figure the flux linkage of phase a will be $\psi_{a1} = 2 \times 10^{-7} \left[I_a \ln(1/r') + I_b \ln(1/D_{ab}) + I_c \ln(1/D_{ca}) \right]$; $\psi_{a2} = 2 \times 10^{-7} \left[I_a \ln(1/r') + I_b \ln(1/D_{bc}) + I_c \ln(1/D_{ab}) \right]$; and $\psi_{a3} = 2 \times 10^{-7} \left[I_a \ln(1/r') + I_b \ln(1/D_{ca}) + I_c \ln(1/D_{ca}) \right]$ Wb/m.

Total flux linkage of phase a will be $\psi_a = (\psi_{a1} + \psi_{a2} + \psi_{a3})/3 = 2 \times 10^{-7} I_a \ln \left[(D_{ab} D_{bc} D_{ca})^{1/3} /r' \right]$. Inductance of phase a, L_a will be equal to $2 \times 10^{-7} \ln(D_{eq}/r')$, where $D_{eq} = (D_{ab} D_{bc} D_{ca})^{1/3}$ H / m is the mutual GMD between the three-phase conductor. From the symmetry, it can be concluded that $L_a = L_b = L_c$. Thus, periodic swapping of the positions of the conductors at regular interval such that each conductor occupies the original position of every other conductor over an equal distance or transposition makes the inductance of each conductor equal after every complete transposition cycle.

3.2.8 INDUCTANCE OF BUNDLED CONDUCTORS

The geometric mean radius (GMR) of two-conductor bundle as shown in Figure 3.8 is given by $D_{s,2b} = \sqrt[4]{(D_s \times d)^2} = \sqrt{D_s \times d}$, where D_s is the GMR of conductor.

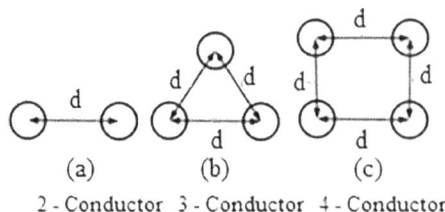

(a)   (b)   (c)

2 - Conductor   3 - Conductor   4 - Conductor

FIGURE 3.8 Bundled conductor.

The GMR of three-conductor bundle is given by $D_{s,3b} = \sqrt[9]{(D_s \times d \times d)^3} = \sqrt[3]{D_s \times d^2}$ and for four-conductor bundle it will be $D_{s,4b} = \left(D_s \times d \times d \times \sqrt{2}d\right)^{4/16} = 1.09\sqrt[3]{D_s \times d^3}$. The inductance of the bundled conductor is then given by $2 \times 10^{-7}\left(\text{GMD}/D_{s,\,nb}\right)$ H/m.

3.2.9 INDUCTANCE OF DOUBLE CIRCUIT THREE-PHASE LINE

It is common practice to build double-circuit three-phase lines so as to increase transmission reliability at somewhat enhanced cost. To achieve low inductance/phase, self GMD (D_S) should be made high and mutual GMD (D_m) should be made low. Therefore, the individual conductors of a phase are to be kept as far apart as possible (for high self GMD), while the distance between phases be kept as low as permissible (for low mutual GMD). Figure 3.9 shows the first section of the transposition cycle of two parallel circuit three-phase lines with vertical spacing.

GMR of conductors of phase a in section 1 is $D_{sa} = \sqrt[4]{r'.n.r'.n} = \sqrt{r'.n}$ and GMR of conductors of phases b and c in section 1 is $D_{sb} = \sqrt[4]{r'.h.r'.h} = \sqrt{r'.h}$ and $D_{sc} = \sqrt[4]{r'.n.r'.n} = \sqrt{r'.n}$. GMR of conductors of phases b and c in section 1 is $D_{sb} = \sqrt[4]{r'.h.r'.h} = \sqrt{r'.h}$ and $D_{sc} = \sqrt[4]{r'.n.r'.n} = \sqrt{r'.n}$. Equivalent GMR is $D_s = \sqrt[3]{D_{sa}.D_{sb}.D_{sc}} = (r')^{1/2}.(n)^{1/3}.(h)^{1/6}$. $\text{GMD} = \sqrt[3]{D_{ab}.D_{bc}.D_{ca}} = \left[\left(\sqrt[4]{D.m.D.m}\right)\right.$ $\left.\left(\sqrt[4]{D.m.D.m}\right)\left(\sqrt[4]{2D.m.2D.m}\right)\right]^{1/3} = 2^{1/6}.D^{1/2}.m^{1/3}.h^{1/6}$. Inductance $L = 2 \times 10^{-7} \ln\left(D_{eq}/D_s\right) = 2 \times 10^{-7} \ln\left[2^{1/6}.(D/r')^{1/2}.(m/n)^{1/3}\right]$ H / Phase / m.

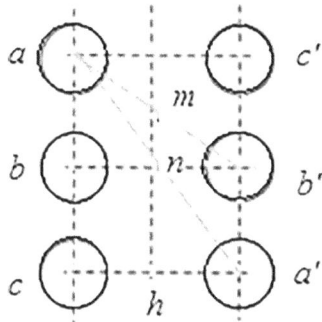

FIGURE 3.9 Double circuit line.

3.3 SHUNT PARAMETER OF TRANSMISSION LINES

Capacitance in a transmission line results due to the potential difference between the conductors. Usually, the capacitance is neglected for the transmission lines that are less than 50 miles (80 km) long. However, the capacitance becomes significant for longer lines with higher voltage. In this section we shall derive the line capacitance of different line configurations.

Let two points P_1 and P_2 be located at distances D_1 and D_2, respectively, from the center of the conductor as shown in Figure 3.10. The conductor is an equipotential surface in which we can assume that the uniformly distributed charge is concentrated at the center of the conductor. The potential difference V_{12} between the points P_1 and P_2 is the work done in moving a unit of charge from P_2 to P_1. Therefore, the voltage drop between the two points can be computed by integrating the field intensity over a radial path between the equipotential surfaces, i.e.,

$$V_{12} = \int_{D_1}^{D_2} E dx = \int_{D_1}^{D_2} \left(q/2\pi x \varepsilon_0\right)dx = \left(q/2\pi x \varepsilon_0\right)\ln(D_2/D_1).$$

E is the electric field intensity and q is the charge carried in coulombs. The potential difference between two conductors say A and B of a group of parallel conductors is $V_{ab} = \left(1/2\pi\varepsilon_0\right)\left[q_a\ln(D_{ab}/r_a) + q_b\ln(r_b/D_{ba}) + q_c\ln(D_{cb}/D_{ca}) + \cdots + q_n\ln(D_{nb}/D_{na})\right].$

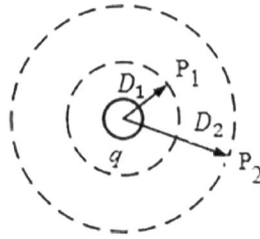

FIGURE 3.10 Electric field of a conductor.

3.3.1 Capacitance of a Straight Conductor

Consider the single-phase line as shown in Figure 3.11 consisting of two round conductors. The separation between the conductors is D. Let us assume that conductor 1 carries a charge of q_1 C/m while conductor 2 carries a charge q_2 C/m. The presence of the second conductor and the ground will disturb field of the first conductor. However, we can assume that the distance of separation between the conductors is much larger compared to the radius of the conductor and the height of the conductor is much larger than D for the ground to disturb the flux. Therefore, the distortion is small and the charge is uniformly distributed on the surface of the conductor. Assuming that the conductor 1 alone has the charge q_1, and the conductor 2 alone has the charge q_2. Then $V_{12} = V_{12(q1)} + V_{12(q2)} = (q_1/2\pi\varepsilon_0)\ln(D/r_1) + (q_2/2\pi\varepsilon_0)\ln(r_2/D)$ V. For a single-phase line let us assume that $q_1 (= -q_2)$ is equal to q. We therefore have

$$V_{12} = \frac{q}{2\pi\varepsilon_0}\ln\frac{D}{r_1} - \frac{q}{2\pi\varepsilon_0}\ln\frac{r_2}{D} = \frac{q}{2\pi\varepsilon_0}\ln\frac{D^2}{r_1 r_2} = \frac{q}{\pi\varepsilon_0}\ln\left(\frac{D}{r}\right) \text{V. So, } C_{12} = \frac{\pi\varepsilon_0}{\ln(D/r)} \text{ F/m} \quad (3.5)$$

The above equation gives the capacitance between two conductors. For the purpose of transmission line modeling, the capacitance is defined between the conductor and neutral. The line-to-line capacitance can be equivalently considered as two equal capacitances in series. The voltage across the lines divides equally between the capacitances such that the neutral point n is at the ground potential. The capacitance of each line to neutral is then given by $C_n = 2C_{12} = 2\pi\varepsilon_0 / \ln(D/r)$ F/m.

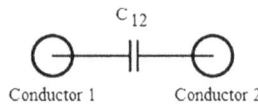

FIGURE 3.11 Single-phase line.

3.3.2 Capacitance of a Three-Phase Line with Asymmetrical Spacing

Consider a three-phase line as shown in Figure 3.12 and assuming that lines are fully transposed, we get for the first section $V_{ab} = (1/2\pi\varepsilon_0)\left[q_{a1}\ln(D_{12}/r) + q_{b1}\ln(r/D_{12}) + q_{c1}\ln(D_{23}/D_{31})\right]$ V, for

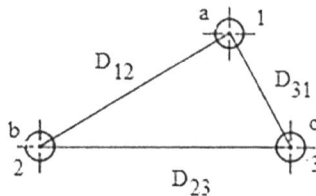

FIGURE 3.12 Three-phase line.

the second section $V_{ab} = (1/2\pi\varepsilon_0)\left[q_{a2}\ln(D_{23}/r) + q_{b2}\ln(r/D_{23}) + q_{c2}\ln(D_{31}/D_{12})\right]$ V and for the third section $V_{ab} = (1/2\pi\varepsilon_0)\left[q_{a1}\ln(D_{31}/r) + q_{b1}\ln(r/D_{31}) + q_{c1}\ln(D_{12}/D_{23})\right]$ V.

Assume $q_{a1} = q_{a2} = q_{a3} = q_a$, and similarly, $q_{b1} = q_{b2} = q_{b3} = q_b$ and $q_{c1} = q_{c2} = q_{c3} = q_c$. This assumption of equal charges/unit length in the three section of the transposition cycle requires, on the other hand, three different values of V_{ab} designated as V_{ab1}, V_{ab2} and V_{ab3} and $V_{ab\,(avg)} = (1/3)(V_{ab1} + V_{ab2} + V_{ab3})$, i.e.,

$$V_{ab} = \frac{1}{6\pi\varepsilon_0}\left(q_a\ln\frac{D_{12}D_{23}D_{31}}{r^3} + q_b\ln\frac{r^3}{D_{12}D_{23}D_{31}} + q_c\ln\frac{D_{12}D_{23}D_{31}}{D_{12}D_{23}D_{31}}\right) = \frac{1}{2\pi\varepsilon_0}\left(q_a\ln\frac{\text{GMD}}{r} + q_b\ln\frac{r}{\text{GMD}}\right)\text{V}$$

(3.6)

Similarly, the voltage V_{ac} is given as $V_{ac} = (1/2\pi\varepsilon_0)\left[q_a\ln(\text{GMD}/r) + q_c\ln(r/\text{GMD})\right]$ V.

$$V_{ab} + V_{ac} = \frac{1}{2\pi\varepsilon_0}\left[2q_a\ln\frac{\text{GMD}}{r} + (q_b + q_c)\ln\frac{r}{\text{GMD}}\right] = \frac{1}{2\pi\varepsilon_0}\left[2q_a\ln\frac{\text{GMD}}{r} - q_a\ln\frac{r}{\text{GMD}}\right]$$

$$= \frac{3}{2\pi\varepsilon_0}q_a\ln\frac{\text{GMD}}{r}\text{V}$$

(3.7)

For a set of balanced three-phase voltages $V_{ab} = V_{an}\angle 0° - V_{an}\angle -120°$ and $V_{ac} = V_{an}\angle 0° - V_{an}\angle -240°$. Therefore, we can write $V_{ab} + V_{ac} = 2V_{an}\angle 0° - V_{an}\angle -120° - V_{an}\angle -240° = 3V_{an}\angle 0°$.

So, $V_{an} = \left[1/2\pi\varepsilon_0\right]q_a\ln(\text{GMD}/r)$ and therefore the capacitance to neutral is given by $C_n = 2\pi\varepsilon_0/\left[\ln(\text{GMD}/r)\right]$ F/m.

For equilateral spacing, $C_n = (q_a/V_{an}) = 2\pi\varepsilon_0/\ln(D/r)$ F/m, where $D_{12} = D_{23} = D_{31} = D$.

3.3.3 Capacitance of a Three-Phase Double Circuit Line with Asymmetrical Spacing

Consider a three-phase double circuit line as shown in Figure 3.13 and assuming that lines are fully transposed, we get

$$V_{ab1} = \left(\frac{1}{2\pi\varepsilon_0}\right)\left[q_a\left(\ln\frac{D}{r} + \ln\frac{m}{n}\right) + q_b\left(\ln\frac{r}{D} + \ln\frac{h}{m}\right)\right] + q_c\left(\ln\frac{D}{2D} + \ln\frac{m}{h}\right)\text{V}$$

(3.8)

$$V_{ab2} = \left(\frac{1}{2\pi\varepsilon_0}\right)\left[q_a\left(\ln\frac{D}{r} + \ln\frac{m}{h}\right) + q_b\left(\ln\frac{r}{D} + \ln\frac{h}{m}\right) + q_c\left(\ln\frac{2D}{D} + \ln\frac{h}{m}\right)\right]\text{V}$$

(3.9)

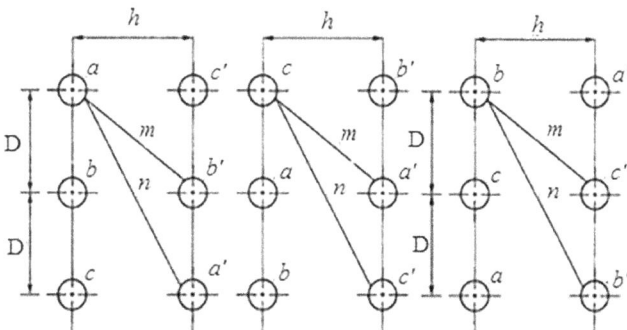

FIGURE 3.13 Double circuit three-phase line.

$$V_{ab3} = \left(\frac{1}{2\pi\varepsilon_0}\right)\left[q_a\left(\ln\frac{2D}{r}+\ln\frac{h}{n}\right)+q_b\left(\ln\frac{r}{2D}+\ln\frac{n}{h}\right)+q_c\left(\ln\frac{D}{D}+\ln\frac{m}{m}\right)\right]\text{V} \qquad (3.10)$$

$$V_{ab} = \frac{1}{3}\left[V_{ab1}+V_{ab2}+V_{ab3}\right] = \left(\frac{1}{6\pi\varepsilon_0}\right)\left[q_a\left(\ln\frac{2D^3}{r^3}+\ln\frac{m^2h}{n^2h}\right)+q_b\left(\ln\frac{r^3}{2D^3}+\ln\frac{n^2h}{m^2h}\right)+q_c\left(\ln\frac{2D^3}{2D^3}+\ln\frac{m^2h}{n^2h}\right)\right.$$

$$= \left(\frac{1}{6\pi\varepsilon_0}\right)\left[q_a\left(\ln\frac{2D^3m^2}{r^3n^2}\right)+q_b\left(\ln\frac{r^3n^2}{2D^3m^2}\right)\right]\text{V} \qquad (3.11)$$

Similarly, $V_{ac} = (1/6\pi\varepsilon_0)\left[q_a\ln\left(2D^3m^2/r^3n^2\right)+q_c\left(\ln r^3n^2/2D^3m^2\right)\right]$ V.

Therefore, $\qquad V_{ab}+V_{ac} = 3V_{an} = (1/6\pi\varepsilon_0)\left[2q_a\ln\left(2D^3m^2/r^3n^2\right)+(q_b+q_c)\left(\ln r^3n^2/2D^3m^2\right)\right] =$
$(1/2\pi\varepsilon_0)\left[q_a\ln 2D^3m^2/r^3n^2\right]$ V.

$$C_{an} = 6\pi\varepsilon_0/\ln\left(2D^3m^2/r^3n^2\right) = 2\pi\varepsilon_0/\ln\left(2^{1/3}Dm^{2/3}/rn^{2/3}\right) = 2\pi\varepsilon_0/\ln\left[2^{1/3}(D/r)(m/n)^{2/3}\right]\text{F/m} \qquad (3.12)$$

3.3.4 CAPACITANCE OF A SINGLE-PHASE LINE TAKING EARTH INTO CONSIDERATION

Consider a single-phase overhead line having two conductors a and b and assume conductors a' and b' as image conductors of conductors a and b, respectively, as shown in Figure 3.14. Let the height of conductors be h meters above the earth and charge of $+q$ coulombs per meter length and $-q$ coulombs per meter length on the conductors A and B, respectively.

$$V_{ab} = \frac{1}{2\pi\varepsilon_0}\left(q_a\ln\frac{D}{r}+q_b\ln\frac{r}{D}+q_a'\ln\frac{\sqrt{D^2+4h^2}}{2h}+q_b'\ln\frac{2h}{\sqrt{D^2+4h^2}}\right)$$

$$= \frac{1}{2\pi\varepsilon_0}\left(q_a\ln\frac{D}{r}+q_b\ln\frac{r}{D}-q_a\ln\frac{\sqrt{D^2+4h^2}}{2h}-q_b\ln\frac{2h}{\sqrt{D^2+4h^2}}\right)$$

$$= \frac{1}{2\pi\varepsilon_0}\left(q_a\ln\frac{2hD}{r\sqrt{D^2+4h^2}}+q_b\ln\frac{r\sqrt{D^2+4h^2}}{2hD}\right)$$

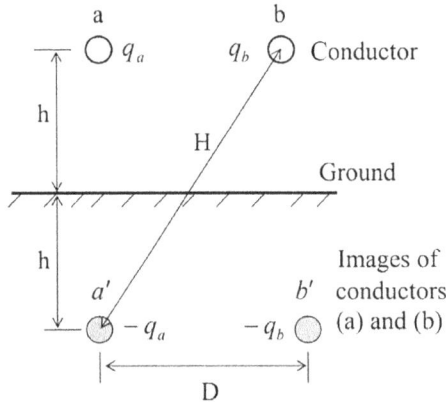

FIGURE 3.14 Single-phase line with images.

$$= \frac{1}{2\pi\varepsilon_0}\left(q_a \ln\frac{2hD}{r\sqrt{D^2+4h^2}} - q_a \ln\frac{r\sqrt{D^2+4h^2}}{2hD}\right) = \frac{1}{2\pi\varepsilon_0}\left(q_a \ln\frac{4h^2D^2}{r^2\left(D^2+4h^2\right)}\right)$$

$$= \frac{q_a}{2\pi\varepsilon_0}\ln\frac{D^2}{r^2\left(1+D^2/4h^2\right)} = \frac{q_a}{\pi\varepsilon_0}\ln\frac{D}{r\left(1+D^2/4h^2\right)^{1/2}}\; \text{V} \tag{3.13}$$

$$C_{ab} = \pi\varepsilon_0\,/\ln D \,/\left[r\left(1+D^2/4h^2\right)^{1/2}\right]\text{F/m} \text{ and } C_n = 2\pi\varepsilon_0\,/\ln D \,/\left[r\left(1+D^2/4h^2\right)^{1/2}\right]\text{F/m}$$

Consider a three-phase overhead line having three conductors as shown in Figure 3.15. With conductor in position 1, B in position 2 and C in position 3, as shown in the figure, and taking earth into consideration, we have

$$V_{ab} = \frac{1}{2\pi\varepsilon_0}\left[q_a\left(\ln\frac{D_{12}}{r}-\ln\frac{h_{12}}{h_1}\right)+q_b\left(\ln\frac{r}{D_{12}}-\ln\frac{h_2}{h_{12}}\right)+q_c\left(\ln\frac{D_{23}}{D_{31}}-\ln\frac{h_{23}}{h_{31}}\right)\right]\text{V} \tag{3.14}$$

Similarly, equations for V_{ab} can be written for the second and third sections of the transposition cycle. If the fairly accurate assumption of constant charge per unit length of the conductor throughout the transmission cycle is made, the average value of three sections of the cycle is given by,

$$V_{ab} = \frac{1}{2\pi\varepsilon_0}\left[q_a\left(\ln\frac{D_{eq}}{r}-\ln\frac{\left(h_{12}h_{23}h_{31}\right)^{1/3}}{\left(h_1h_2h_3\right)^{1/3}}\right)+q_b\left(\ln\frac{r}{D_{eq}}-\ln\frac{\left(h_{12}h_{23}h_{31}\right)^{1/3}}{\left(h_1h_2h_3\right)^{1/3}}\right)\right]\text{V}; \text{ where } D_{eq}=\sqrt[3]{d_{12}d_{23}d_{31}}$$

$$\tag{3.15}$$

The equation for the average value of the voltage V_{ac} can be determined in the same way. Since, $V_{ab}+V_{ac}=3V_{an}$ and $q_a+q_b+q_c=0$, so the capacitance to neutral of the three-phase line taking earth into consideration will be $C_n = 2\pi\varepsilon_0\left[1/\left[\ln\left(D_{eq}/r\right)-\ln\sqrt[3]{\left(h_{12}h_{23}h_{31}/h_1h_2h_3\right)}\right]\right]$ F/m.

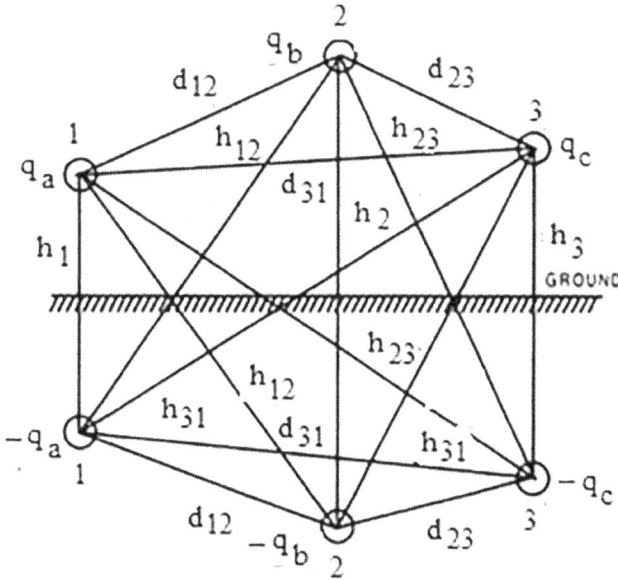

FIGURE 3.15 Three-phase line with images.

3.3.5   Effect of Earth on the Calculation of Capacitance

Earth affects the calculation of capacitance of three-phase lines as its presence alters the electric field lines. Usually, the height of the conductors placed on transmission towers is much larger than the spacing between the conductors. Therefore, the effect of earth can be neglected for capacitance calculations, especially when balanced steady-state operation of the power system is considered. However, for unbalanced operation when the sum of the three line currents is not zero, the effect of earth needs to be considered. However, the presence of earth increases the capacitance by 0.15%–2% due to the term $\ln \sqrt[3]{\left(h_{12}h_{23}h_{31} / h_1 h_2 h_3\right)}$ in the denominator.

3.3.6   Skin Effect and Proximity Effect

Skin Effect – The phenomena arising due to unequal distribution of alternating electric current over the entire cross section of the conductor is referred as the skin effect. Skin effect results in increased effective resistance but reduced effective internal reactance of the conductor. Skin depth is given by $\delta = \sqrt{\left(1 / \pi f \mu \sigma\right)}$, where f is the frequency, μ, the permeability and σ ($=1 / \rho$), the conductivity of the conductor. Skin effect attenuates the higher frequency components of a signal more than the lower frequency components. The conductor resistance affects the attenuation of traveling waves due to lightning and switching operations, as well as radio-frequency energy generated by corona.

Proximity Effect – The AC current in two round, parallel wires is not distributed uniformly around the conductors. The magnetic fields from each wire affect the current flow in the other, resulting in a non-uniform current distribution, which in turn increases the apparent resistance of the conductors. In parallel round wires, we call this the proximity effect. If each carries a current in the same direction, the halves of the conductors in close proximity are cut by more magnetic flux than the remote halves. Consequently, the current distribution is not even throughout the cross section, and a greater proportion is being carried by the remote halves. If the currents are in opposite directions, the halves close proximity will carry the greater density of current.

Both skin effect and proximity effects depend upon the conductor size, frequency, distance between conductors and permeability of conductor material. The greater the spacing between phases of a transmission line, greater will be the inductance of the line. However, increase in distance between phases of the transmission line reduces the capacitance. Greater the radius of the conductor of the transmission, lesser the inductance but greater will be the capacitance.

3.4   LEARNING OUTCOME

An electric transmission line primarily has three circuit parameters which are resistance that causes power loss in a transmission line; inductance that arises due to change in flux linkage that contributes to induced voltage that causes loss in a transmission line; and capacitance that is the result of the potential difference between the conductors, the effect of which is predominant of increased voltage. The frequency of the AC voltage produces effect on the conductor resistance due to the non-uniform distribution of the current. The performance of transmission line depends on these parameters of the line. Once evaluated, the line parameters are used to model the transmission line and to perform design calculations.

In this chapter, readers have been able to learn all the above aspects.

4 Corona and Sag

4.1 CORONA

When alternating potential between two parallel conductor increases beyond a certain limit, a point is reached where a pale violet glow appears on the conductor surface. This is called corona and is accompanied by hissing sound as corona discharges generate positive and negative ions which are alternately attracted and repelled by the periodic reversal of polarity of the ac excitation. Their movement gives rise to sound-pressure waves at frequencies of twice the power frequency and its multiples, in addition to the broadband spectrum which is the result of random motions of the ions. The luminous envelope surrounding the conductors is composed of air which has become ionized and has become conducting due to effect of high electrostatic stress. Corona is accompanied by power loss and there is flow of non-sinusoidal current due to corona that causes non-sinusoidal voltage drop. The effect of corona is equivalent to increase in effective diameter of the conductor. If the spacing between conductors is small enough, the corona may bridge the conductors and cause flash over. So, the spacing of the conductors is maintained large enough compared to the line diameter for reducing the coronal effect. For, this bundled conductor is chosen.

If a high-voltage DC is applied between two conductors, there is a difference in appearance of the positive and negative conductors. The positive conductors have a uniform glow, while the glow around the negative conductors is more spotty.

The disruptive critical voltage is the minimum phase to neutral voltage at which the electric field intensity at the surface of the conductor exceeds the critical value and generates corona. The disruptive critical value of air is 30 KV/cm at 25°C temperature and pressure of 76 cm of mercury. In case of two parallel conductors, the maximum potential gradient occurs at the surface of the conductors, and if V is the phase-neutral potential, then potential gradient at the conductor surface is given by: $g_{\max} = V / r \left[\log_e(d / r)\right]$ V/cm.

In order that corona is formed, the value of g must be made equal to the breakdown strength of air. If V_d is the phase-neutral potential required, under such condition, $g_0 = V_d / r \left[\log_e(d / r)\right]$ V/cm. g_0 is the breakdown strength of air at 76 cm of mercury and 25°C and is equal to 30 kV/cm (max) or 21.2 kV/cm (rms). This gives $V_d = g_0 r \left[\log_e(d / r)\right]$ V. The expression for disruptive voltage V_d is under standard conditions, i.e., at 76 cm of Hg and 25°C. However, if these conditions vary, the air density also changes, thus altering the value of g_0. The value of g_0 is proportional to density of air and is therefore proportional to the barometric pressure and inversely proportional to the absolute temperature. Taking the relative density at θ_0°C and 76 cm of mercury as 1.0, the relative density δ at a barometric pressure p cm and a temperature θ is $(p / 76)\left[(273 + \theta_0) / (273 + \theta)\right]$. Taking $\theta_0 = 20$°C, δ will be equal to $\left[(3.86 \ p) / (273 + \theta)\right]$. Taking air density factor δ into consideration, disruptive critical voltage $V_d = g_0 \delta r \ \log_e(d / r)$.

Under standard conditions, the value of $\delta = 1$. Correction must also be made for the surface condition of the conductor. This is accounted for by multiplying the above expression by irregularity or surface factor m_0. Then, $V_d = m_0 g_0 \delta r \ \log_e(d / r)$. $m_0 = 1.0$ for smooth conductor and 0.93–0.98 for rough conductor and 0.87–0.8 for stranded conductors.

Visual corona does not occur when electric intensity becomes equal to critical value but starts at higher value of phase to neutral voltage. Visual critical voltage V_C is the minimum phase-neutral voltage at which corona glow appears all along the line conductors and can be given by $V_C = g_0 \ \delta \ m_v \ r \left[1 + \left(0.03 / \sqrt{\delta r}\right)\right]\log_e(d / r)$. m_v is the roughness factor having a value of 1·0 for polished conductors and 0.72–0.82 for rough conductors.

DOI: 10.1201/9781003231240-4

Corona power loss according to Peek's formula is: $P_C = (244/\delta)(f+25)\left[\sqrt{(r/d)}\right](V-V_d)^2 10^{-5}$ KW/km/phase under fair weather condition, where f is the system frequency, r is the radius of conductor in meter, d is the spacing between the conductor in meter, V = Phase voltage = $V_L/\sqrt{3}$ and V_d is the Disruptive critical voltage. This formula is valid when corona loss is predominant and the ratio of V and V_d is greater than 1.8. The limitations of applying this formula are as follows: (i) frequency should be within 25–120 Hz; (ii) r must be greater 0.25 cm and (iii) applicable to fair weather and humidity should not be very low. When the ratio of V and V_d is less than 1.8 the coronal power loss $P_C = 21 \times 10^{-6} \times f \ V^2 F \times \left[1/(\log_{10} d/r)^2\right]$, where F is a factor that depends on the condition V/V_d.

Coronal power loss is dependent on the following: (i) frequency (directly proportional); (ii) system voltage (greater the potential difference, greater is the electric field and greater is corona loss). In the region near the disruptive critical voltage, the rate of increase of corona power loss with increase of system voltage is small but when the difference is large, corona loss increases at a very fast rate with the increase of system voltage. (iii) Conductivity of air (higher conductivity leads to greater corona loss); (iv) rain and dust (increases corona); (v) conductor radius (greater the diameter of conductor, lesser the surface field intensity); (vi) density of air (higher the density, lesser the corona loss); and (vii) conductor surface.

There are in general two types of corona discharge from transmission-line conductors: (i) Pulse less or Glow Corona; and (ii) Pulse Type or Streamer Corona. Both these give rise to energy loss, but only the pulse-type of corona gives interference to radio broadcast in the range of 0.5–1.6 MHz.

Advantages of corona are as follows: (i) due to corona formation, the surrounding air of the conductor becomes conducting and hence virtual diameter of conductor increases. This increase in diameter reduces the electrostatic stress between the conductors. (ii) Corona reduces the effects of transients by surge. The disadvantages of corona are as follows: (i) corona is accompanied by loss of energy; and (ii) ozone formation may cause corrosion of conductor due to chemical reaction.

Using bundled conductors, Corona can be reduced. When power is being transferred at very high voltages using a single conductor, the voltage gradient around it is high and there are high chances of formation of corona, especially in bad weather conditions. However, using several conductors instead of one in close proximity forming bundled conductors leads to reduction in voltage gradient and hence possibility of corona formation.

4.2   SAG

The difference in levels between point of support and the lowest point on the conductor is called sag. If sag is too low, the wire gets subjected to an extra tension. There are some other factors present like temperature, wind pressure, etc., due to which conductor gets subjected to stress. If sufficient sag is not kept, then conductor cannot sustain such stress and there is possibility of mechanical failure. The minimum sag and less tension cannot be satisfied simultaneously. If the sag is large, and the line becomes heavily loaded, then the sag will further increase and breach the safety clearances. Similarly, if the sag is low, then when the line contracts in the winter, low sag will indicate a high tension, and as a result of this contraction, the line may snap.

The exact shape of the line under sag is that of catenary (Latin chain). Except for lines with very very long span and large sag, it is sufficiently accurate to assume that the shape of line is that of parabola ($y = ax^2$). At any point of the conductor, tension acts horizontally. It has two components; one acts horizontally and other acts vertically. Sag at any span = sag at basic span $\times \left[(\text{Span length})^2/(\text{Basic span})^2\right]$.

4.2.1   SAG WHEN THE SUPPORTS ARE AT EQUAL LEVELS AND SPAN IS SMALL

When the supports of the conductor are at equal levels as shown in Figure 4.1, span is small and the shape of the conductor sag is a parabola:

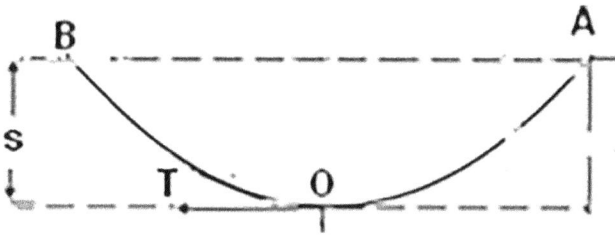

FIGURE 4.1 Supports at same level.

l = length of span, i.e., horizontal distance between supports in m. w = weight/length in kg/m. T = tension in kg. s = sag at mid span in m. $y = s$, when $x = l / 2$; since the equation of parabola is considered as $y = ax^2$, $s = a(l / 2)^2$ or $a = 4s / l^2$ and this gives $y = 4s(x / l)^2$. Consider the equilibrium of half-line OA and assuming that conductor is almost horizontal and taking moment about A, $Ts = (wl / 2)(l / 4)$, or $s = wl^2 / 8T$. So, $a = 4s / l^2 = w / 2T$ and $y = wx^2 / 2T$.

4.2.2 Sag When the Supports Are at Unequal Levels and Span Is Small

When the supports are at different levels as shown in Figure 4.2, span is small and the shape of the conductor sag is a parabola:

Let a section of conductor be suspended between different levels B and C. The curve BOCA is the complete parabola with A and B at the same level. The actual lone BOC is a part of the complete parabola. Let l be the actual span (horizontal distance between B and C), l_c be the span of the complete parabola and x_1, y_1 be the coordinates of point C. $x_1 = l - (l_c / 2)$. The equation of parabola $y = ax^2$ is valid for the curve BOCA as well as BOC. $s = wl_c^2 / 8T$. Substituting the coordinates of point C, $y_1 = s - h = 4s(x_1)^2 / l_c^2$, or $\left[(x_1)^2 / (s - h)\right] = l_c^2 / 4s$. This gives $l_c = l + 2TH / wl$. Since h, T, l and w are known, l_c can be found out.

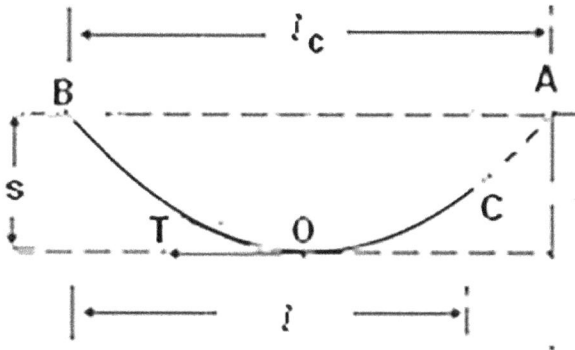

FIGURE 4.2 Supports at different level.

4.2.3 Sag When the Supports Are at Unequal Levels and Span Is Large

When the supports are at different levels, span is large and the shape of the conductor sag is forming catenary:

Let a conductor pulled between supports A and B is as shown in Figure 4.3 with a certain tension and O is the lowest point of the conductor. This is at a distance x from support B and $(2l - x)$ from support A where $2l$ is the length of the span. The two supports A and B are at different levels, and the difference being h. Take any point P on the conductor.

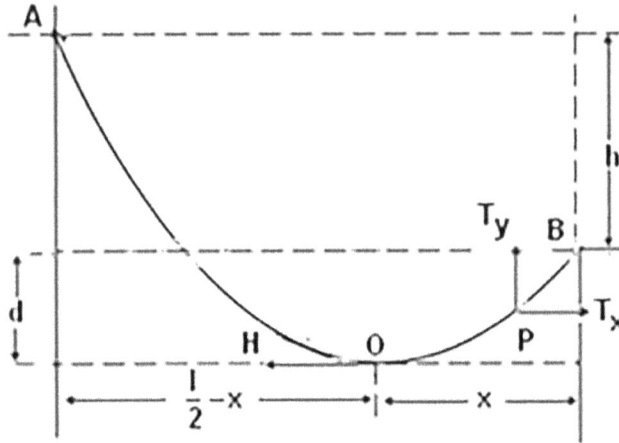

FIGURE 4.3 Supports at different level.

Let H = Horizontal component of tension, T = Conductor tension at point P, Tx = Horizontal component of tension at P, Ty = Vertical component of tension at P, s = length of the conductor OP, ds = small length of the conductor, dx = Horizontal component of ds, dy = Vertical component of ds, d = sag at the lowest point with reference to support B, $d + h$ = sag at the lowest point with reference to point A, w = weight of conductor per unit length and ws = weight of conductor of length OP, which acts downwards and at the center of length OP.

$$H = Tx \text{ and } Ty = ws; \tan\varphi = dy \,/\, dx = Ty \,/\, Tx = ws \,/\, H; \tag{4.1}$$

$$ds = \sqrt{(dx)^2 + (dy)^2}, \text{ or } ds \,/\, dx = \sqrt{1 + (dy \,/\, dx)^2} = \sqrt{1 + (ws \,/\, H)^2} \tag{4.2}$$

$$\int dx = \int 1 / \left[1 + (ws \,/\, H)^2\right]: \text{ or } x + C_1 = (H \,/\, w)\sinh^{-1}(ws \,/\, H), \text{ where } C_1 \text{ is a constant} \tag{4.3}$$

At the condition $x = 0$, $s = 0$ and $C_1 = 0$ and the equation will be $x = (H \,/\, w)\sinh^{-1}(ws \,/\, H)$, or $s = (H \,/\, w)\sinh(wx \,/\, H)$

$$dy \,/\, dx = ws \,/\, H = \sinh (wx \,/\, H); \text{ integrating } y = (H \,/\, w)\cosh(wx \,/\, H) + C_2$$

At O, $y = 0$, when $x = 0$; $0 = (H \,/\, w)\cosh 0 + C_2 = (H \,/\, w) + C_2$, i.e. $C_2 = -(H \,/\, w)$ and $y = H \,/\, w\left[\cosh(wx \,/\, H) - 1\right]$, which is the equation of catenary.

$$y = \frac{H}{w}\left[\cosh\left(\frac{wx}{H}\right) - 1\right] \cong \frac{H}{w}\left[1 + \frac{w^2 x^2}{2H^2} + \ldots + \ldots - 1\right] = \frac{w^2 x^2}{2H}; \text{ For } x = \frac{l}{2}; d = \frac{w^2 l^2}{8H} \tag{4.4}$$

$$T = (T_x)^2 + (T_y)^2 = H^2 + (ws)^2 = H^2 + H^2 \sinh^2(wx \,/\, H)$$

$$= H^2 \cosh^2(wx \,/\, H); \text{Thus } T = H\cosh(wx \,/\, H) \tag{4.5}$$

Since $\cosh(wx \,/\, H) = \left[1 + (w^2 x^2 \,/\, 2H^2) + \ldots\right]; T \cong H$. If the supports are at the same level, and the span is l, or half span is l, the above expressions reduce to $s = (H \,/\, w)\sinh(wl \,/\, 2H)$; $d = (H \,/\, w)\cosh\left[(wl \,/\, 2H) - 1\right] \cong wl^2 \,/\, 8H$; $T = H\cosh(wl \,/\, 2H) \cong H$

4.2.4 Effect of Ice Coating

Ice coating affects the conductor by (i) increasing the weight per meter and (ii) increasing the projected surface area subject to wind pressure. A conductor with diameter d is coated with ice of thickness t as shown in Figure 4.4. Hence overall diameter of the coated conductor is $D = d + 2t$. The area of the coated conductor $= (\pi/4)d^2$. The area of ice covering $= (\pi/4)(D^2 - d^2)$. If D and d are in meter and taking a length of 1 m, weight of ice covering $=$
$$w_i = 915\left[(\pi/4)(D^2 - d^2)\right] = 915\left[(\pi/4)(D^2 - d^2)\right] = 915\left[(\pi/4)((d+2t)^2 - d^2)\right] = 915\left[\pi t(d+t)\right]$$
kg/m; since density of ice $= 915\,\text{kg/m}^3$. This weight acts vertically downward.

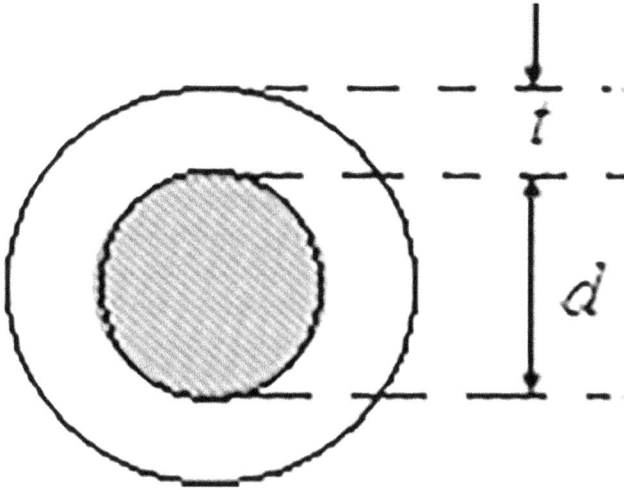

FIGURE 4.4 Ice coated conductor.

4.2.5 Effect of Wind

The wind flows horizontally as shown in Figure 4.5 and hence the wind pressure on the conductor is considered to be acting perpendicularly to the conductor and it affects only the transverse loading on the conductor. Thus, the force due to wind acts at right angles to the projected surface of the conductor.

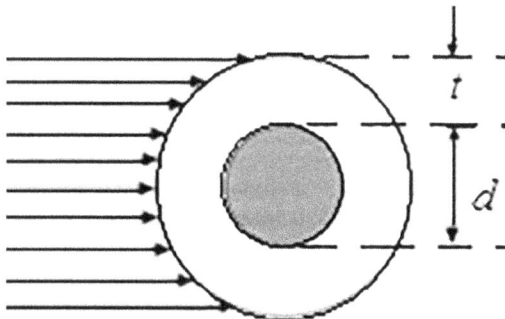

FIGURE 4.5 Effect of wind.

The wind force w_w = wind force per unit length in kg/m = wind pressure per unit area × projected surface area per unit length = wind pressure $\times \left[(d + 2t) \times 1 \right] = p(d + 2t)$. Hence the conductor is acted upon by the two additional for one vertically downwards w_i and other in horizontal direction w_w.

The total force acting upon the conductor is the vector sum of the horizontal and vertical forces. If w is the weight of the conductor, then the total vertical force is $w + w_i$ and the horizontal force is w_w. The total weight $w_t = \sqrt{(w + w_i)^2 + w_w^2}$. This is shown in Figure 4.6. The conductor adjusts itself in a plane of angle θ, where $\tan\theta = w_w / (w + w_i)$. Hence, it is necessary to calculate the sag direction of an angle θ measured with respect to vertical direction. Hence the sag is called the slant sag. This is calculated by considering total weight w_t. Slant sag is $S = w_t l^2 / 8T$. The vertical sag is $S \cos\theta$.

4.2.6 Effect of Sag in Transmission Line

It Reduces Excessive Tension: While erecting an overhead line, it is very important that the conductors are under safe tension. If the conductors are too much stressed between the supports (towers, utility poles), then the stress on the conductors may reach to an unsafe level and the conductor may break due to excessive pressure (i.e., tension). To permit safe tension in the conductors, the conductors (i.e., the transmission lines) are not fully stretched but are allowed having a dip or sag. If there is too much sag in a transmission line, it will increase the amount of conductor used, increasing the cost more than is necessary. When a transmission line sag excessively, it is liable of causing power failure.

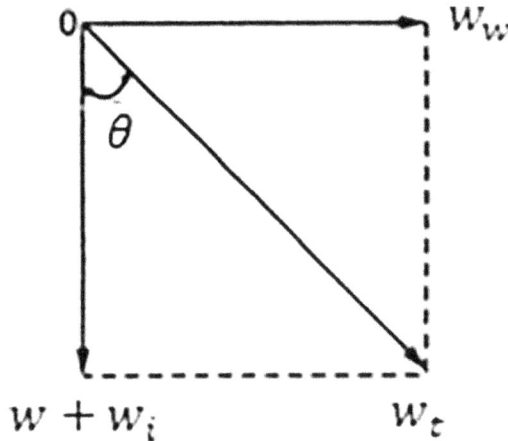

FIGURE 4.6   Force diagram.

4.2.7 Stringing Chart

The process of pulling conductors between supports of the towers is known as stringing of the lines. Stretching conductors as tightly as possible with safety helps in reducing the height of the towers and thus cost. Stringing necessary clearance should be kept between the lowest point of the conductor and the ground. The stringing is done by either measuring the tension correctly while pulling at the supports or by measuring the sag accurately. While stringing, weather conditions may vary, particularly temperature, and it is necessary to know "temperature – tension" and "sag – temperature curve for the conductors. For normal or average span used for transmission line work for particular voltage range, temperature sag and temperature tension relations are worked out for different temperatures and worst loading conditions to which line will be subjected and the chart is called *Stringing Chart*. As the temperature increases tension between two points where the conductors are fixed decreases; however, increasing temperature causes increase of sag.

4.2.8 SAG TEMPLATE

It is a very important tool with the help of which the position of towers on the profile is decided so that they conform to minimum clearances, as per I.E. Rules required to be maintained between the line conductor to ground, telephone lines, buildings, streets, navigable canals, power lines, etc. A sag template is specific for the particular line voltage, the conductor used and the applicable design conditions. The template consists of a set of parabolic curves that consists of: (i) cold or up-lift curve showing sag of conductor at minimum temperature (−2.5°C and still wind); (ii) hot or maximum sag curve showing maximum sag of conductor at maximum temperature and still wind; (iii) ground clearance curve; and (iv) tower footing curve.

4.2.9 VIBRATIONS OF CONDUCTOR

The conductors are supported by the string insulators at each tower. Under widely varying atmospheric conditions like strong wind velocities; the conductors can vibrate in the vertical plane. Such vibrations can have different frequencies, amplitudes and modes. There are three types of vibrations:

- **Aeoline Vibrations** – They have the frequency range of 8–40 Hz with amplitudes varying between 2 and 5 cm. Thus, these vibrations are high-frequency low-amplitude vibrations. This is caused by aerodynamic forces generated as the wind blows across the conductor. The wind velocities of about 2–40 km/hour can generate these vibrations. The frequency of aeoline vibration is given by $(2V_p / d)10^3$, where V_p is the velocity of the wind perpendicular to the conductor in kmph and d is the diameter of the conductor in meter. This vibration is harmful from suspension point of view and travel along the length of the conductor. This kind of vibration is the major cause of fatigue failure of conductor strands or of items associated with the support, use and protection of the conductor. Aeoline vibrations can be restricted by using stock bridge type dampers. The shape of damper is known as a "dog bone".
- **Galloping or Dancing of Conductors** – It means low-frequency high-amplitude vibrations usually occur during modest wind in areas of relatively flat terrain under freezing rain and icing of conductors. The flat terrain provides winds that are uniform and of a low turbulence. When a conductor is iced, it presents an unsymmetrical cross section with the windward side having less ice accumulation than the leeward side of the conductor. When the wind blows across such a surface, there is an aerodynamic lift as well as a drag force due to the direct pressure of the wind. This gives rise to torsional modes of oscillation of complete span of conductors. It causes the longitudinal movement of conductors through the conductor clamps. The frequency of galloping is about 0.25–1.5 Hz with the amplitudes of about 6 m. Since galloping is caused by the wind blowing over conductors which are not circular, the conductors are designed to be circular to prevent gallop-type vibrations. There are three modes of galloping: (i) one loop, (ii) two loop and (iii) three loop. Galloping cannot be restricted. Hoop Spacers, inter-phase Spacers, Rotating Clamp Spacers and Pendulum de-tuners take the form of weights applied at different locations on the span. Wind Dampers are used to reduce galloping.

- **Wake-Induced Oscillation** – It occurs generally in a bundle conductor, and is similar to aeolian vibration and galloping occurring principally in flat terrain with winds of steady velocity and low turbulence. The frequency of the oscillation does not exceed 3 Hz but may be of sufficient amplitude to cause clashing of adjacent sub-conductors, which are separated by about 50 cm. Wind speeds for causing wake-induced oscillation are normally in the range of 25–65 km/hour. Fatigue failure to spacers is one of the chief causes for damage to insulators and conductors. Wake-induced oscillation, also called "flutter instability", is caused when one conductor on the windward side aerodynamically shields the leeward conductor. The conductor spacing to diameter ratio in the bundle is also critical. If the spacing bundle spacing is less than $15d$, where d being the conductor diameter, a tendency to oscillate is created while for $B/d > 15$ the bundle is found to be more stable. As mentioned earlier, the electrical design, such as calculating the surface voltage gradient on the conductors, will depend upon these mechanical considerations.

The various factors affecting the vibrations are as follows: (i) atmospheric conditions like icing, rain and high wind; (ii) span of conductor; (iii) tension of conductor; (iv) conductor configuration; (v) types of clamp used and (vi) height of towers, etc. When the wind energy imparted to the conductor achieves a balance with the energy dissipated by the vibrating conductor, steady amplitudes for the oscillations occur. A damping device helps to achieve this balance at smaller amplitudes of aeolian vibrations than an undamped conductor. The damper controls the intensity of the wave-like properties of travel of the oscillation and provides an equivalent heavy mass which absorbs the energy in the wave.

4.3   LEARNING OUTCOME

Corona is produced due to ionization of air, surrounding the conductor of overhead transmission lines, which causes non-sinusoidal voltage drop accompanied by power loss and is undesirable and uneconomical.

Sag and tension, the vertical bending in shape of the conductor due to its weight, are related to laying of transmission line to accommodate the change in length due to differential temperature and heat generation. In addition, ice- and wind-loading functionally increase line weight and, along with thermal loading on the conductor, cause additional line sag or stretching. The amount of sag is a function of the conductor size, material and span distance. To satisfy the National Electric Safety Code, minimum mid-span ground clearance is required for a particular voltage level for transmission and distribution of electricity. A Sag Template is a very important tool with the help of which the position of towers on the profile is decided so that they conform to the limitations of vertical and wind loads on any particular tower and minimum clearances.

In this chapter, readers have been able to learn all the above aspects.

5 Cable

Cables are laid underground through cable trenches for transmission/distribution of power. Unlike in overhead lines, air does not form part of the insulation, and the conductor must be completely insulated. Thus, cables are much more costly than overhead lines. Also, unlike for overhead lines where tappings can easily be given, cables must be connected through cable boxes which provide the necessary insulation for the joint. Cables are of two types in terms of voltage level: (i) Low-voltage power and control cables pertain to electrical cables that typically have a voltage grade of 0.6/1 kV or below and (ii) Medium/High voltage cables pertain to cables used for electric power transmission at medium and high voltage (usually from 1 to 33 kV are medium voltage cables and those over 50 kV are high-voltage cables).

Different parts of cable are as follows: (i) Conductor – Multilayer stranded copper (usually annealed – is the process of gradually heating and cooling the conductor material to make it more malleable and less brittle and surface coated for giving insulation) or aluminum. Stranding gives flexibility/bendability. Aluminum conductors have a cross-sectional area approximately 1.6 times larger than copper, but is half the weight of the copper. The reduction of weight per unit length of aluminum cable makes the cable handling during laying easier. (ii) Conductor Screen – A semiconducting tape consists of either lapped copper tape or metallic foil which acts as an interface between the conductor and the insulation. Conductor screen "smoothes" out the surface irregularities of the conductor and maintains a uniform electric field. Screen also minimizes electrostatic stresses (for MV/HV power cables) and allows for operation at a higher temperature. (iii) Filler and Binding Tape – The three cores of a three core cables are laid up with polymer compound or non-hygroscopic fillers like polypropylene fillers and a binder tape with an overlap to provide a circular shape of the cable. (iv) Insulation – usually Polyvinyl Chloride (PVC), Ethylene Propylene Rubber (EPR) or cross-linked polyethylene (XLPE) type. Insulation must have high resistivity and dielectric strength, low thermal coefficient, low viscosity at working temperature, high tensile strength, low absorption of moisture, high mechanical strength, tough and flexible. (v) Insulation Screen – A semiconducting material that has a similar function as the conductor screen (i.e., control of the electric field for MV/HV power cables). (vi) Conductor Sheath – usually made of lead alloy, to keep electromagnetic radiation in, and also provides a path for fault and leakage currents (sheaths are earthed at one cable end). Lead sheaths provide better earth fault capacity. (vii) Armor – For mechanical protection of the conductor bundle. Steel wire armor or braid is typically used. Tinning or galvanizing is used for rust prevention. (viii) Outer Sheath – First outer lair of the cable for overall mechanical, weather, chemical and electrical protection.

Parts of a single core cable are shown in Figure 5.1.

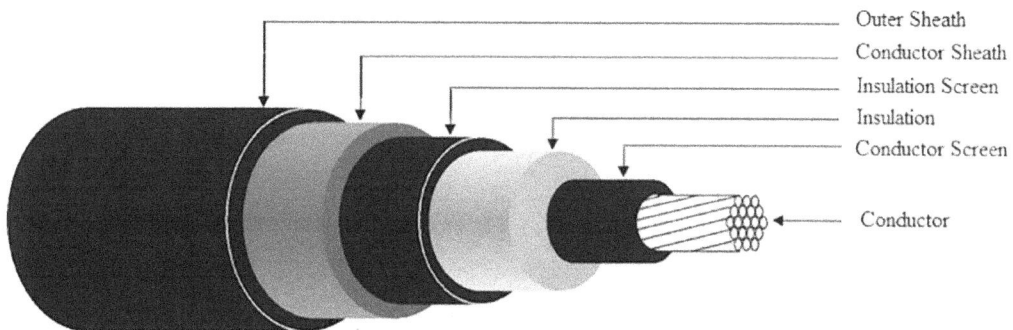

FIGURE 5.1 Parts of a single-core cable.

DOI: 10.1201/9781003231240-5

The cables are classified in terms of the insulation provided in the cable. Initially, Paper Insulated Lead Covered (PILC) cable was used. Paper has little insulation value alone; however, when impregnated with a high grade of mineral oil, it provides satisfactory insulation for high-voltage cables. The oil has a high dielectric strength, and tends to prevent breakdown of the paper insulation. However, for use over the years, the insulation property deteriorates. PVC insulated cables are used as they are cheap and durable but there is a fire hazard for PVC insulated cable as the same when burned produces toxic, black smoke. This has caused restriction in use of PVC cable. Presently XLPE cables are extensively used along with EPR cable [11] as XLPE is superior to EPR cable. The basic material XLPE is low density polyethylene.

The advantage of the cable system is that there is less likelihood of breakdown due to lightning or storms. Thus, the cable system provides better reliability for continuity of supply. In the case of large cities to protect the look, cable distribution is often done. Single-core cables are usually used for short interconnecting circuits or where very high load and fault levels are required.

Inside buildings or industrial plants cables are generally installed on racks in air fixed to the walls or supported from ceiling. Racks may be ladder or perforated type cable trays. The vertical distance between the two racks are kept to be minimum 0.3 m and the clearance between the first cable and the wall (if racks are mounted on wall) is kept 25 mm. The width of the rack should not exceed 0.75 m in order to facilitate installation of cables. While laying of cable sharp bends are avoided. Larger bending radii are kept within 15/20 D, where D is the diameter of the cable. Underground cabling system for power supply is done in trench, and after that, the cable is covered with soil. The trench is covered with bricks or concrete slab and then again filled with soil. The depth of the trench is generally kept within 1.25–1.50 m from ground level and the width of the trench is kept 1.0 m.

5.1 INSULATION RESISTANCE OF CABLE

Let "d" is the conductor diameter in centimeter, and the internal sheath diameter is "D" cm as shown in Figure 5.2.

If ρ is the specific resistance of the insulating material in Ω cm, taking a small section at a distance "x" from the center of the cable, and of thickness "dx", the cross section for l cm length of the cable is $l \times 2\pi x$ cm^2 and the insulation resistance "dR" can be expressed as $dR = \rho[dx\,/\,2\pi x l]$. The insulation R therefore will be

$$R = \int_{d/2}^{D/2} \rho[dx\,/\,2\pi x l] = \rho\,/\,2\pi l \log_e(D\,/\,d) \tag{5.1}$$

The insulation resistance of cable is inversely proportional to the length of the cable. This is because as the length increases, the leakage current from the conductor through the insulation increases and this reduces the insulation resistance of the cable.

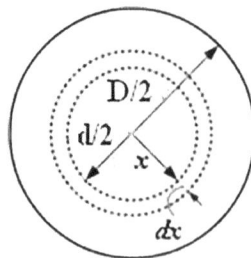

FIGURE 5.2   Single-core cable.

5.2 CAPACITANCE OF A SINGLE-CORE CABLE

A single-core cable can be considered to be equivalent of two long coaxial cylinders. The conductor or core of the cable is the inner cylinder and the outer cylinder is represented by lead sheath which is at ground potential. Let "d" is the conductor diameter in centimeter, and the internal sheath diameter is "D" cm. Let the charge per meter axial length of the cable be Q coulombs and $\varepsilon = \varepsilon_0 \varepsilon_r$ be the relative permittivity of the insulation material between core and lead sheath. Consider a cylinder of radius x meters and axial length l m. The surface area of this cylinder will be $2\pi x$ m^2. Electric flux density at any point on the cylinder is $D_x = (Q/2\pi x) C/m^2$ and the electric intensity at that point $E_x = Q/2\pi \varepsilon x = Q/2\pi \varepsilon_0 \varepsilon_r x$ V/m. The potential difference

$$V = \int_{d/2}^{D/2} E_x \, dx = \int_{d/2}^{D/2} (Q/2\pi\varepsilon_0\varepsilon_r x) \, dx = (Q/2\pi\varepsilon_0\varepsilon_r)\log_e D/d \tag{5.2}$$

Capacitance of the cable is $C = Q/V = 2\pi\varepsilon_0\varepsilon_r / \log_e D/d$.

5.3 DIELECTRIC STRESS OF A SINGLE CORE CABLE

Electric intensity at that point $E_x = Q/2\pi\varepsilon x = Q/2\pi\varepsilon_0\varepsilon_r x$ V/m. By definition, electric field intensity is equal to Potential Gradient [g]. Hence $g = Q/2\pi\varepsilon_0\varepsilon_r x$ and since $V = Q/2\pi\varepsilon_0\varepsilon_r \log_e D/d$ and so substituting Q it can be written as $g = V/(x\log_e D/d)$. The electrostatic stress is maximum when $x = d/2$ and minimum when $x = D/2$; thus $g_{max} = 2V/(d\log_e D/d)$ and $g_{min} = 2V/(D\log_e D/d)$. The ration of g_{max} and g_{min} is equal to D/d.

5.4 MOST ECONOMICAL CONDUCTOR SIZE IN A CABLE

For given values of V and D, the most economical conductor diameter will be one for which g_{max} has a minimum value. The value of g_{max} will be minimum when

$$\frac{d}{dd}\left[\frac{2V}{d\log_e D/d}\right] = 0 \text{ or, } \log_e D/d + d\left(\frac{d}{D}\right)\left(\frac{-D}{d^2}\right)$$

$$= 0 \text{ or, } \log_e D/d = 1 \text{ or, } \frac{D}{d} = e = 2.718 \tag{5.3}$$

Most economical conductor diameter is $d = D/2.718$.

5.5 GRADING OF CABLES

The process of achieving uniform electrostatic stress in the dielectric of cables is known as grading of cables. This can be done in two ways:

Capacitance Grading – by using two or more number of insulating materials of different dielectric strength, one with higher value of ε_r (dielectric constant), being placed near to the core of the cable. If three grading is done as shown in Figure 5.3, then $g_{1max} = Q/\pi\varepsilon_0\varepsilon d$; $g_{2max} = Q/\pi\varepsilon_0\varepsilon_1 d_1$ and $g_{3max} = Q/\pi\varepsilon_0\varepsilon_1 d_2$. If $g_{1max} = g_{2max} = g_{3max} = g_{max}$, then $1/\varepsilon_1 d = 1/\varepsilon_2 d_1 = 1/\varepsilon_3 d_2$.

$$V_1 = \int_{d/2}^{d_1/2} g_1 \, dx = \frac{Q}{2\pi\varepsilon_0\varepsilon_1}\log_e d_1/d = \frac{g_{max}}{2}d\log_e d_1/d;$$

$$V_2 = \frac{g_{max}}{2}d_1 \log_e \frac{d_2}{d_1} \text{ and } V_3 = \frac{g_{max}}{2}d_2 \log_e \frac{D}{d_2} \tag{5.4}$$

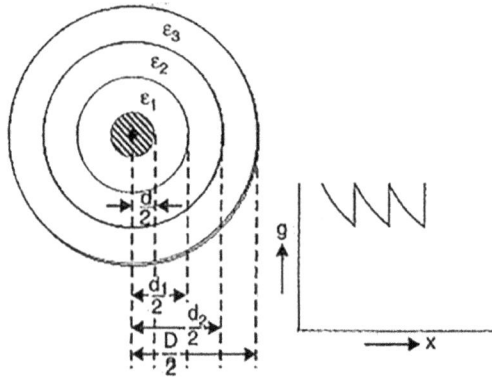

FIGURE 5.3 Capacitance grading.

Thus, $V = V_1 + V_2 + V_3 = (g_{max} / 2)\left[d \log_e (d_1 / d) + d_1 \log_e (d_2 / d_1) + d_2 (\log_e D / d_2)\right]$.

Intersheath Grading – It is a method of ensuring that the voltage gradient across the insulation of a cable does not become so steep as to cause the failure of the insulation by proving metallic intersheath between successive layers of the same dielectric materials. This is shown in Figure 5.4.

The electrostatic stress at any point x from the center of the cable to inner sheath is $g_1 = (V - V_1) / (x \log_e d_1 / d)$ and the maximum stress g_{1max} will occur at the conductor surface where $x = d / 2$, i.e., $g_{1max} = (V - V_1) / \left[(d / 2) \log_e (d_1 / d)\right]$.

Similarly potential difference between intersheath 1 and intersheath 2 is $V_1 - V_2$ and $g_{2max} = (V_1 - V_2) / \left[(d_1 / 2) \log_e (d_2 / d_1)\right]$. The potential difference between intersheath 2 and outermost sheath V_2 is only as potential of intersheath is maintained at V_2 with respect to earth. Thus, $g_{3max} = V_2 / \left[((d_2 / 2)) \log_e (D / d_2)\right]$. Choosing proper values of V_1 and V_2, g_{1max}, g_{2max}, etc. can be made equal and hence uniform distribution of stress can be obtained. Let $d_1 / d = d_2 / d_1 = D / d_2 = \alpha$ and $g_{1max} = g_{2max} = g_{3max}$

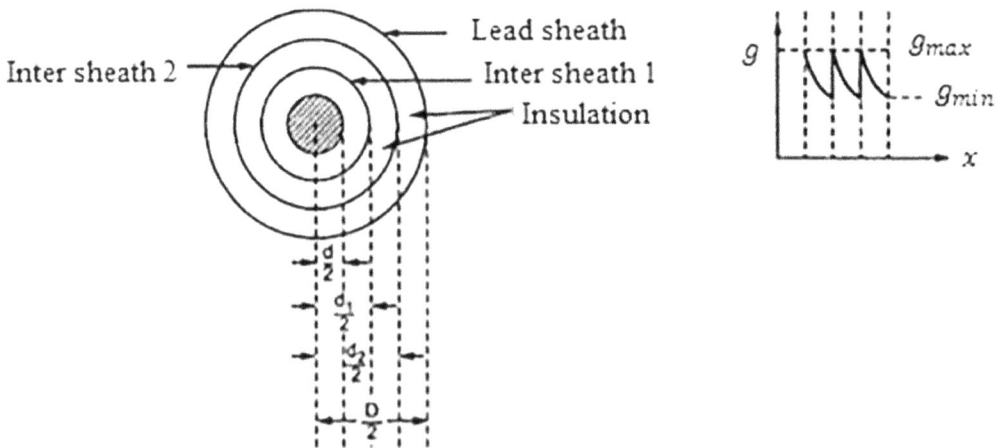

FIGURE 5.4 Intersheath grading.

$$\frac{V-V_1}{\left[\left(\dfrac{d}{2}\right)\log_e \alpha\right]} = \frac{V_1-V_2}{\left[\left(\dfrac{d_1}{2}\right)\log_e \alpha\right]} = \frac{V_2}{\left[\left(\dfrac{d_2}{2}\right)\log_e \alpha\right]}, \text{ i.e. } \frac{V_2}{d_2}$$

$$= \frac{V_1-V_2}{d_1} = \frac{V-V_1}{d} \text{ or } V_2 = \alpha(V_1 - V_2) \text{ or } V_2 = \frac{\alpha}{1+\alpha}V_1 \tag{5.5}$$

From $(V_1 - V_2)/d_1 = (V - V_1)/d$ we get $V_1 - V_2 = \alpha(V - V_1)$ or $(1+\alpha)V_1 = \alpha V + V_2$ or, $V_1 = [\alpha(1+\alpha)]/(\alpha^2 + \alpha + 1) V$

$$V_1 = \frac{\left[1+\dfrac{1}{\alpha}\right]}{\left[1+\dfrac{1}{\alpha}+\dfrac{1}{\alpha^2}\right]} \text{ Volt and } V_2 = \frac{\alpha}{1+\alpha}V_1 = \frac{1}{\left[1+\dfrac{1}{\alpha}+\dfrac{1}{\alpha^2}\right]} \text{ Volt} \tag{5.6}$$

$$g_{1\max} = \frac{V-V_1}{\left[\left(\dfrac{d}{2}\right)\log_e \dfrac{d_1}{d}\right]} = \frac{V}{\alpha^2\left[1+\dfrac{1}{\alpha}+\dfrac{1}{\alpha^2}\right]\dfrac{d}{2}\log_e \alpha} = \frac{V}{\left[1+\alpha+\alpha^2\right]\dfrac{d}{2}\log_e \alpha} \tag{5.7}$$

$$\frac{d_1}{d} \times \frac{d_2}{d_1} \times \frac{D}{d_2} = \alpha^3 \text{ i.e. } \frac{D}{d} = \alpha^3 \text{ or, } \log\frac{D}{d} = 3\log \alpha \text{ or, } \log\alpha$$

$$= \frac{1}{3}\log\frac{D}{d} \text{ and } g_{1\max} = \frac{V}{\dfrac{1}{3}\left[1+\alpha+\alpha^2\right]\dfrac{d}{2}\log\dfrac{D}{d}} \tag{5.8}$$

If intersheath is used, the electrostatic stress is reduced.

5.6 POWER FACTOR OF A SINGLE-CORE CABLE

Power loss in the cable can occur by (i) the conductor current passing through the resistance of the conductor – conductor loss (known as copper loss), (ii) dielectric losses caused by the voltage across the insulation and (iii) sheath losses caused by the induced currents in the sheath, and intersheath losses caused by circulating currents in loops formed between sheaths of different phases. The dielectric loss is voltage dependent, while the rest of the losses are current dependent.

For a perfect dielectric, the power factor is zero. Since the cable is not a perfect dielectric, the power factor is not zero. The current leads the voltage by an angle of less than 90° as shown in Figure 5.5, and hence there is a power loss. If V is the applied voltage, C is the capacitance of cable, θ is the phase angle between voltage current called the power factor angle and δ is the loss angle of the dielectric.

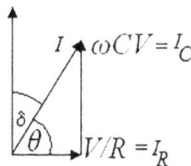

FIGURE 5.5 Power factor angle.

Charging current $I_C = \omega CV$, which increases as the working voltage and the cable length increases (hence, it is not recommended to transmit power through a long-distance cable. This also creates overvoltage problem) and the leakage current $I_R = V / R$. Power loss (dielectric loss) is $VI\cos\theta = VI\sin\delta = V(I_C / \cos\delta)\sin\delta = V\omega CV\tan\delta = \omega CV^2\tan\delta \cong \omega CV^2\delta$. Since δ is very small, so $\tan\delta$ can be taken equal to δ. So, power factor of the cable is $\cos\theta = \sin\delta$. Thus, dielectric loss is proportional to square of the voltage, and hence for low voltage, although the loss can be neglected but for higher voltage, it is of great importance. Due to increased dielectric loss, the temperature for the dielectric increases, which may damage the insulation.

5.7 CAPACITANCE OF A THREE CORE CABLE

The capacitance of a three-core belted cable arises due to (i) phase conductors that are nearer to each other and to the earthed sheath and (ii) the phase conductors that are separated by a dielectric of permittivity much greater than that of air as in case of an overhead line. Figure 5.6 shows the system of capacitances in a three-core belted cable used for a three-phase system. Since the potential difference exists between pairs of conductors, which gives the core to core capacitance C_C and the potential difference between each conductor and earthed sheath gives core to earth capacitance C_S. The three C_C are delta connected and three C_S are star connected, the sheath forming the star point as shown in Figure 5.6. It can be assumed that all C_C are equal, likewise all C_S are equal. The three delta connected C_C can be converted into equivalent star connected capacitance, $C_{eq} = 3C_C$. The star point may be assumed to be zero potential and if the sheath is also at zero potential, the capacitance of each core to neutral is $C_N = C_{eq} + C_S = 3C_C + C_S$.

Permissible current of a conductor of cable or current capacity is $I = \sqrt{t / [nR(S + G)]}$ where n = number of core per phase, R = conductor resistance at 60°C in Ω/m, S = thermal resistance of the cable between the cores and the sheath in Ω/m, G = thermal resistance between sheath and earth and t = temperature rise between the conductor and soil.

The thermal resistance of the cable depends on the geometry of the cable and the thermal resistivity K of the insulating material of the cable. "Thermal Ohm" is defined as the thermal resistance which requires a temperature difference of 1°C to produce heat flow of 1 joule/s. The thermal resistance T_r of a single core cable is given by $T_r = (K / 2\pi)\ln(D / d)$, where D is the sheath diameter, d is the diameter of the conductor and K varies between 750 at low voltages to 550 at high voltages.

Power loss in AC cables will be nI^2R_θ (W), where n = number of cores, R_θ = ohmic a.c. resistance of the conductor at θ°C. The AC resistance of XLPE cables is greater than the equivalent PILC cable, meaning that conductor losses in XLPE cables will be higher than for equivalent PILC cables.

Decreasing the cross-sectional area of any conductor increases its resistance (and hence its losses) and this is one argument against using the ability of XLPE to run at a higher temperature as a reason to reduce the conductor cross-sectional area. Dielectric losses are proportional

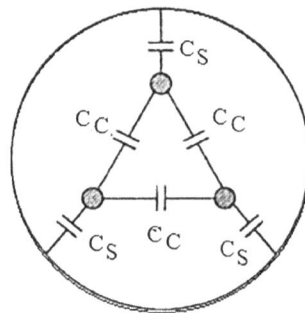

FIGURE 5.6 Capacitance of cable.

to the capacitance, frequency, phase voltage and power factor and are given by $D = n\omega CV^2 \tan\delta$ 10^{-6} (W/km), where C =capacitance to neutral (in μF/km), V =phase to neutral voltage (in V) and $\tan\delta$ =dielectric power factor.

5.8 LEARNING OUTCOME

The power distribution through cable ensures safety and aesthetics and provides the uninterruptable power supply including other advantage which includes reduced risks of fault due to external factors like rain, wind and adverse climatic conditions. More and more cable distribution is being used for integrated township, in densely populated urban areas, in factories and even to supply power from the overhead posts to the consumer premises.

In this chapter, readers have been able to learn all the above aspects.

6 Characteristics and Performance of Transmission Line

6.1 TRANSMISSION LINES

The main function of the transmission line is to transmit power from sending end over required distance economically maintaining voltage, current, power and power factor at the receiving end. The receiving end requirements are dictated by load conditions. The difference between sending end voltage and receiving end voltage is defined as regulation of transmission line. Lower the voltage regulation, better it is. *Regulation* of transmission line is the ratio of difference of sending and receiving end voltage to receiving end voltage, i.e., $(V_S - V_R)/V_R$. *Efficiency* of transmission line is the ratio of difference of sending and receiving end power to receiving end power, i.e., $(P_S - P_R)/P_R$.

Transmission line constants, i.e. resistance, inductance and capacitance, may be taken as lumped or distributed. If the values are considered per unit length of the line and the variation of these are considered point to point, they are distributed parameter. If the constants are assumed for the whole length of the lines and then considered at suitable places for solving the circuit, they can be taken as lumped constants. If V_s is the sending end voltage, I_s is the sending end current, V_R is the receiving end voltage, I_R is the receiving end current, then $V_s = AV_R + BI_R$; $I_s = CV_R + DI_R$, where A–D are called transmission parameters, through which sending end voltage and current is expressed by receiving end voltage and current. A is the ratio of sending end voltage and the open circuit receiving end voltage and is dimensionless. B is the ratio of sending end voltage and short circuit receiving end current and unit is Ω. C is the ratio of sending end current and the open circuit receiving end voltage and unit is Ω. D is the ratio of sending end current and short circuit receiving end current and is dimensionless.

6.2 GENERAL RELATIONS FOR THE ANALYSIS OF A TRANSMISSION LINE

Transmission lines are classified into three types: short transmission lines, medium transmission lines and ling transmission lines.

6.2.1 SHORT TRANSMISSION LINE APPROXIMATION

The lines which have length between 80 and 100 km are termed as short transmission lines. The series parameters are represented through lumped parameters and the effect of shunt capacitance is neglected. The simple equivalent circuit of a short line is shown in Figure 6.1.

The sending end voltage and current for this approximation are given by $V_s = V_R + ZI_R$; $I_s = I_R$. The phasor diagram for the short line is shown in Figure 6.2 for the lagging power factor condition.

Thus, ABCD parameters are given by $\begin{bmatrix} 1 & Z \\ 0 & 1 \end{bmatrix}$

From this figure it can be written as $V_s^2 = (V_R \cos\theta_r + IR)^2 + (V_R \sin\theta_r + IX)^2 = V_R^2 + I^2 (R^2 + X^2) + 2V_R I(R\cos\theta_r + X\sin\theta_r)$. So, $V_s = V_R \left[1 + (2IR/V_R)\cos\theta_r + (2IX/V_R)\sin\theta_r + I^2(R^2 + X^2)/V_R^2\right]^{1/2}$. Expanding binomially and retaining first-order terms, V_s can be expressed as $V_s = V_R \left[1 + (IR/V_R)\cos\theta_r + (IX/V_R)\sin\theta_r\right] \cong V_R + I(R\cos\theta_r + X\sin\theta_r)$. Since, *Regulation* is $(V_S - V_R)/V_R$, so, it will be $(IR\cos\theta_r + IR\sin\theta_r)/V_R$.

DOI: 10.1201/9781003231240-6

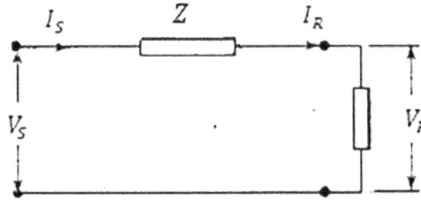

FIGURE 6.1 Representation of short transmission line.

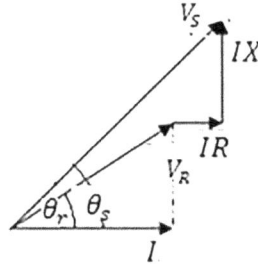

FIGURE 6.2 Phasor representation.

6.2.2 MEDIUM TRANSMISSION LINE APPROXIMATION

For medium transmission lines, line length is considered between 100 and 250 km. The parameters are lumped parameter representations but the effect of shunt capacitance is not neglected. It can be placed totally in load end or half of the value of the capacitance of the line at each of the two ends of the line or can be placed at the center of the line.

End Condenser Method – In this representation the lumped series impedance is placed in the middle while the shunt admittance is placed at the load end as shown in Figure 6.3.

$$\text{ABCD parameters are given by } \begin{bmatrix} 1+YZ & Z \\ Y & 1 \end{bmatrix} \text{ where } Z = R + jX$$

Nominal-π Representation – In this representation the lumped series impedance is placed in the middle while the shunt admittance is divided into two equal parts and placed at the two ends as shown in Figure 6.4.

$I_s = I_1 + I_2 = I_1 + I_3 + I_R = I_R + V_R Y / 2 + V_S Y / 2$ and $V_s = V_R + (I_R + V_R Y / 2)Z = V_R (1 + YZ / 2) + I_R Z$. So, $I_s = I_R + V_R Y / 2 + (Y / 2) \left[V_R (1 + YZ / 2) + I_R Z \right] = V_R Y (1 + YZ / 4) + I_R (1 + YZ / 2)$.

$$\text{Thus, ABCD parameters are given by } \begin{bmatrix} 1+YZ/2 & Z \\ Y(1+YZ/4) & 1+YZ/2 \end{bmatrix}$$

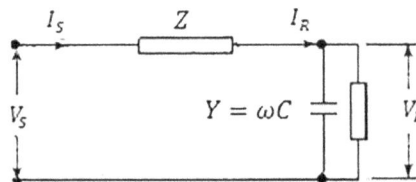

FIGURE 6.3 End condenser representation.

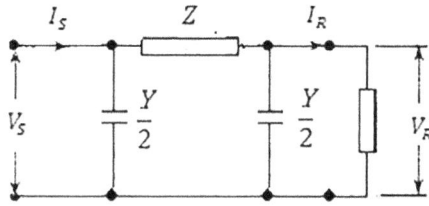

FIGURE 6.4 π representation.

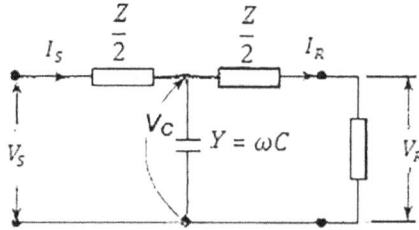

FIGURE 6.5   T representation.

Nominal-T Representation – In this representation the lumped shunt admittance is placed in the middle and the series impedance is divided into two equal parts and placed at the two ends as shown in Figure 6.5.

$V_C = V_R + I_R Z / 2$ and $I_s = I_R + V_M Y = I_R + V_R Y + I_R ZY / 2$. This gives $V_s = V_C + I_S Z / 2$. Substituting for V_C and I_S gives $V_s = V_R + I_R Z / 2 + (Z / 2)[I_R(1 + YZ / 2) + YV_R] = V_R (1 + YZ / 2) + I_R Z(1 + YZ / 4)$

Thus, ABCD parameters are given by $\begin{bmatrix} 1 + YZ / 2 & Z(1 + YZ / 4) \\ Y & 1 + YZ / 2 \end{bmatrix}$.

Nominal-π Representation and Nominal-T Representation with the above constants are not equivalent to each other.

6.2.3   LONG TRANSMISSION LINE

For long transmission lines, the line length is above 250 km. The parameters are distributed parameter representations and the effect of shunt capacitance is not neglected. The single-line diagram of a long transmission line is shown in Figure 6.6.

The length of the line is l. Let us consider a small strip dx that is at a distance x from the receiving end. The voltage and current at the end of the strip are V_x and I_x, respectively, and the beginning of the strip are $V_x + dV_x$ and $I_x + dV_x$, respectively. The voltage drop across the strip is then dV_x. Since

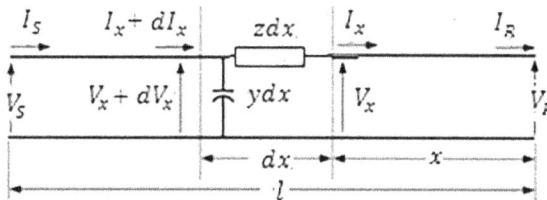

FIGURE 6.6   Long line representation.

the length of the strip is dx, the series impedance and shunt admittance are zdx and ydx. It is to be noted here that the total impedance and admittance of the line are $Z = zl$ and $Y = yl$. z = series impedance per unit length per phase, y = shunt admittance per unit length, per phase to neutral

$$dV_x = I_x zdx, \text{ or } \frac{dV_x}{dx} = zI_x; dI_x = V_x ydx, \text{ or } \frac{dI_x}{dx} = yV_x, \text{ so } \frac{d^2V_x}{dx^2} = \frac{dI_x}{dx} z = yzV_x \quad (6.1)$$

This is a linear differential equation whose general solution is $V_x = A_1 e^{\gamma x} + A_2 e^{-\gamma x}$ where $\gamma = \sqrt{yz}$ = propagation constant

$$\frac{dV_x}{dx} = A_1 \gamma e^{\gamma x} - A_2 \gamma e^{-\gamma x} = zI_x; \text{ so } I_x = \frac{A_1}{Z_c} e^{\gamma x} - \frac{A_2}{Z_c} e^{-\gamma x}, \text{ where}$$

$$Z_c = \text{characteristic impedance} = \sqrt{z/y} \quad (6.2)$$

The constants A_1 and A_2 can be evaluated by using the end conditions, i.e., when $x = 0$, $V_x = V_R$ and $I_x = I_R$; then, $V_R = A_1 + A_2$ and $I_R = (A_1 - A_2)/Z_C$. So, $A_1 = (V_R + Z_C I_R)/2$ and $A_2 = (V_R - Z_C I_R)/2$. Thus, $V_x = (V_R + Z_C I_R)e^{\gamma x}/2 + (V_R - Z_C I_R)e^{-\gamma x}/2$ and $I_x = (V_R/Z_C + I_R)e^{\gamma x}/2 + (V_R/Z_C - I_R)e^{-\gamma x}/2$.

Hence, $V_x = V_R(e^{\gamma x} + e^{-\gamma x})/2 + Z_C I_R(e^{\gamma x} - e^{-\gamma x})/2 = V_R \cosh\gamma x + Z_C I_R \sinh\gamma x$; $I_x = V_R(e^{\gamma x} - e^{-\gamma x})/2Z_C + I_R(e^{\gamma x} + e^{-\gamma x})/2 = I_R \cosh\gamma x + (V_R/Z_C)\sinh\gamma x$. Also note that for $x = l$ we have $V_x = V_s$ and $I_x = I_s$.

Therefore, replacing x by l and substituting the values of A_1 and A_2, we get, $V_s = V_R \cosh\gamma l + Z_C I_R \sinh\gamma l$ and $I_s = V_R \sinh\gamma l/Z_C + I_R \cosh\gamma l$

The ABCD parameters of the long transmission line can then be written as

$$\begin{bmatrix} \cosh\gamma l & Z_C \sinh\gamma l \\ \sinh\gamma l/Z_C & \cosh\gamma l \end{bmatrix}$$

If $\gamma = \alpha + j\beta$, where α = attenuation constant and β = phase constant, $\cosh\gamma l = 1 + (\gamma^2 l^2/\angle 2) + (\gamma^4 l^4/\angle 4) + \cdots\cdots = 1 + (YZ/2)$ and $\sinh\gamma l = \gamma l + (\gamma^3 l^3/\angle 3) + (\gamma^5 l^5/\angle 5) + \cdots\cdots = \sqrt{YZ}(1 + YZ/6)$, then $A = D = 1 + YZ/2$; and $B = Z(1 + YZ/6)$ and $C = Y(1 + YZ/6)$.

Equivalent-π Representation of a Long Line – The equivalent-π of a long transmission line is shown in Figure 6.7.

In this the series impedance is denoted by Z' while the shunt admittance is denoted by Y' and the ABCD parameters are defined as $A = D = Y'Z'/2 + 1$ and $B = Z'$ and $C = Y'(Y'Z'/4 + 1)$. $Z' = Z_C \sinh\gamma l = (\sqrt{z/y})\sinh\gamma l = Z(\sinh\gamma l)/\gamma l$.

Since, $1 + Y'Z'/2 = \cosh\gamma l = 1 + Y'Z_C \sinh\gamma l/2$

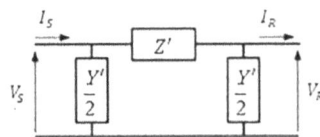

FIGURE 6.7 Equivalent-π representation.

$$\text{so } \frac{1}{2}Y' = \frac{1}{Z_C}\frac{(\cosh\gamma l - 1)}{\sinh\gamma l} = \frac{1}{Z_C}\tanh\left(\frac{\gamma l}{2}\right) = \left(\sqrt{y/z}\right)\tanh\left(\frac{\gamma l}{2}\right)$$

$$= \frac{yl}{2}\left[\frac{\tanh\left(\dfrac{\gamma l}{2}\right)}{\gamma l/2}\right] = \frac{Y}{2}\left[\frac{\tanh\left(\dfrac{\gamma l}{2}\right)}{\gamma l/2}\right] \tag{6.3}$$

For the T-equivalent of a long transmission line $Z'/2 = Z/2\left[\tanh(\gamma l/2)/\gamma l/2\right]$ and $Y' = Y$ $(\sinh\gamma l)/\gamma l$.

6.3 POWER FLOW THROUGH A TRANSMISSION LINE

$$V_s = AV_R + BI_R; \text{which gives } I_R = (1/B)V_s - (A/B)V_R \text{ and } I_s$$

$$= CV_R + DI_R; \text{which gives } I_s = (D/B)V_s - (1/B)V_R$$

Let the transmission line constants be written as $A = |A|\angle\alpha$, $B = |B|\angle\beta$, $A = |D|\angle\alpha$. Then $I_R = |1/B||V_s|\angle(\delta - \beta) - |A/B||V_R|\angle(\alpha - \beta)$ and $I_s = |D/B||V_s|\angle(\alpha + \delta - \beta) - |1/B||V_R|\angle(-\beta)$.

The receiving and sending end power $S_R = P_R + jQ_R = \bar{V}_R\bar{I}_R^*$ and $S_S = P_S + jQ_S = \bar{V}_S\bar{I}_S^*$. This gives $S_R = |1/B||V_s||V_R|\angle(\beta - \delta) - |A/B||V_R|^2\angle(\beta - \alpha)$ and $S_S = |D/B||V_s|^2\angle(\beta - \alpha) - |1/B||V_s||V_R|\angle(\beta + \delta)$.

Receiving end active power is $P_R = |1/B||V_s||V_R|\cos(\beta - \delta) - |A/B||V_R|^2\cos(\beta - \alpha)$ and reactive power is $Q_R = |1/B||V_s||V_R|\sin(\beta - \delta) - |A/B||V_R|^2\sin(\beta - \alpha)$.

Sending end active power $P_S = |D/B||V_s|^2\cos(\beta - \alpha) - |1/B||V_s||V_R|\cos(\beta + \delta)$ and reactive power is $Q_S = |D/B||V_s|^2\sin(\beta - \alpha) - |1/B||V_s||V_R|\sin(\beta + \delta)$.

P_R will be maximum when $\beta = \delta$. S_R and S_S are each composed of two-phase components, one a constant phasor and the other a phasor of fixed magnitude but variable angle. The loci of S_R and S_S will be a circle and is called power circle diagram. Figure 6.8 shows the receiving end power circle diagram.

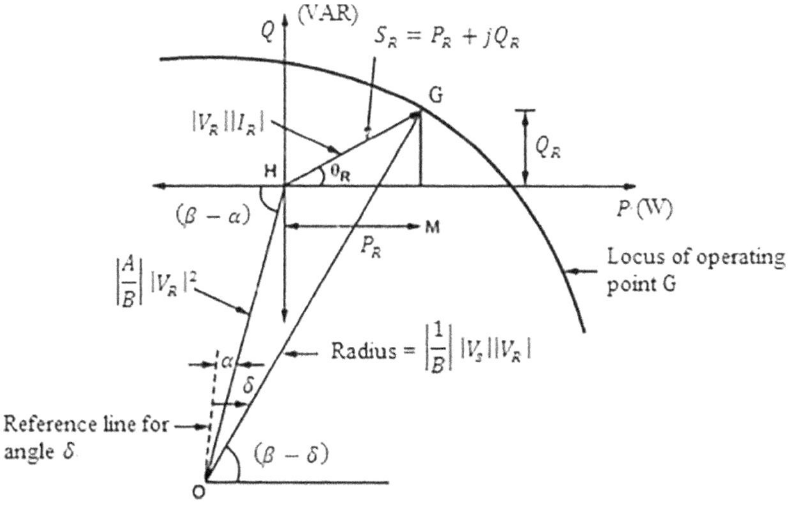

FIGURE 6.8 Receiving end power circle diagram.

The receiving end maximum active and reactive power will be $P_{R\,(\text{Max})} = |1/B||V_s||V_R| - |A/B||V_R|^2 \cos(\beta - \alpha)$ and $Q_{R\,(\text{Max})} = -|A/B||V_R|^2 \sin(\beta - \alpha)$. Thus the load must draw this much leading Mega Volt Amperes Reactive (MVAR) in order to receive the maximum real power. Normally the resistance of a transmission line is small compared to its reactance, so that $\theta = \tan^{-1} X/R = 90°$; where $Z = R + jX$. The receiving end power $P_{R\,(\text{Max})} = |1/B||V_s||V_R| - |A/B||V_R|^2 \cos(\beta - \alpha)$ and $Q_{R\,(\text{Max})} = -|A/B||V_R|^2 \sin(\beta - \alpha)$. The receiving end active and reactive power can be approximated as $P_R = |1/B||V_s||V_R| - |A/B||V_R|^2 \cos(\beta - \alpha) \approx |V_s||V_R| \sin\delta/X$ and $Q_R = -|A/B||V_R|^2 \sin(\beta - \alpha) \approx [|V_s||V_R|]\cos\delta/X - |V_R|^2/X$. Since, $\cos\delta \approx 1$ as δ is normally small $Q_R = (|V_R|/X)(|V_s| - |V_R|)$.

For $R \approx 0$, the real power transferred to the receiving end is proportional to $\sin\delta \approx \delta$, while the reactive power is proportional to the voltage drop across the line. The real power received is maximum for $\delta = 90°$ and has a value $(|V_s||V_R|)/X$. Maximum real power transferred for a given line can be increased by raising its voltage level. It is from this consideration voltage levels are progressively pushed up to transmit large amount of power.

6.4 TRAVELLING WAVE EQUATION OF A TRANSMISSION LINE

We know $= \cosh \gamma l$, $B = Z_C \sinh \gamma l$ and $C = \sinh \gamma l / Z_C$, where $\gamma = \alpha + j\beta$. Now,

$$V_x = \left(\frac{V_R + Z_C I_R}{2}\right) e^{\gamma x} + \left(\frac{V_R - Z_C I_R}{2}\right) e^{-\gamma x}$$

$$= \left|\frac{V_R + Z_C I_R}{2}\right| e^{\alpha x} e^{j(\beta x + \phi_1)} + \left|\frac{V_R - Z_C I_R}{2}\right| e^{-\alpha x} e^{-j(\beta x - \phi_2)} \tag{6.4}$$

where $\phi_1 = \angle(V_R + Z_C I_R)$ and $\phi_2 = \angle(V_R - Z_C I_R)$. The instantaneous voltage $V_x(t)$ can be written as

$$v_x(t) = Re\left[\sqrt{2}\left|\frac{V_R + Z_C I_R}{2}\right| e^{\alpha x} e^{j(\beta x + \phi_1)} + \sqrt{2}\left|\frac{V_R - Z_C I_R}{2}\right| e^{-\alpha x} e^{-j(\beta x - \phi_2)}\right] = v_{x1} + v_{x2} \tag{6.5}$$

So, the instantaneous voltage consists of two terms each of which is a function of two variables – time and distance. Thus, they represent traveling wave. So, $v_{x1} = \sqrt{2}|(V_R + Z_C I_R)/2| e^{\alpha x} \cos(\omega t + \beta x + \phi_1)$, which implies that at any instant of time t, v_{x1} is sinusoidally distributed along the distance from the receiving end with amplitude increasing exponentially (as $\alpha > 0$) with distance. This wave is traveling toward the receiving end and is the incident wave. Line losses cause its amplitude to decrease exponentially in going from the sending end to the receiving end.

This wave $v_{x2} = \sqrt{2}|(V_R - Z_C I_R)/2| e^{-\alpha x} \cos(\omega t - \beta x + \phi_2)$ is traveling toward the sending end and is the reflected wave. If the load impedance $Z_L = V_R/I_R = Z_C$, i.e., the line is terminated in its characteristic impedance, the reflected wave is zero. A line terminated in its characteristic impedance is called the infinite line. Z_C is generally called the surge impedance. $Z_C = (\sqrt{l/c})$. The term "surge impedance loading" (SIL) is often used to indicate the nominal capacity of the line. The surge impedance is the ratio of voltage and current at any point along an infinitely long line. The term SIL or natural power is a measure of power delivered by a transmission line when terminated by surge impedance and is given by SIL $= V_0^2/Z_C$, where V_0 is the rated voltage of the line. At SIL $Z_C = V/I$ and hence $V = V_R e^{-j\beta x}$ and $I = I_R e^{-j\beta x}$.

6.5 CHARACTERIZATION OF A LONG LOSSLESS LINE

For a lossless line, the line resistance is assumed to be zero. The characteristic impedance then becomes a pure real number and it is often referred to as the surge impedance. The propagation constant becomes a pure imaginary number. Defining the propagation constant as $\gamma = j\beta$ and replacing l by; $V = V_R \cos \beta x + jZ_C I_R \sin \beta x$ and $I = jV_R \sin \beta x / Z_C + I_R \cos \beta x$.

The term SIL is often used to indicate the nominal capacity of the line. The surge impedance is the ratio of voltage and current at any point along an infinitely long line. The term SIL or natural power is a measure of power delivered by a transmission line when terminated by surge impedance and is given by $SIL = V_0^2 / Z_C$ where V_0 is the rated voltage of the line. At SIL : $Z_C = V_R / I_R$ and hence $V = V_R e^{-j\beta x}$ and $I = I_R e^{-j\beta x}$. This implies that as the distance x changes, the magnitudes of the voltage and current in the above equations do not change. The voltage then has a flat profile all along the line. Also as Z_C is real, V and I are in phase with each other all throughout the line. The phase angle difference between the sending end voltage and the receiving end voltage is then $\theta = \beta_l$.

6.6   VOLTAGE AND CURRENT CHARACTERISTICS OF AN SINGLE MACHINE INFINITE BUS SYSTEM

For the analysis presented below we assume that the magnitudes of the voltages at the two ends are the same. The sending and receiving voltages are given by $V_S = |V_S| \angle \delta$ and $V_R = |V_R| \angle 0$, where δ is angle between the sources and is usually called the *load angle*.

As the total length of the line is l, we replace x by l to obtain the sending end voltage as $V_S = |V_S| \angle \delta = (|V_R| + Z_C I_R) e^{j\theta} / 2 + (|V_R| - Z_C I_R) e^{-j\theta} / 2 = |V_R| \cos \theta + jZ_C I_R \sin \theta$. This gives $I_R = (|V_S| \angle \delta - |V_R| \cos \theta) jZ_C \sin \theta$. So, $V = [|V_S| \angle \delta \sin \beta x + |V_R| \sin (\theta - \beta x)] / \sin \theta$ and $I = j[|V_S| \angle \delta \cos \beta x + |V_R| \cos (\theta - \beta)] / Z_C \sin \theta$. When the system is unloaded, the receiving end current is zero ($I_R = 0$). Therefore, we can rewrite $V_S = |V_S| \angle \delta = |V_R| \cos \theta$.

6.7   KELVIN'S LAW

It states that the most economical size of the conductor will be when the sum of the annual charge on the capital investment and the annual charge due to loss of energy in transmission would be minimum. The capital cost and cost of energy wasted in the line is based on the size of the conductor. If conductor size is big, then due to its lesser resistance, the running cost (cost of energy due to losses) will be lower while the conductor may be expensive. For smaller size conductor, its cost is less but running cost will be more as it will have more resistance and hence greater losses. The cost of energy loss is inversely proportional to the conductor cross section while the fixed charges (cost of conductor, interest and depreciation charges) are directly proportional to area of cross section of the conductor. The annual charges on the cost of conductor for distribution is $= P_1 + P_2 A$, where A is the cross-sectional area of the conductor.

The cost of energy wasted is P_3 / A. By Kelvin's law, $P_1 + P_2 A + (P_3 / A)$ should be minimum and will be when $A = \sqrt{P_3 / P_2}$. In the plot, the lowest point X on the total annual cost curve gives the most economical size of the conductor and it is the intersection two curves, i.e., annual charges on the cost of conductor for distribution, which is a straight line since the same is proportional to conductor area, and cost of energy wasted, which is a hyperbola since the same is inversely proportional to the conductor area. Graphical illustration of Kelvin's Law is shown in Figure 6.9.

Limitations of Kelvin's Law – (i) The amount of energy loss cannot be determined accurately as the annual load factor of the line which is the ratio of average load over a period and maximum load over that period or the ratio of number of KWh generated per year and Maximum demand $\times 24 \times 365$; (ii) the cost of energy losses also cannot be determined accurately; (iii) the cost of conductor and the rates of interest change frequently; (iv) if economical conductor size is selected, then voltage drop may be beyond the acceptable limits; (v) the economical conductor size may not have the desired mechanical strength; and (vi) due to problem of corona and leakage currents, the economical size of the conductor cannot be used for extra high voltages.

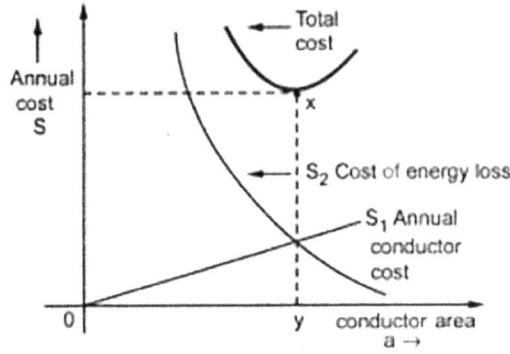

FIGURE 6.9 Graphical illustration of Kelvin's law.

Modified Kelvin's Law: the most economical size is one for which annual energy loss cost is equal to the annual cost of interest and depreciation for the part of initial investment which is proportional to the area of conductor, i.e., $A = \sqrt{P_3 / (P.r_2)}$, where r is the rate of depreciation.

6.8 LEARNING OUTCOME

A transmission line is a four-terminal passive, linear and bilateral network that can be represented by ABCD parameters. ABCD constants provide a straightforward means of writing equations in a more concise form and are very convenient in problems involving network reduction. The model used to calculate voltages, currents and power flows depends on the length of the line. The circuit parameters and voltage and current relations can be developed for "short", "medium" and "long" lines. The long transmission line equations are valid for a line of any length. The approximations for the short and medium length lines make analysis easier. These are required for performance measurement of transmission line. Circle diagrams are used because of their instructional value in showing the maximum power which can be transmitted by a line and also in showing the effect of the power factor of the load or the addition of capacitors.

In this chapter, readers have been able to learn all the above aspects.

7 Insulators for Overhead Lines

7.1 BASICS OF INSULATORS FOR OVERHEAD LINES

Electrical Insulator for overhead lines provide insulation by preventing unwanted flow of electric current to the earth from its supporting points. In transmission and distribution system, the overhead conductors are generally supported by supporting towers or poles. The towers and poles both are properly grounded. So there must be *insulator* between tower or pole body and current-carrying conductors to prevent the flow of current from conductor to earth through the grounded supporting towers or poles. The insulators are connected to the cross arm of the tower and the power conductor passes through the clamp of the insulator. Properties of insulating material are as follows: (i) it must be mechanically strong enough to carry tension and weight of conductors; (ii) it must have very high dielectric strength to withstand the voltage stresses in High-Voltage system; (iii) it must possess high Insulation Resistance to prevent leakage current to the earth; (iv) the insulating material must be free from unwanted impurities; (v) it should not be porous; (vi) there must not be any entrance on the surface of electrical insulator so that the moisture or gases can enter in it; and (vii) the physical as well as electrical properties must be less effected by changing temperature.

Insulators are substances that permit very less current flow through them. Dielectrics are also insulators. But, more specifically, they are materials that can be polarized. When an external voltage is applied to the dielectric, the nucleus of the atoms is attracted to the negative side and the electrons are attracted to the positive side. Hence, the material gets polarized. This is a key feature of a dielectric. Thus, a dielectric can be defined as an insulator that can be polarized. Thus all dielectrics are insulators, but all insulators are not dielectrics. A dielectric can thus store charge.

7.2 MATERIALS OF INSULATOR

Porcelain is the most commonly used material for overhead insulator in present days. The porcelain is aluminum silicate. The aluminum silicate is mixed with plastic kaolin, feldspar and quartz to obtain final hard and glazed porcelain insulator material. The surface of the insulator should be glazed enough so that water should not be traced on it. Porcelain also should be free from porosity since porosity is the main cause of deterioration of its dielectric property. It must also be free from any impurity and air bubble inside the material which may affect the insulator properties. Dielectric strength of Porcelain is 6.5 kV/mm, tensile strength is 2 kg/mm^2 and composite strength is 30 kg/mm.

Glass insulator has become popular in transmission and distribution system. Annealed tough glass is used for insulating purpose. Glass insulator has numbers of advantages over conventional porcelain insulator since glass has very high dielectric strength, high resistivity and low coefficient of thermal expansion compared to porcelain. Since glass is transparent, it is not heated up in sunlight as porcelain. Glass has very long service life because mechanical and electrical properties of glass are not affected by ageing. Glass is cheaper than porcelain. However, moisture can easily be condensed on glass surface and hence air dust will be deposited on the wet glass surface which will provide path to leakage of current in the system. Also casting in irregular shape is difficult, and irregular internal cooling causes irregular internal strains. Polymers like glass fiber reinforced epoxy resin rod shaped core or silicone rubber or Ethylene Propylene Diene Monomer can be used for making polymer insulators, which are also called composite insulators.

DOI: 10.1201/9781003231240-7

7.3 TYPES OF INSULATORS

Pin Insulator – High tension (HT) and low tension (LT) Pin Insulators are used for voltages between 11 and 33 kV. Beyond operating voltage of 33 kV, the pin type insulator becomes too bulky and hence uneconomical as shown in Figure 7.1. Pin insulator can be of one piece or multi-piece construction. Multi-piece construction makes these insulators less vulnerable to damage.

In 11 kV system one part type insulator is generally used where whole pin insulator is one piece of properly shaped porcelain or glass. As the leakage path of insulator is through its surface, it is desirable to increase the vertical length of the insulator surface area for lengthening leakage path. To obtain lengthy leakage path, one, two or more rain sheds or petticoats are provided on the insulator body. These rain sheds or petticoats are designed so that during raining the outer surface of the rain shed becomes wet but the inner surface remains dry and non-conductive. So, there will be discontinuations of conducting path through the wet pin insulator surface.

Post-insulator – This insulator is more or less similar to Pin insulator but former is suitable for higher voltage application. Post insulator has higher numbers of petticoats and has greater height. This type of insulator can be mounted on supporting structure horizontally as well as vertically. The insulator is made of one piece of porcelain but has fixing clamp arrangement in both top and bottom end.

Suspension Disc Insulators – This insulator is shown in Figure 7.2 and is the most widely used model for transmission and distribution lines where a number of insulators are connected in series to form a string. The line conductor is carried by the bottom most insulator. Each disc is provided with a metal cap at the top and a metal pin under. A string of any number of units can be built according to the line operating voltage. The number of discs in a string depends on the line voltage and the atmospheric conditions. As the current-carrying conductors are suspended from supporting structure by suspension string, the height of the conductor position is always less than the total height of the supporting structure. Therefore, the conductors may be safe from lightening. Suspension type insulators are cheaper than pin type insulator. If any one string is damaged, it can be replaced by a sound one.

Strain Insulators – The insulator is as shown in Figure 7.3. When suspension string is used to sustain extraordinary tensile load of conductor it is referred as strain insulator. When there is a dead end or there is a sharp corner in transmission line, the line has to sustain a

FIGURE 7.1 Pin insulator.

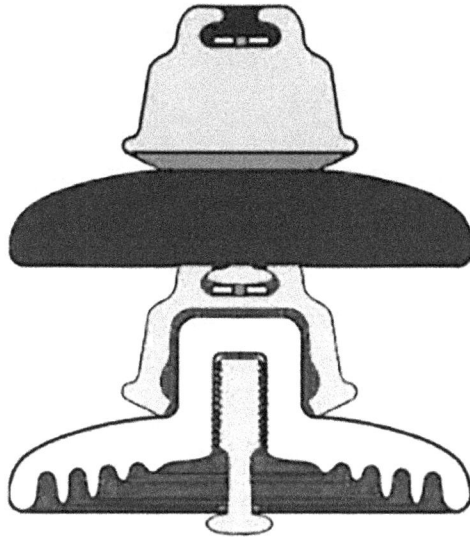

FIGURE 7.2 Suspension disc insulator.

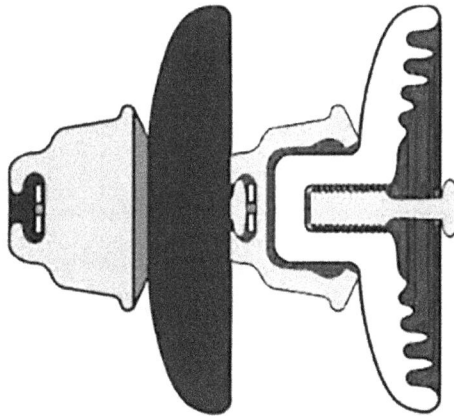

FIGURE 7.3 Strain insulator.

great tensile load of conductor or strain and a strain insulator is used. They are used to take the tension of the conductors at the line terminals, at angle towers and at road crossings. The string is placed in horizontal plane. Two or three strings of insulators in parallel can be used when the tension in conductor is very high.

Stay Insulator – The insulator is used for low-voltage lines. This is shown in Figure 7.4.

Shackle Insulator or Spool Insulator – This insulator is shown in Figure 7.5. In early days, the shackle insulators were used as strain insulators. But nowadays, they are used for low-voltage distribution lines. It can be used in both horizontal and vertical positions.

Solid Core Station Post-insulators – Solid core station post-insulators are comparatively lighter and have smaller diameters than other types of post-insulators of similar voltage class. These are ideally suitable for compact installations which result in slim and neat appearance and afford better harmony with other modern apparatus in any EHV/UHV substation. Solid core station post-insulators have superior antipollution performance, low RIV, better arc resistibility and lower deflection under cantilever load.

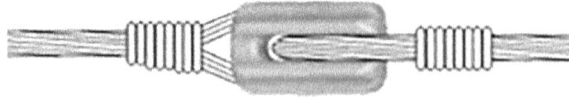

FIGURE 7.4   Stay insulator.

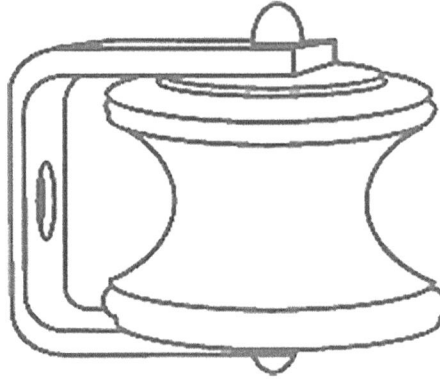

FIGURE 7.5   Shackle insulator.

Long Rod Insulators – Long rod insulators are absolutely puncture-proof, possess high arc resistibility and are free from cement growth trouble. Long rod insulators can be used at suspension and tension locations both as single and multiple strings as per system requirements. Standard long rod insulators have high strength in compression and tension, provide higher leakage distance with superior self-cleaning characteristics and are lighter when compared to disc insulator strings of similar voltage class.

Solid Core Line Post-insulators – Solid core line post-insulators can be used as support for conductors on cross-arms of transmission and distribution line poles. These insulators are recommended for polluted zones and areas prone to vandal damages. Even if the insulator shed is damaged, there will be little decrease in flashover voltage and power supply will not be disturbed.

7.4   VOLTAGE DISTRIBUTION ON INSULATOR STRING

Voltage distribution across different units of an insulator string can be found considering C as the capacitance between each insulator unit and C' as the capacitance between each unit and ground. This is shown in Figure 7.6. Let $C'/C = m$ or $C' = mC$.

Applying Kirchhoff's current to node 1, $I_2 = I_1 + i_1$; or $\omega CV_2 = \omega CV_1 + \omega C'V_1 = \omega CV_1 + \omega mCV_1 = \omega CV_1(1+m)$ i.e. $V_2 = V_1(1+m)$.

Applying Kirchhoff's current to node 2, $I_3 = I_2 + i_2$; or $\omega CV_3 = \omega CV_2 + \omega C'(V_1 + V_2) = \omega CV_2 + \omega mC(V_1 + V_2)$ i.e. $V_3 = mV_1 + V_2(1+m) = mV_1 + V_1(1+m)(1+m) = V_1(1+3m+m^2)$.

Applying Kirchhoff's current to node 3, $I_4 = I_3 + i_3$; or $\omega CV_4 = \omega CV_3 + \omega C'(V_1 + V_2 + V_3) = \omega CV_3 + \omega mC(V_1 + V_2 + V_3)$ i.e. $V_4 = V_1(1+3m+m^2) + \left[V_1 + V_1(1+m) + V_1(1+3m+m^2)\right] = V_1(1+6m+5m^2+m^3)$.

If there are five discs, then $I_5 = I_4 + i_4$; or $\omega CV_5 = \omega CV_4 + \omega C'(V_1 + V_2 + V_3 + V_4) = \omega CV_3 + \omega mC(V_1 + V_2 + V_3 + V_4)$ i.e. $V_5 = V_1(1+6m+5m^2+m^3) + mV_1\left[1+(1+m)+(1+3m+m^2)+(1+6m+5m^2+m^3)\right] = (1+10m+15m^2+7m^3+m^4)$.

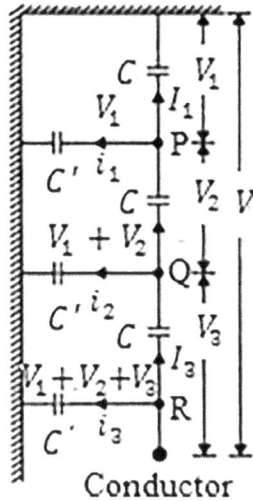

FIGURE 7.6 Voltage distribution.

7.5 STRING EFFICIENCY

The ratio of voltage across the whole string to the product of number of discs and the voltage across the disc nearest to the conductor is called as string efficiency: String efficiency = Voltage across the string/(number of discs × voltage across the disc nearest to the conductor). The voltage distribution across an insulator string is not uniform. The units nearest to the line end are stressed to their maximum allowable value while those near to the tower end are considerable less stressed. Greater the string efficiency, more uniform is the voltage distribution. String efficiency becomes 100% if the voltage across each disc is exactly the same, but this is an ideal case and impossible in practical scenario. However, for DC voltages, insulator capacitances are ineffective and voltage across each unit would be the same. This is why string efficiency for DC system is 100%. If the total voltage across the string is distributed across each unit equally, then the voltage across the power conductor is reduced. So the efficiency of the string increases. This V_1 can be obtained by reducing the value of m. If m approaches zero, the voltage across each unit will be equalized. Inequality in voltage distribution increases with the increase in the number of discs in a string. Therefore, shorter strings are more efficient than longer string insulators.

7.5.1 Methods of Improving String Efficiency

- **Using Longer Cross Arms** – It is clear from the above mathematical expression of string efficiency that the value of string efficiency depends upon the value of m. The lesser the value of m, the greater is the string efficiency. As the value of m approaches to zero, the string efficiency approaches to 100%. The value of m can be decreased by reducing the shunt capacitance. To decrease the shunt capacitance, the distance between the insulator string and the tower should be increased, i.e., longer cross-arms should be used. However, there is a limit in increasing the length of cross-arms due to economic considerations.
- **Grading of Insulator Discs** – In this method, voltage across each disc can be equalized by using discs with different capacitances. For equalizing the voltage distribution, the top unit of the string must have minimum capacitance, while the disc nearest to the conductor must have maximum capacitance. The insulator discs of different dimensions are so chosen that each disc has a different capacitance. They are arranged in such a way that the capacitance increases progressively toward the bottom.

- **By Using a Guard or Grading Ring** – A guard ring or grading ring is basically a metal ring which is electrically connected to the conductor surrounding the bottom unit of the string insulator. The guard ring introduces capacitance between metal links and the line conductor which tends to cancel out the shunt capacitances. As a result, nearly same charging current flows through each disc and, hence, improving the string efficiency.

In general case, if the string has of insulator is "n" units, E is the maximum voltage across the string, e_1, e_2, ...e_n denotes the voltage across the insulator unit starting from the conductor position and $C'/C = m$, the voltage distribution across the unit x is given by $e_x = 2E\sinh(1/2\sqrt{m})\cosh$ $\left[\sqrt{m}(n-x+1/2)\right]/\sinh(n\sqrt{m})$. The largest voltage across the insulator unit 1 nearest to the conductor is given by e_j

$$e_j = E \frac{2\sinh\left(\frac{1}{2}\sqrt{m}\right)\cosh\left[\sqrt{m}\left(n-\frac{1}{2}\right)\right]}{\sinh(n\sqrt{m})} \tag{7.1}$$

The string efficiency is given by

$$\frac{2\sinh(n\sqrt{m})}{2n\sinh\left(\frac{1}{2}\sqrt{m}\right)\cosh\left[\sqrt{m}\left(n-\frac{1}{2}\right)\right]} \tag{7.2}$$

Creepage Distance – It is the shortest length between two metallic fittings of insulator along the surface of insulator. In the string of insulator for creepage length calculation the metallic portion between consecutive insulator discs is not taken.

Flash Over Distance – A disruptive discharge external to the insulator, connecting those parts which normally have the operating voltages between them, is called flashover. Flashover distance is the shortest distance through air between the conductor and the metallic part.

Flash Over Voltage – The voltage at which air around the insulator breaks down and flash over takes place.

Puncture Voltage – A disruptive discharge passing through the solid insulating parts of an insulator is called puncture and the voltage at which the insulator breaks down and current flows through the insulator is called puncture voltage. Before puncture there is a flash over.

Safety factor – Flash over voltage / working voltage

The insulator should have good mechanical and dielectric strengths to withstand the load and operating or flashover voltages, respectively. However, it should be free from pores or voids, which may damage the insulator. For this, the following three tests are performed

- **Impulse Voltage Withstand Test** – A standard 1.2/50 µs impulse wave of peak value equal to the specified value of the impulse withstand voltage corrected for atmospheric conditions is applied across the insulator. Five such impulse voltage waves are applied to the insulator. If there is no flashover or puncture, the insulator is considered to have passed the test.
- **Dry and Wet Power-Frequency Voltage Withstand Test** – A voltage of about 75% of the test voltage shall be applied and then increased gradually to reach the test voltage in a time not less than 5 seconds. The test voltage shall be maintained at this value for 1 minute.

The insulator shall not flashover or puncture during the application of the test voltage. Before the commencement of the wet power-frequency voltage withstand test, the insulator is kept exposed to the artificial rain for at least 1 minute before application of voltage and then throughout the test.

Puncture Test – The insulators, after having been cleaned and dried, shall be completely immersed in a tank containing a suitable insulating medium to prevent surface discharges on them. If the tank is made of metal, its dimensions shall be such that the shortest distance between any part of the insulator and the side of the tank is not less than 1.5 times the diameter of the largest insulator shed. The immersion medium shall be at about room temperature. The test voltage is applied and raised as rapidly as is consistent with its value being indicated by the measuring instrument to the specified puncture voltage.

Dry and Wet Flashover Voltage Withstand Test – In the flashover test, voltage is applied between the electrodes of the insulator mounted in the manner in which it is to be used. Gradually the applied voltage is increased until the surrounding air breaks down. This voltage is known as flashover voltage, and must be greater than that of the minimum specified voltage. The insulator must be capable of withstanding the minimum specified voltage for 1 minute. In the wet flashover test the insulator is sprayed throughout the test with artificial rain.

7.6   LEARNING OUTCOME

Overhead Transmission Lines are not insulated but is carrying current at extremely high voltage. It is suspended from the transmission towers through insulator that also works as protectors. Electrical design of overhead insulators dictates the number of insulator discs, vertical or V-shaped string arrangement and phase-to-tower clearance. The insulator must withstand transient overvoltages due to lightning and switching surges, even when insulators are contaminated by fog or in the rainy season or industrial pollution. Mechanical design focuses on the strength of the conductors, insulator strings and support structures besides tensile and oscillatory forces of the conductors.

In this chapter, readers have been able to learn all the above aspects.

8 Overvoltages and Insulation Requirements

8.1 TYPES OF SYSTEM TRANSIENTS AND SOME BASIC FEATURES

A transient occurs in the power system when the network changes from one steady state into another and can be defined as an instantaneous change in the state leading to a burst of energy for a limited time which is termed as a transient event. The causes can be both external – for instance, the case when lightning hits the ground in the vicinity of a high-voltage transmission line or when lightning strikes a substation directly, and internal – the result of a switching action, with the aftermath being sequential and affecting the other parts too. Transients may be of two types: (i) Impulsive and (ii) Oscillatory. An impulsive transient is a sudden, non-power frequency change in the steady-state condition of voltage, current or both, which are unidirectional in polarity (primarily either positive or negative) caused by say lightning. Impulsive transients are not usually transmitted far from the source of where they enter the power system. However, in some cases, they may propagate for some distance along distribution utility lines. An oscillatory transient is a sudden, non-power frequency change in the steady-state condition of voltage, current or both, which include both positive and negative polarity values. Oscillatory transients with a primary frequency component greater than 500kHz and a typical duration measured in microseconds (or several cycles of the principal frequency) are considered high-frequency transients. Oscillatory transients with a primary frequency component between 5 and 500kHz with duration measured in the tens of microseconds (or several cycles of the principal frequency) is termed a medium-frequency transient. Oscillatory transients with a primary frequency component less than 5kHz and a duration from 0.3 to 50 μs is considered a low-frequency transient.

8.2 TRANSIENTS ON A TRANSMISSION LINE

In a series circuit of resistance R Ω and inductance L H, the transient current i is given by the solution of the differential equation: $e(t) = Ri + Ldi/dt$ or, $di/dt = [e(t) - Ri]/L = [(e(t)/R) - i]/(L/R) = (i_s(t) - i)/T = -i_t/T$. At every instant the rate of change of current is equal to the difference i_t between the steady-state current $i_s(t)$ which would be reached if the electro motive force (e.m.f.) remained at its instantaneous value and the actual current $i_s(t)$ is divided by the time constant T.

Transients happen in response to a sudden change either when input voltage/current abruptly changes its magnitude and frequency/phase or a switch alters the circuit. Let us consider the short circuit transient of a transmission line having resistance R and inductance L supplied by an ac source voltage v, such that $v = V_m \sin(\omega t + \alpha)$. To analyze the short circuit transients, the following assumptions are generally made: (i) the supply to the line is a constant voltage source; (ii) the short circuit occurs when the line is unloaded; and (iii) the line capacitance is negligible and the line can be represented by a lumped RL series circuit.

If the short circuit takes place at $t = 0$, the parameter α controls the instant on the voltage wave when short circuit occurs. It can be said that the current after short circuit is composed of two parts as under: $i = i_s + i_t$, where i_s is the steady-state current $= \left[\left(\sqrt{2} \ V \right)/|Z| \right] \sin(\omega t + \alpha - \theta); Z = \sqrt{\left(R^2 + \omega^2 L^2 \right)} \tan^{-1}(\omega L/R)$. i_t is the transient current and it is such that $i(0) = i_s(0) + i_t(0) = 0$ being an inductive circuit; it decays corresponding to the time constant $L/R = -i_s(0)e^{-(R/L)t} = \left[\left(\sqrt{2} \ V \right)/|Z| \right] \sin(\theta - \alpha)e^{-(R/L)t}$.

DOI: 10.1201/9781003231240-8

Thus, the short circuit current is given by $\left[\left(\sqrt{2}\,V\right)/|Z|\right]\sin(\omega t+\alpha-\theta)+\left[\left(\sqrt{2}\,V\right)/|Z|\right]$ $\sin(\theta-\alpha)e^{-(R/L)t}$.

The first part is called symmetrical short circuit current and the second part is called DC off-set current, which causes the total short circuit current to be unsymmetrical till transient decays. The maximum momentary short circuit current i_{mm} corresponds to first peak and is equal to $i_{mm}=\left[\left(\sqrt{2}\,V\right)/|Z|\right]\sin(\theta-\alpha)+\left[\left(\sqrt{2}\,V\right)/|Z|\right].$

Since transmission line resistance is small, $\theta=90°$ and $i_{mm}=\left[\left(\sqrt{2}\,V\right)/|Z|\right]\cos\alpha+\left[\left(\sqrt{2}\,V\right)/|Z|\right].$ This has the maximum possible value for $\alpha=0$, i.e., short circuit occurring when the voltage wave is going through zero. Thus $i_{mm\,(max)}=2\left[\left(\sqrt{2}\,V\right)/|Z|\right]=$ twice the maximum of symmetrical short circuit current (Doubling Effect).

In general, a surge is a transient wave of current, voltage or power in an electric circuit. Any disturbance on Transmission Lines or system such as sudden opening or closing of line, a short circuit or a fault results in the development of overvoltages or over-currents at that point. This disturbance propagates as a traveling wave to the ends of the line or to a termination. As the waves travel along the line their wave shapes and magnitudes are also modified. This is called distortion. The study of traveling waves helps in knowing the voltages and currents at all points in a power system. The establishment of a potential difference between the conductors of an overhead transmission line is accompanied by the production of an electrostatic flux, while the flow of current along the conductor results in the creation of a magnetic field. The electrostatic fields are due, in effect, to a series of shunt capacitors while the inductances are in series with the line.

Short and medium transmission lines are studied by their equivalent lumped parameter T or π model and these models are only useful to study and analyze the steady-state response of the transmission line. To study the transient behavior, these models are not useful and are very important to consider the line parameters like shunt capacitance and inductance that is distributed. A circuit with distributed parameter has a finite velocity of electromagnetic field propagation. When a transmission line is suddenly connected to a voltage source by the closing of a switch the whole of the line in not energized at once, i.e., the voltage does not appear instantaneously at the other end. This is due to the presence of distributed constants (inductance and capacitance in a loss-free line).

So, for accurate modeling of the transmission line for transient analysis the consideration that the parameters are not lumped but is distributed throughout line is necessary and hence the long transmission line model is considered. The single-line diagram of a long transmission line is shown in Figure 8.1. The length of the line is l.

Let us consider a small strip Δx that is at a distance x from the receiving end as shown in Figure 8.1. The voltage and current at the end of the strip are V and I, respectively, and the beginning of the strip are $V+\Delta V$ and $I+\Delta I$, respectively. The voltage drop across the strip is then ΔV and equals to $Iz\,\Delta x$. So, $\Delta V/\Delta x=Iz$. Since the length of the strip is Δx, the series impedance and shunt admittance are $z\Delta x$ and $y\Delta x$. It is to be noted here that the total impedance and admittance of the line are $Z=zl$ and $Y=yl$, respectively.

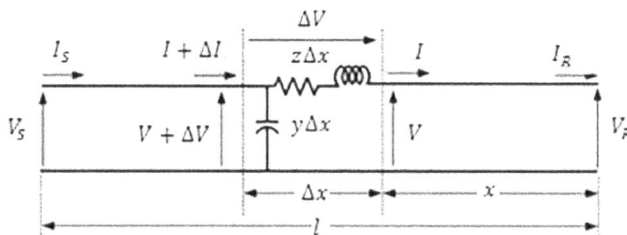

FIGURE 8.1 Long transmission line.

As $\Delta x \to 0$, $dV/dx = Iz$. Now for the current through the strip, applying Kirchhoff's current law (KCL) we get $\Delta I = (V + \Delta V) y \Delta x = Vy\Delta x + \Delta Vy\Delta x$. The second term of the expression is the product of two small quantities and therefore can be neglected. For $\Delta x \to 0$ we then have

$$\frac{dI}{dx} = Vy; \quad \frac{d}{dx}\left(\frac{dV}{dx}\right) = z\frac{dI}{dx}; \text{so} \quad \frac{d^2V}{dx^2} - yzV = 0 \tag{8.1}$$

The roots of the above equation are located at $\pm\sqrt{yz}$. Hence the solution will be of the form $V = A_1 e^{x\sqrt{yz}} + A_2 e^{-x\sqrt{yz}}$.

$$\frac{dV}{dx} = A_1\sqrt{yz}e^{x\sqrt{yz}} - A_2\sqrt{yz}e^{-x\sqrt{yz}} \text{ and } I = \frac{1}{z}\left(\frac{dV}{dx}\right) = \frac{A_1}{\sqrt{z/y}}e^{x\sqrt{yz}} - \frac{A_2}{\sqrt{z/y}}e^{-x\sqrt{yz}} \tag{8.2}$$

$Z_C = \sqrt{z/y} = \sqrt{zl/yl} = \sqrt{Z/Y} = \sqrt{(R + j\omega L)/(G + j\omega C)}$, which is called Characteristic Impedance. For lossless line, $R = 0$ and $G = 0$, $Z_C = \sqrt{L/C}$ and $\gamma = \sqrt{yz}$, the Propagation Constant.

Then V and I can be written in terms of the characteristic impedance and propagation constant as $V = A_1 e^{\gamma x} + A_2 e^{-\gamma x}$ and $I = (A_1/Z_C)e^{\gamma x} - (A_2/Z_C)e^{-\gamma x}$. Let us assume that $x=0$. Then $V=V_R$ and $I=I_R$. Then, $V_R = A_1 + A_2$ and $I_R = (A_1/Z_C) - (A_2/Z_C)$. So, $A_1 = (V_R + Z_C I_R)/2$; $A_2 = (V_R - Z_C I_R)/2$. With A_1 and A_2 as indicated, voltage at and current at a distance x can be stated as

$$V_x = (V_R + Z_C I_R)e^{\gamma x}/2 + (V_R - Z_C I_R)e^{-\gamma x}/2 \quad \text{and} \quad I_x = (V_R/Z_C + I_R)e^{\gamma x}/2 - (V_R/Z_C - I_R)e^{-\gamma x}/2.$$

Let $\gamma = \alpha + j\beta$, where α = attenuation constant (It causes a signal amplitude to decrease along a transmission line. The natural units of the attenuation constant are Nepers/meter.) and β = phase constant (it represents the change in phase per meter along the path traveled by the wave at any instant). Thus, $V_x = \left[|V_R + Z_C I_R|e^{\alpha x}e^{j(\beta x + \varphi_1)}\right]/2 + \left[|V_R - Z_C I_R|e^{-\alpha x}e^{-j(\beta x - \varphi_2)}\right]/2$, where $\varphi_1 = \angle(V_R + Z_C I_R)$ and $\varphi_2 = \angle(V_R - Z_C I_R)$

The instantaneous voltage is $v_x(t) = Re\left[\sqrt{2}|(V_R + Z_C I_R)/2|e^{\alpha x}.e^{j(\omega t + \beta x + \varphi_1)} + \sqrt{2}|(V_R - Z_C I_R)/2|e^{-\alpha x}.e^{j(\omega t - \beta x + \varphi_2)}\right]$.

The instantaneous voltage consists of two terms, each of which is a function of two variables – time and distance. Thus, they represent two traveling wave, i.e., $v_x = v_{x1} + v_{x2}$ where $v_{x1} = \sqrt{2}|(V_R + Z_C I_R)/2|e^{\alpha x}\cos(\omega t + \beta x + \varphi_1)$ and $v_{x2} = \sqrt{2}|(V_R - Z_C I_R)/2|e^{-\alpha x}\cos(\omega t - \beta x + \varphi_2)$. At any instant of time t, v_{x1} is sinusoidally distributed along the distance from the receiving end with amplitude increasing exponentially with distance. After time Δt, the distribution advances in distance phase by $\omega\Delta t/\beta$. Thus, this wave is traveling toward the receiving end from the sending end and is the incident wave. Line losses cause its amplitude to decrease exponentially in going from sending end to receiving end.

At any instant of time t, v_{x2} is sinusoidally distributed along the distance and this wave is traveling from the receiving to the sending wave and is the reflected wave.

At any point along the line, the voltage is the sum of incident wave and traveling wave. The same is for the current wave. The ratio of the amplitude of incident wave and reflected wave is called the reflection coefficient. It is equal to $\sqrt{(Z_L - Z_C)/(Z_L + Z_C)}$, where Z_L is the load impedance. If $Z_L = Z_C$, the reflection coefficient is zero and there will be no reflected voltage wave, and in that case, i.e., when the line is terminated at characteristic impedance, it is called infinite line.

At any time, the voltage or current varies sinusoidally along the line with respect to x, the space coordinate. A complete voltage or current cycle along the line corresponds to a change of 2π rad in the angular argument βx. The corresponding line length is defined as wavelength. If β is expressed in rad/m, then wavelength $\lambda = 2\pi/\beta$ m. For a typical power transmission line, g, the shunt conductance per unit length is zero. Also, resistance per unit length of the line $r \ll \omega L$, so $\gamma = \alpha + j\beta = \sqrt{yz} = \sqrt{j\omega C (r + j\omega L)} = j\omega\sqrt{LC}\left[1 - j(r/2\omega L)\right]^{1/2}$; $\alpha = (r/2)\sqrt{C/L}$ and $\beta = \omega\sqrt{LC}$.

Now, time for a phase change of 2π is $1/f$ s, where $f = \omega/2\pi$ is the frequency in cycles/sec. During this time the wave travels a distance equal to λ, i.e., wavelength. So, velocity of propagation of wave is $v = \lambda/(1/f) = f\lambda$ m/s. Voltage Standing Wave Ratio is defined as the ratio of the maximum voltage to the minimum voltage in standing wave pattern along the length of a transmission line structure.

When a transmission line is terminated with impedance Z_L, which is not equal to the characteristic impedance of the transmission line, Z_C, not all of the incident power is absorbed by the termination. Part of the power is reflected back so that phase addition and subtraction of the incident and reflected waves create a voltage standing wave pattern on the transmission line. The voltage standing wave ration VSWR $= V_{max}/V_{min} = (V_i + V_r)/(V_i - V_r)$; V_i is the incident voltage and V_r is the reflected voltage wave magnitude. The reflection coefficient is defined as V_r/V_i. Voltage standing wave ration can have any value from 1 to *infinity*.

8.3   SWITCHING SURGE

A transient wave is set up in the transmission line due to switching is called Traveling wave which travels from the sending end of a transmission line to the other end. The transmission lines may not have physical inductor and capacitor elements but the effects of inductance and capacitance exist in a line. Therefore, when the switch is closed the voltage will build up gradually over the line conductors. This phenomenon is usually called as the voltage wave, which is traveling from transmission line's sending end to the other end. And similarly, the gradual charging of the capacitances happens due to the associated current wave.

Consider a lossless transmission line with a wave after time t has traveled through a distance x. Consider the wave traveled a distance dx in a time dt. An electrostatic flux is associated with the electromagnetic flux and voltage wave with the current wave. Consider the charge between conductors of a line up to a distance x is $q = VCx$, where V is voltage, C is capacitance, x is distance traveled by wave and q is charge. The current that flows through the conductor is determined by the rate at which the charge flows into and out of the line as $I = dq/dt = dVCx/dt = VC(dx/dt) = VCv$. In a similar way, the electromagnetic flux linkages are created around the conductor due to current wave, $\psi = ILx$, where ψ is electromagnetic flux and L is inductance. Consider the voltage is determined as $V = d\psi/dt = dILx/dt = IL(dx/dt) = ILv$. So, $V/I = ILv/VCv$ and $V/I = \sqrt{L/C} = Z_0$, where Z_0 is the characteristic impedance or natural impedance value of transmission line. The characteristic impedance of a lossless line is a real quantity. It has the characteristics of resistance and the dimensions of ohm. Therefore, it is also called characteristic resistance or surge resistance. The velocity of traveling wave is $v = 1/\sqrt{LC}$.

For an overhead transmission line $L = 2 \times 10^{-7} \ln(D/r)$H/m and $C = 2\pi\varepsilon_0/\ln(D/r)$ F/m, where D is the distance between the centers of the conductors and r is the radius of the conductor and $D \gg r$. Hence $Z_0 = \sqrt{L/C} = 60\ln(D/r)$

$$v = \cfrac{1}{\sqrt{\left[2\times10^{-7}\ln(D/r)\right]\left[2\pi\varepsilon_0/\ln(D/r)\right]}} = \cfrac{1}{\sqrt{4\pi\left(10^{-7}\right)\varepsilon_0}} = \cfrac{1}{\sqrt{4\pi\left(10^{-7}\right)\dfrac{1}{36\pi}\left(10^{-9}\right)}} = 3\times10^8\,\text{m/s}$$

Since the product of L and C is the same for all overhead lines, it follows that the velocity of propagation is also the same. This velocity is the same as the velocity of light. In the aforesaid analysis, lossless line is considered, i.e., resistance is equal to zero. In practice, the velocity will be from 5% to 10% less. Normally a velocity of approximately 285 m/μs is obtained.

The velocity of propagation over the cables will be smaller than that over the overhead lines because in case of overhead lines $\varepsilon_r = 1$, while for cables $\varepsilon_r > 1$ the cable core is surrounded by insulations and sheath, being the dielectric constant. For cable $L = 2 \times 10^{-7} \ln(R/r) H /$ and $C = 2\pi\varepsilon_0\varepsilon_r /\ln(R/r)$ F/m, where R is the radius of the cable and r is the radius of the conductor. So, $Z_0 = \sqrt{L/C} = 60\ln(R/r)/\sqrt{\varepsilon_r}$.

The velocity of wave propagation in case of cables can be given as $v = 3 \times 10^8 /\sqrt{\varepsilon_r}$ m/s where ε_r varies from 2.5 to 4 in case of cables. A value of 500 Ω is usually assumed for the surge impedance of an overhead line while a value of 50 Ω is assumed for the surge impedance of a cable.

8.4 LIGHTNING SURGE

Lightning is an electrical discharge between cloud and the earth, between clouds or between the charge centers of the same cloud. There are two main ways in which the lightning may strike the power system. They are (i) Direct stroke and (ii) Indirect stroke. In direct stroke, the lightning discharge is directly from the cloud to an overhead line. From the line, current path may be over the insulators down to the pole to the ground. The overvoltage set up due to the stroke may be large enough to flashover this path directly to the ground. The direct stroke can be of two types: (i) Stroke A, where the lightning discharge is from the cloud to the subject equipment (e.g., overhead lines). The cloud will induce a charge of opposite sign on the tall object. When the potential between the cloud and line exceeds the breakdown value of air, the lightning discharge occurs between the cloud and the line. (ii) Stroke B, where the lightning discharge occurs on the overhead line as a result of stroke A between the clouds.

Indirect stroke results from eletrostatically induced charges on the conductors due to the presence of charged cloud. If a positively charged cloud is above the line, it induces a negative charge on the line by electrostatic induction. This negative charge however will be only on that portion on the line right under the cloud and the portion of the line away from it will be positively charged. The induced positive charge leaks slowly to earth. When the cloud discharges to earth or to another cloud, negative charge on the wire is isolated as it cannot flow quickly to earth over the insulator. The result is that negative charge rushes along the line in both directions in the form of traveling wave. The majority of the surges in a transmission lines are caused by indirect lightning stroke. Accordingly, the more frequent threat of lightning hazards will become. Furthermore, the more serious destructive consequence would be than ever before.

Surges caused due to lightning discharge are generally of non-oscillatory in nature with very steep wave front and have gradually decreasing tail. The wave is designated by the maximum or crest value of the voltage and the time at which it occurs and the time at which the voltage reduces to half its maximum value. In general, the voltage wave due to lightning is a unidirectional impulse of nearly double exponential in shape which can be expressed as $e = E(\epsilon^{-at} - \epsilon^{-bt})$. For a 1/50 μs wave form $a = 0.0139$ and $b = 6.1$. For a 1.2/50 μs wave form $a = 0.0143$ and $b = 4.87$.

The lightning surge causes traveling waves similar to that of switching surge but of very great intensity as the value of the maximum voltage is very high and the wave has very steep wave front. This results in a very large current which passes quickly along the lines and should be provided with path to ground to avoid damage.

8.5 LIGHTNING ARRESTERS

There are several types of lightning arresters in general use. They differ only in constructional details but operate on the same principle, providing low resistance path for the surges to the round.

Rod Gap Arrester – It is a very simple type of arrestor that consists of two 1.5 cm rods, which are bent at right angles with a gap in between. One rod is connected to the line circuit and the other rod is connected to earth. The distance between gap and insulator must be not less than one-third of the gap length so that the arc may not reach the insulator and damage it.

Generally, the gap length is so adjusted that breakdown should occur at 80% of spark-voltage in order to avoid cascading of very steep wave fronts across the insulators. This type of arrestor is simplest, cheapest and most rugged one. Limitations: (i) After the surge is over, the arc in the gap is maintained by the normal supply voltage, leading to short circuit on the system; (ii) the rods may melt or get damaged due to excessive heat produced by the arc; (iii) the climatic conditions (e.g., rain, humidity, temperature) affect the performance of rod gap arrester; and (iv) the polarity of the surge also affects the performance of this arrester.

Horn Gap Arrester – It consists of a horn-shaped metal rods A and B separated by a small air gap. The horns are so constructed that distance between them gradually increases toward the top. The horns are mounted on porcelain insulators. One end of horn is connected to the line through a resistance and choke coil L while the other end is effectively grounded. The resistance helps in limiting the follow current to a small value. The choke coil is so designed that it offers small reactance at normal power frequency but a very high reactance at transient frequency. Thus, the choke does not allow the transients to enter the apparatus to be protected. The gap between the horns is so adjusted that normal supply voltage is not enough to cause an arc across the gap. The time of operation of this type arrestor is comparatively high.

Multi Gap Arrester – It consists of a series of metallic (generally alloy of zinc known as the non-arcing metal) cylinders insulated from one another and separated by small intervals of air gaps between the main gap and the arc extermination gap cluster. The first cylinder in the series is connected to the line. When the lightning strikes, the gap of the multiple-gap lightning arrester is punctured causing formation of multiple short arcs within the arc extinguishing gap cluster.

Expulsion Type Lightning Arrester – It consists of a rod gap as mentioned above in series with a second gap enclosed within the tube having fiber lining inside the tube. The gap in the fiber tube is formed by two electrodes. The upper electrode is connected to rod gap and the lower electrode to the earth. There is a vent at the lower end of the tube for releasing gases after the arc is formed and extinguished between the spaces in the electrodes due to surge voltage.

Valve Type Lightning Arrester – It consists of two assemblies: (i) series spark gaps, each gap having two electrodes with a fixed gap spacing, and (ii) non-linear resistor discs (made of material such as thyrite or metrosil) in series. This type of arrestor is known as non-linear surge diverter. The impulse ratio (which is defined as the ratio of breakdown due to an impulse of specified shape to the breakdown voltage at power frequency) is nearly unity.

A lightning arrestor is specified with (i) Rated voltage – the voltage at which the leakage current flowing through the arrester is limited up to its specified rated value. Consider a solidly earthed system having line to line nominal voltage "M" Kv. Line to line highest system voltage will be 1.1 M Kv. Line to earth highest system voltage will be 1.1 $M/\sqrt{3}$ Kv. Line to earth peak system voltage will be $\sqrt{2}(1.1\ M)/\sqrt{3}$ Kv, which is the voltage rating of the arrestor. For non-effectively earthed system, the rated voltage will be $1.4\left[\sqrt{2}(1.1\ M)/\sqrt{3}\right]$ Kv. (ii) Power frequency spark overvoltage – the root-mean-square (rms) value of the lowest power frequency sinusoidal voltage that will cause sparkover when applied across the terminals of an arrester. (iii) Residual (Discharge) voltage – the voltage that appears between the line and earth terminal of the arrester during the passage of current; and (iv) Nominal discharge current – surge current that flows through the arrester after spark over. This current is 1.5, 2.5,5 and 10 KA.

Surge and Transient Voltage Surge are temporary rise in voltage and current on an electrical circuit. Typical rise time is in the 1–10 μs range. The surge absorbers are a protective device which can reduce the steepness of the wave front of a surge and absorbs energy contained in the traveling wave. The surge diverters divert the surge to ground but the surge absorbers absorbs the energy contained in the surge. A capacitor connected between line and earth can reduce the steepness of wave front. Similarly, as the impedance of capacitor is inversely proportional to the frequency, it gives protection against low-voltage high-frequency waves. Thus, it can be considered that the lines are connected to the earth so far as discharge at high frequency is considered. A pure capacitor cannot dissipate the energy in wave front or in high-frequency discharge but it just reflects the energy away from the equipment to be protected while the energy is dissipated in the resistance of line conductors and earthing resistances. With a series combination of resistor and capacitor, a part of energy is dissipated in the series combination in addition to prevent it from approaching the equipment.

The Ferranti surge absorbers consist of inductive coil magnetically coupled but not electrically to a metal shield, which is earthed and the steel tank containing it. The coil is enclosed in cylindrical boiler plate tank and is vacuum impregnated. It is similar to transformer with short circuited secondary. The inductor is primary while the metallic acts as a short-circuited secondary dissipater. Whenever a traveling wave is incident on the surge absorber, energy is transformed by mutual inductance between the coil and dissipater. Because of the series inductance the steepness of the wave is also reduced.

ERA surge absorber is having a gap G and an expulsion gap E. When a high-frequency wave reaches the inductor L, a high voltage is induced across it and causes the gap G to break down and so the resistor R and the expulsion gap E are included in the circuit. An incoming wave is thus flattened by the inductor L and resistor R and its amplitude is reduced by the expulsion gap E.

8.6 INSULATION COORDINATION IN POWER SYSTEM

This was introduced to arrange the electrical *insulation levels* of different components in the power system including transmission network, in such a manner that the failure of insulator, if occurs, is confined to the place where it would result in the least damage of the system, easy to repair and replace, and results in least disturbance to the power supply. When any overvoltage appears in the power system, there may be a chance of failure of its insulation system. Probability of failure of insulation is high at the weakest insulation point nearest to the source of overvoltage. In power system and transmission networks, insulation is provided to the all equipment and components. For proper understanding the insulation coordination we have to understand, some basic terminologies of the electrical power system.

Nominal System Voltage – It is the phase to phase voltage of the system for which the system is normally designed, such as 11, 33, 66, 132, 220 and 400 kV.

Maximum System Voltage – It is the maximum allowable power frequency voltage which may occur for long time during no load or low load condition of the power system such as 12, 36, 72.5, 145, 245 and 420 kV. It is also measured in phase to phase manner.

Factor of Earthing – It is the ratio of the highest rms phase to earth power frequency voltage on a sound phase during an earth fault to the rms phase to phase power frequency voltage which would be obtained at the selected location without the fault. This ratio characterizes, in general terms, the earthing conditions of a system as viewed from the selected fault location.

Insulation Level – Every electrical equipment has to undergo different abnormal transient overvoltage situation in different times during its total service life period. The equipment may have to withstand lightning impulses, switching impulses and/or short duration power frequency overvoltages. Depending upon the maximum level of impulse voltages and short duration power frequency overvoltages that one power system component can withstand, the insulation level of high-voltage power system is determined.

Power system may suffer from different level of transient voltage stresses, switching impulse voltage and lightning impulse voltage. The maximum amplitude the transient overvoltages reach the components can be limited by using protecting device like lightning arrestor in the system. If we maintain the insulation level of all the power system components above the protection level of

protective device, then ideally there will be no chance of breakdown of insulation of any component. Since the transient overvoltage reaches at the insulation after crossing, the surge protective
devices will have amplitude equals to protection level voltage and protection level voltage < impulse
insulation level of the components. This can be termed as Insulation Coordination.

8.7 BEWLEY LATTICE DIAGRAM

Bewley's lattice diagram is a graphical method that has been widely used for determining value of
a wave in transient analysis. When a traveling wave on a transmission line reaches a junction with
another line, or a termination, then part of the incident wave is reflected back, and a part of it is
transmitted beyond the junction or termination. The incident wave, the reflected wave and the transmitted wave are formed in accordance with Kirchhoff's laws. They must also satisfy the differential
equation of the line. This convenient diagram devised by Bewley shows at a glance the position
and direction of motion of every incident, reflected and transmitted wave on the system at every
instant of time. In a power system network that has many junctions and terminations, the number of
transmitted and reflected waves, initiated by a single incident wave, increases as the wave meets different junctions and it becomes difficult to find the transmitted and reflected waves. With the use of
Bewley's lattice diagram, it is possible to find the position and direction of motion of every incident,
reflected and transmitted wave on the system at every instant of time.

Let us consider a line connected to a source of constant voltage V is connected across the lossless transmission line of length l meters as shown in Figure 8.2. The load at the receiving end of the
line is one-third of the surge impedance of the line, i.e., $(Z_0/3)$. The impedance of the generator is
one-half of the surge impedance of the surge impedance of the line, i.e., $(Z_0/2)$.

The procedure for developing lattice diagram is as follows:

i. Find out the sending end voltage and current.
Sending end current $= V / [(Z_0/2) + Z_0 + (Z_0/3)] = 6V/11Z_0$
Sending end voltage $= V - [(6V/11Z_0)(Z_0/2)] = 8V/11$
ii. Find out the reflection coefficients at the sending end and at the receiving or load end.
Reflection coefficient at the load end $k_1 = (Z_L - Z_0)/(Z_L + Z_0) = [(Z_0/3) - Z_0]/[(Z_0/3) + Z_0] = -1/2$.
Reflection coefficient at the generator end $k_2 = [(Z_0/2) - Z_0]/[(Z_0/2) + Z_0] = -1/3$
iii. Find out the time for the wave to reach at the end of the line, which is equal to $l/v = t$s,
where v is the velocity of the wave propagation. The lattice diagram is shown in Figure 8.3.

The horizontal line of the Lattice Diagram represents the line and two vertical lines at the ends on
which equal intervals of time t are marked. The diagram begins at the top left corner at the source
and proceeds along the line.

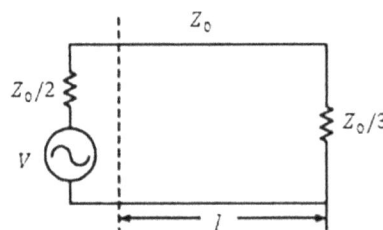

FIGURE 8.2 Line configuration.

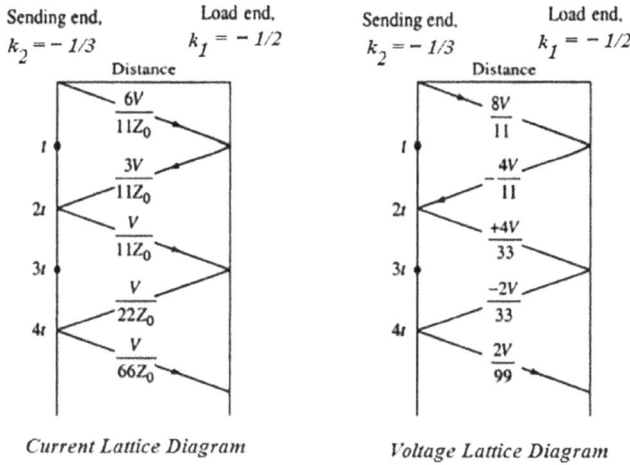

Current Lattice Diagram

Voltage Lattice Diagram

FIGURE 8.3 Lattice diagram.

8.8 LEARNING OUTCOME

Insulation requirements are essential in high voltage (HV) system and equipment design. Overvoltages in a power system include temporary overvoltages, switching impulse and lightning impulse overvoltages; all are having different peak magnitude and shape of waveforms, although all are transient in nature. Any kind of overvoltage generates stress to the insulation of the electrical equipment and likely to cause damage resulting outage of power supply. Overvoltage caused by surges can result in sparkover and flashover between phase and ground at the weakest point in the network. Overvoltage protective device like surge arrestors or lightning arrestors are designed to withstand a certain level of transient overvoltage beyond which the devices drain the surge energy to the ground and therefore maintain the level of transient overvoltage up to a specific level.

In this chapter, readers have been able to learn all the above aspects.

9 Electrical Fault Analysis

9.1 FAULT

A fault is an abnormal behavior of power system. A fault can also be defined as the departure of the voltage and current from the normal or rated value. Fault can be either transient or persistent in nature or a fault may be symmetrical or unsymmetrical in nature. A symmetrical fault is a direct short circuit in three terminals with or without involving ground and of two types – line to line to line to ground fault and line to line to line fault. For a symmetrical fault, the system remains balanced. Hence, these kinds of faults are called symmetrical or a balanced fault in the fault analysis. The unsymmetrical faults are: line to ground fault (LG fault), double line to ground fault and line to line to ground fault. In these kinds of faults, balanced state of the network is disturbed, and hence they are called unsymmetrical faults. Again, LG fault, double line to ground fault, line to line to ground fault and line to line to line fault are called shunt fault.

Analysis of fault is required to select proper switchgear and setting of relays for protection, including ascertaining reliability and stability of power system. Fault level is the maximum fault current that can flow into a zero-impedance fault. The fault level defines the value for symmetrical condition, which is the ration of Nominal MVA to Fault MVA and is equal to $1 / Z_{\mathrm{pu}}$.

9.2 SYMMETRICAL COMPONENTS

An unbalanced three-phase system can be resolved into three balanced systems in the sinusoidal steady state. This method of resolving an unbalanced system into three balanced phasor system has been proposed by C. L. Fortescue. This method is called resolving symmetrical components of the original phasors or simply symmetrical components. A system of three unbalanced phasors can be resolved in the following three symmetrical components: (i) Positive Sequence: A balanced three-phase system with the same phase sequence as the original sequence; (ii) Negative sequence: A balanced three-phase system with the opposite phase sequence as the original sequence; and (iii) Zero Sequence: Three phasors that are equal in magnitude and phase.

Figure 9.1 depicts a set of three unbalanced phasors that are resolved into the three sequence components mentioned above. In this the original set of three phasors are denoted by V_a, V_b and V_c, while their positive, negative and zero-sequence components are denoted by the subscripts 1, 2 and 0, respectively. This implies that the positive, negative and zero-sequence components of phase are denoted by V_{a1}, V_{a2} and V_{a0}, respectively. Note that, just like the voltage phasors given in (a), we can also resolve three unbalanced current phasors into three symmetrical components.

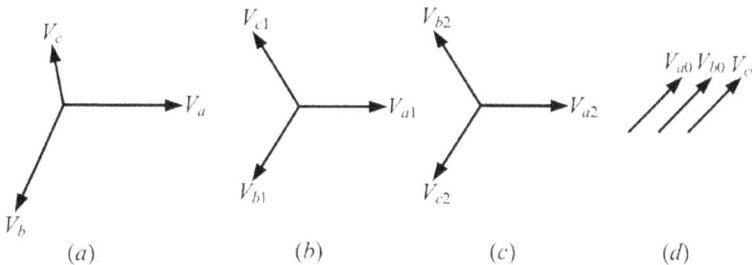

FIGURE 9.1 Representation of (a) an unbalanced network, its (b) positive sequence, (c) negative sequence and (d) zero sequence.

DOI: 10.1201/9781003231240-9

Any arbitrary set of three phasors, say V_a, V_b, V_c, can be represented as a sum of the three sequence sets: $V_a = V_{a0} + V_{a1} + V_{a2}$; $V_b = V_{b0} + V_{b1} + V_{b2}$; $V_c = V_{c0} + V_{c1} + V_{c2}$. V_{a0}, V_{b0} and V_{c0} is the zero-sequence set. V_{a1}, V_{b1} and V_{c1} is the positive sequence set. V_{a2}, V_{b2} and V_{c2} is the negative sequence set. Only three of the sequence values are unique; V_{a0}, V_{a1}, V_{a2}; the others are determined as by considering an operator $\alpha = e^{j120°} = 1\angle 120°$, $\alpha^2 = e^{j240°} = e^{-j120°} = \alpha^*$, $\alpha + \alpha^2 + \alpha^3 = 0$, or $1 + \alpha + \alpha^2 = 0$, $\alpha^3 = 1$.

By definition $V_{a0} = V_{b0} = V_{c0}$; for a phase sequence abc (positive sequence) $V_a = V_a$, $V_b = \alpha^2 V_a$, and $V_c = \alpha V_a$ and for a phase sequence acb (negative sequence) $V_a = V_a$, $V_b = \alpha V_a$ and $V_c = \alpha^2 V_a$. Thus $V_a = V_{a1} + V_{a2} + V_{a0} = V_{a1} + V_{a2} + V_{a0}$; $V_b = V_{b1} + V_{b2} + V_{b0} = \alpha^2 V_{a1} + \alpha V_{a2} + V_{a0}$; and $V_c = V_{c1} + V_{c2} + V_{c0} = \alpha V_{a1} + \alpha^2 V_{a2} + V_{a0}$. Thus,

$$
\begin{bmatrix} V_a \\ V_b \\ V_c \end{bmatrix} = \begin{bmatrix} 1 & 1 & 1 \\ \alpha^2 & \alpha & 1 \\ \alpha & \alpha^2 & 1 \end{bmatrix} \begin{bmatrix} V_{a1} \\ V_{a2} \\ V_{a0} \end{bmatrix} ; V_p = A V_s; \text{ where } A = \begin{bmatrix} 1 & 1 & 1 \\ \alpha^2 & \alpha & 1 \\ \alpha & \alpha^2 & 1 \end{bmatrix} ; \begin{bmatrix} V_a \\ V_b \\ V_c \end{bmatrix}
$$

$$
= V_p = \text{ Vector of original phasors;} \quad \begin{bmatrix} V_{a1} \\ V_{a2} \\ V_{a0} \end{bmatrix} = V_s = \text{ Vector of symmetrical components}
$$

Thus $V_s = A^{-1} V_p = \dfrac{1}{3} \begin{bmatrix} 1 & \alpha & \alpha^2 \\ 1 & \alpha^2 & \alpha \\ 1 & 1 & 1 \end{bmatrix} V_p$

V_{a1}, V_{a2} and V_{a0} can be determined as $V_{a1} = (1/3)\left(V_a + \alpha V_b + \alpha^2 V_c\right)$, $V_{a2} = (1/3)\left(V_a + \alpha^2 V_b + \alpha V_c\right)$ and $V_{a0} = (1/3)(V_a + V_b + V_c)$. The symmetrical component transformations though given above in terms of voltages hold for any set of phasors and therefore automatically apply for a set of currents. Thus $I_p = A I_s$ and $I_s = A^{-1} I_p$.

Consider the balanced Y-connected load that is shown in Figure 9.2. Certain observations can now be made regarding a three-phase system with neutral return. The sum of the three line voltage is always zero, i.e., $V_{abo} = (1/3)(V_{ab} + V_{bc} + V_{ca}) = 0$. On the other hand, the sum of phase voltages (line to neutral) may not be zero so that their zero-sequence component V_{ao} may exist. Since the sum of three line currents equals the current in the neutral, we have $I_{ao} = (1/3)(I_a + I_b + I_c) = (1/3)(I_n)$, i.e., the current in the neutral is three times the zero-sequence line current. If there is no neutral

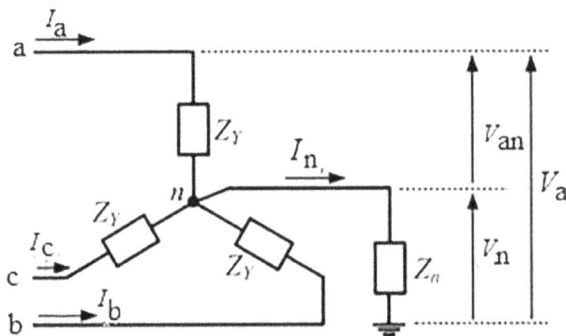

FIGURE 9.2 Star connected load with neutral return.

$I_{ao} = (1/3)(I_n) = 0$, i.e., in the absence of neutral connection, the zero-sequence line current is always zero.

The symmetrical component transformation is *power invariant*, which means the sum of powers of the three symmetrical components equals the three-phase power. Total complex power in a three-phase circuit is given by $S = V_a I_a^* + V_b I_b^* + V_c I_c^* = [V_{abc}][I_{abc}]^* = V_p^T I_p^* = [AV_s]^T[AI_s]^* = V_s^T A^T A^* I_s^*$.

$$A^T A^* = \begin{bmatrix} 1 & \alpha^2 & \alpha \\ 1 & \alpha & \alpha^2 \\ 1 & 1 & 1 \end{bmatrix} \begin{bmatrix} 1 & 1 & 1 \\ \alpha & \alpha^2 & 1 \\ \alpha^2 & \alpha & 1 \end{bmatrix} = 3\begin{bmatrix} 1 & 0 & 0 \\ 0 & 1 & 0 \\ 0 & 0 & 1 \end{bmatrix}$$

$$= 3U; \text{So } S = 3V_s^T U I_s^* = 3V_s^T I_s^* = 3V_{a1}I_{a1}^* + 3V_{a2}I_{a2}^* + 3V_{a0}I_{a0}^*$$

S is the sum of symmetrical component of powers. We can write $[V_{abc}] = [Z_{abc}][I_{abc}]$, where $[Z_{abc}]$ is 3×3 matrix, in which the diagonal elements are the self-impedances and the off-diagonal elements are mutual impedances. Thus, $[A][V_{120}] = [Z_{abc}][A][I_{120}]$ or $[V_{120}] = [A]^{-1}[Z_{abc}][A][I_{120}] = [Z_{120}][I_{120}]$ where $[Z_{120}] = [A]^{-1}[Z_{abc}][A]$.

9.3 SEQUENCE IMPEDANCE AND SEQUENCE NETWORKS OF POWER SYSTEM

When a numerical analysis of power system under fault is done, knowledge of sequence impedance of power system elements is required. Power system elements, have a three-phase symmetry – transmission lines, transformers and synchronous machines, because of which when currents of a particular sequence is passed through these elements, voltage drops of the same sequence appear, i.e., the elements possess only self-impedances to sequence elements. Each element can therefore represented by three decoupled sequence networks – on single phase basis, pertaining to positive, negative and zero sequence. EMFs are only involved in a positive sequence network of synchronous machine.

9.3.1 SEQUENCE IMPEDANCE AND SEQUENCE NETWORKS OF SYNCHRONOUS MACHINE

Figure 9.3 shows an unloaded synchronous machine (generator or motor) grounded through a reactor (impedance Z_n). E_a, E_b and E_c are the induced emfs of the three phases. If a fault occurs at machine terminals, currents I_a, I_b and I_c will flow in the lines. Whenever the fault involves the ground, current $I_n = I_a + I_b + I_c$ flows to neutral from ground via Z_n.

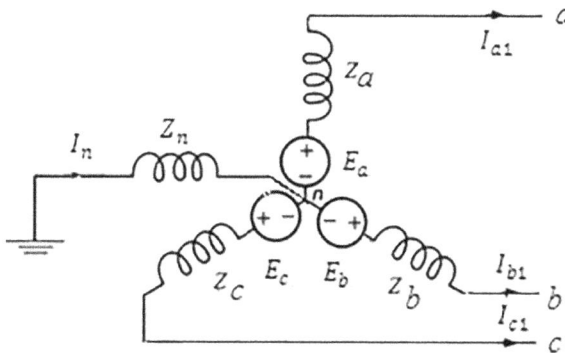

FIGURE 9.3 Three-phase synchronous generator with grounded neutral.

Whenever an unsymmetrical fault occurs at machine terminal, the unbalanced current flowing through the machine can be resolved into their symmetrical components, I_{a1}, I_{a2} and I_{a0}. Since a synchronous machine is designed with a symmetrical windings, it has induced emfs of positive sequence only, i.e., no negative or zero-sequence voltages are induced in it. If the machine short circuit takes place on unloaded condition, the terminal voltage constitutes the positive sequence voltage; on the other hand, if the short circuit occurs on loaded conditions, the voltage behind appropriate reactance (subtransient, transient or synchronous) constitutes the positive sequence voltage. Z_n does not appear as $I_n = 0$. Figure 9.4 shows the three-phase positive sequence network of a synchronous machine including single phase model. The reference bus for a positive sequence network is at neutral potential. Further, since no current flows from ground to neutral, the neutral is at ground potential.

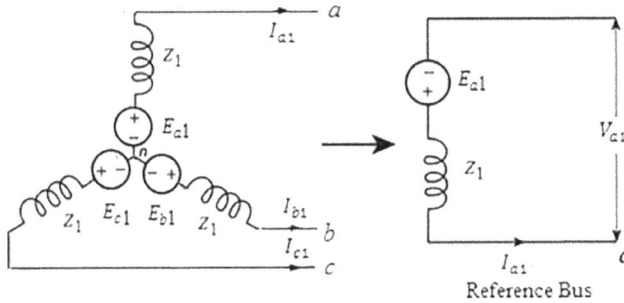

FIGURE 9.4   Positive sequence network of synchronous machine.

The positive sequence voltage can be expressed as $V_{a1} = E_{a1} - I_{a1}Z_1$, where V_{a1} is positive sequence terminal voltage, E_{a1} is the positive sequence induced voltage, I_{a1} is the positive sequence armatrure current and Z_1 is the positive sequence impedance. The armature reaction field caused by positive sequence currents rotates at synchronous speed in the same direction as the rotor, i.e., it is stationary with respect to field excitation. The machine equivalently offers a direct axis reactance whose value reduces from sub-transient reactance (X_d'') to transient reactance (X_d') and finally to steady state or synchronous reactance (X_d).

A synchronous machine has zero negative sequence induced voltages. With the flow of negative sequence currents in the stator a rotating field is created which rotates in the opposite direction to that of positive sequence field and, therefore, at double the synchronous speed with respect to rotor. Currents at double the stator frequency are therefore induced in rotor field and damper winding. In sweeping over the rotor surface, the negative sequence mmf is alternately presented with reluctances of direct and quadrature axis. The negative sequence impedance presented by the machine with consideration given to damper windings is defined as $Z_2 = j(X_d'' + X_q'')/2$. Negative sequence network model of a synchronous machine, on a three-phase and single-phase basis are shown in Figure 9.5. The negative sequence voltage can be expressed as $V_{a2} = -I_{a2}Z_{21}$, where V_{a2}, I_{a2} and Z_2 is the negative sequence voltage, current and impedance, respectively.

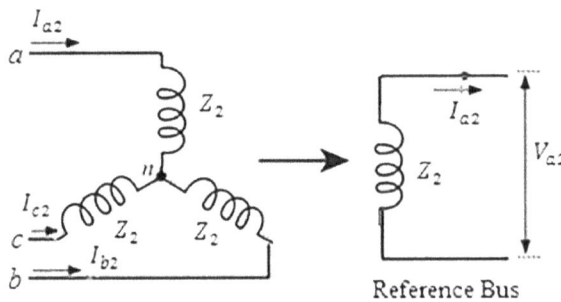

FIGURE 9.5   Negative sequence network of synchronous machine.

The flow of zero-sequence currents creates three mmfs which are in time phase but are distributed in space phase by $120°$. The resultant air gap field caused by zero-sequence currents is therefore zero. Hence, the rotor windings present leakage reactance only to the flow of zero-sequence currents $(Z_{0g} < Z_2 < Z_1)$. The current flowing in the impedance Z_n between neutral and ground is $I_n = 3I_{a0}$. The zero-sequence voltage of terminal a with respect to ground, the reference bus, is therefore $V_{a0} = -3Z_n I_{a0} - Z_{0g}I_{a0} = -(3Z_n + Z_{0g})Z_{0g}$. Thus, the zero-sequence impedance of the machine is $(3Z_n + Z_{0g})$. The current flowing in the impedance Z_n between neutral and ground is $I_n = 3I_{a0}$. Zero-sequence network models on a three- and single-phase basis are shown in Figure 9.6.

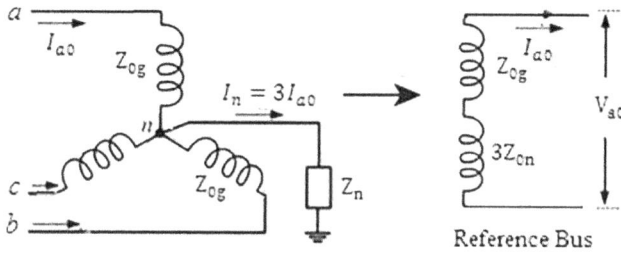

FIGURE 9.6 Zero-sequence network of synchronous machine.

9.3.2 Sequence Impedance and Sequence Networks of Transformer

The positive sequence impedance of a transformer equals the leakage impedance. Since the transformer is a static device, the leakage impedance does not change, if the phase sequence is altered from RYB to RBY. Therefore, the negative sequence impedance of transformer is the same as the positive sequence impedance. The zero-sequence impedance of the transformer depends on the winding type (star or delta) and also on the type of earth connection. The zero-sequence equivalent circuits of three-phase transformers deserve special attention. The different possible combinations of the primary and the secondary windings in Y and Δ alter the zero-sequence network.

If either one of the neutrals of a Star-Star bank is ungrounded, zero-sequence current cannot flow in either winding (as the absence of a path through one winding prevents current in the other). An open circuit exists for zero-sequence current between two parts of the system connected by the transformer bank.		
Star-Star Bank with both neutral grounded: In this case, a path through transformer exists for the zero-sequence current. Hence zero-sequence current can flow in both sides of the transformer provided there is closed path for it to flow. Hence the points on the two sides of the transformer are connected by the zero-sequence impedance of the transformer.		

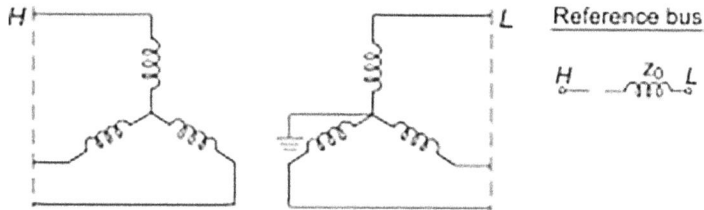

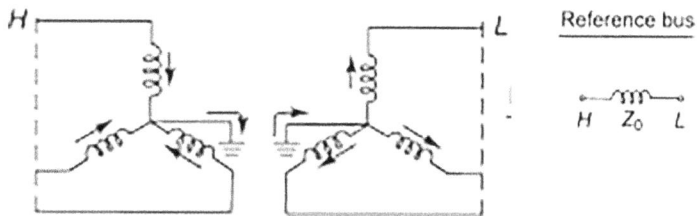

Star-Delta Bank with grounded Star: there is path for zero-sequence current to ground through the star as the corresponding induced current can circulate in the delta. The equivalent circuit must provide for a path from lines on the star side through zero-sequence impedance of the transformer to the reference bus. However, an open circuit must exist between line and the reference bus on the delta side. If there is an impedance Zn between neutral and ground, then the zero-sequence impedance must include 3Zn along with zero-sequence impedance of the transformer.

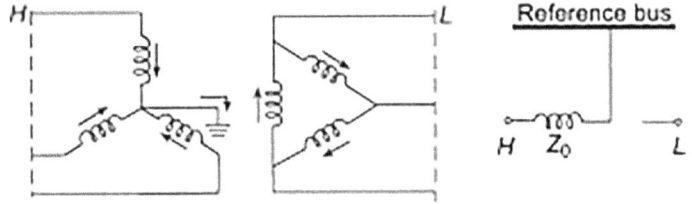

Star-Delta Bank with ungrounded Star: In this case, there is no path for zero-sequence current. The zero-sequence impedance is infinite and is shown by an open circuit.

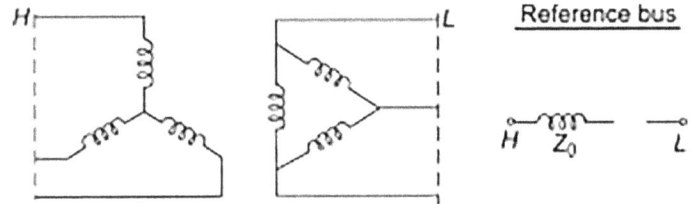

Delta-Delta Bank: In this case, there is no return path for zero-sequence current. The zero-sequence current cannot flow in lines, although it can circulate in the delta windings.

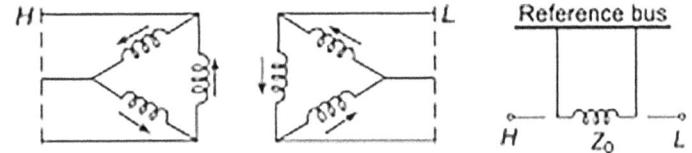

The zero-sequence components

Y-connected, with no connection from the neutral to ground or to another neutral point in the circuit, no zero-sequence currents can flow, and hence the impedance to zero-sequence current is infinite.

This is represented by an open circuit between the neutral of the Y-connected circuit and the reference bus, if the neutral of the Y-connected circuit is grounded through zero impedance, a zero-impedance path (short circuit) is connected between the neutral point and the reference bus.

If an impedance Zn is connected between the neutral and the ground of a Y-connected circuit, an impedance of 3Zn must be connected between the neutral and the reference bus (because all the three zero-sequence currents ($3I_{a0}$) flow through this impedance to cause a voltage drop of $3I_{a0} Z_0$), as shown in the figure Zero-sequence equivalent networks of Y-connected load.

A delta-connected circuit can provide no return path; its impedance to zero-sequence line currents is therefore infinite. Thus, the zero-sequence network is open at the delta-connected circuit, as shown in the figure Zero-sequence equivalent networks of delta-connected load. However, zero-sequence currents can circulate inside the delta-connected circuit.

9.3.3 Sequence Circuits for Symmetrical Transmission Line

The schematic diagram of a transmission line is shown in Figure 9.7. In this diagram the self-impedance of the three phases are denoted by Z_a, Z_b and Z_c, while that of the neutral wire is denoted by Z_n. The self-impedances of the conductors are same, i.e., $Z_a = Z_b = Z_c$

$V_A - V_a = V_{Aa}$, the voltage drop in Z_a and $Z_n = Z_a I_a + Z_n(I_a + I_b + I_c)$; $V_B - V_b = V_{Bb}$, the voltage drop in Z_b and $Z_n = Z_b I_b + Z_n(I_a + I_b + I_c)$ and $V_C - V_c = V_{Cc}$, the voltage drop in Z_c and $Z_n = Z_c I_c + Z_n(I_a + I_b + I_c)$.

$$\begin{bmatrix} V_{Aa} \\ V_{Bb} \\ V_{Cc} \end{bmatrix} = \begin{bmatrix} Z_a + Z_n & Z_n & Z_n \\ Z_n & Z_b + Z_n & Z_n \\ Z_n & Z_n & Z_c + Z_n \end{bmatrix} \begin{bmatrix} I_a \\ I_b \\ I_c \end{bmatrix}; \text{ or, } V_p = Z I_p \qquad (9.1)$$

Applying symmetrical component transformation $AV_s = ZAI_s$ or, $V_s = A^{-1}ZAI_s = Z_s I_s$, where $Z_s = A^{-1}ZA =$ symmetrical components impedance matrix.

$$Z_s = \frac{1}{3} \begin{bmatrix} 1 & \alpha & \alpha^2 \\ 1 & \alpha^2 & \alpha \\ 1 & 1 & 1 \end{bmatrix} \begin{bmatrix} Z_a + Z_n & Z_n & Z_n \\ Z_n & Z_b + Z_n & Z_n \\ Z_n & Z_n & Z_c + Z_n \end{bmatrix} \begin{bmatrix} 1 & 1 & 1 \\ \alpha^2 & \alpha & 1 \\ \alpha & \alpha^2 & 1 \end{bmatrix}$$

$$= \begin{bmatrix} \frac{1}{3}(Z_a + Z_b + Z_c) & \frac{1}{3}(Z_a + \alpha^2 Z_b + \alpha Z_c) & \frac{1}{3}(Z_a + \alpha Z_b + \alpha^2 Z_c) \\ \frac{1}{3}(Z_a + \alpha Z_b + \alpha^2 Z_c) & \frac{1}{3}(Z_a + Z_b + Z_c) & \frac{1}{3}(Z_a + \alpha^2 Z_b + \alpha Z_c) \\ \frac{1}{3}(Z_a + \alpha^2 Z_b + \alpha Z_c) & \frac{1}{3}(Z_a + \alpha Z_b + \alpha^2 Z_c) & \frac{1}{3}(Z_a + Z_b + Z_c) + 3Z_n \end{bmatrix}$$

$$= \begin{bmatrix} Z_{11} & Z_{12} & Z_{10} \\ Z_{21} & Z_{22} & Z_{20} \\ Z_{01} & Z_{02} & Z_{00} \end{bmatrix} \qquad (9.2)$$

So,

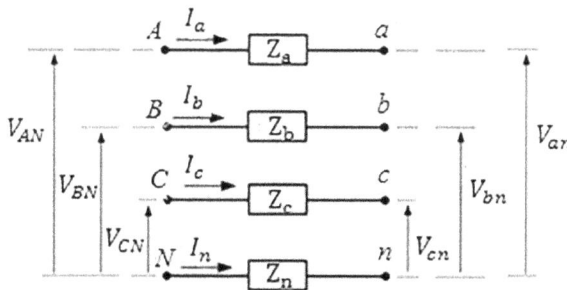

FIGURE 9.7 Transmission line with lumped parameter representation.

$$\begin{bmatrix} V_{Aa1} \\ V_{Aa2} \\ V_{Aa0} \end{bmatrix} = \begin{bmatrix} Z_{11} & Z_{12} & Z_{10} \\ Z_{21} & Z_{22} & Z_{20} \\ Z_{01} & Z_{02} & Z_{00} \end{bmatrix} \begin{bmatrix} I_{a1} \\ I_{a2} \\ I_{a0} \end{bmatrix} \tag{9.3}$$

$Z_{11} = Z_{22} = (Z_a + Z_b + Z_c)/3$; $Z_{00} = (Z_a + Z_b + Z_c)/3 + 3Z_n$; $Z_{12} = Z_{20} = Z_{01} = (Z_a + \alpha^2 Z_b + \alpha Z_c)/3$
$Z_{10} = Z_{21} = Z_{02} = (Z_a + \alpha Z_b + \alpha^2 Z_c)/3$. If $Z_a = Z_b = Z_c = Z_l$, then Z_s will be a diagonal matrix with diagonal elements are Z_l, Z_l and $Z_l + 3Z_n$.

9.4 UNSYMMETRICAL FAULTS

Three Types of Unsymmetrical Faults are as follows: (i) Single line to ground (LG) fault, (ii) Line to line fault and (iii) Double line to ground fault. Assumptions for simplifying the fault calculations are as follows: (i) The power system is balanced before the fault occurs such that of the three sequence networks only the positive sequence network is active. Also as the fault occurs, the sequence networks are connected only through the fault location. (ii) The fault current is negligible such that the pre-fault positive sequence voltages are same at all nodes and at the fault location. (iii) All the network resistances and line charging capacitances are negligible. (iv) All loads are passive except the rotating loads which are represented by synchronous machines.

9.4.1 LINE TO GROUND FAULT

Consider a three-phase system with an earthed neutral. Let a single LG fault occur on the red phase as shown in Figure 9.8. It is clear from this figure that at the fault point F, the currents out of the power system and the LG voltage are constrained as $I_b = 0$, $I_c = 0$ and $V_a = Z_f I_a$.

The symmetrical components of the fault currents are

$$\begin{bmatrix} I_{a1} \\ I_{a2} \\ I_{a0} \end{bmatrix} = \frac{1}{3} \begin{bmatrix} 1 & \alpha & \alpha^2 \\ 1 & \alpha^2 & \alpha \\ 1 & 1 & 1 \end{bmatrix} \begin{bmatrix} I_a \\ 0 \\ 0 \end{bmatrix}, \text{ i.e., } I_{a1} = I_{a2} = I_{a0} = \frac{1}{3} I_a \tag{9.4}$$

Since $V_a = Z_f I_a$, so $V_{a1} + V_{a2} + V_{a0} = Z_f I_a = 3Z_f I_{a1}$. This indicates a series connection of sequence networks through an impedance $3Z_f$. The positive sequence current $I_{a1} = V_f / [(Z_1 + Z_2 + Z_0) + 3Z_f]$ and the fault current $I_a = 3I_{a1} = 3V_f / [(Z_1 + Z_2 + Z_0) + 3Z_f]$. The voltage of line b to ground under fault condition $V_b = \alpha^2 V_{a1} + \alpha V_{a2} + V_{a0} = \alpha^2 [E_a - I_a Z_1 / 3] + \alpha(-I_a Z_2 / 3) + (-I_a Z_0 / 3)$. This gives $V_b = E_a [3\alpha^2 Z_f + Z_2(\alpha^2 - \alpha) + Z_0(\alpha^2 - \alpha)] / [(Z_1 + Z_2 + Z_0) + 3Z_f]$. The equivalent of the sequence network of the single LG fault is shown in Figure 9.9.

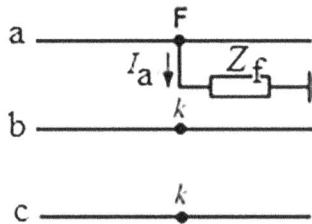

FIGURE 9.8 Single line to ground fault.

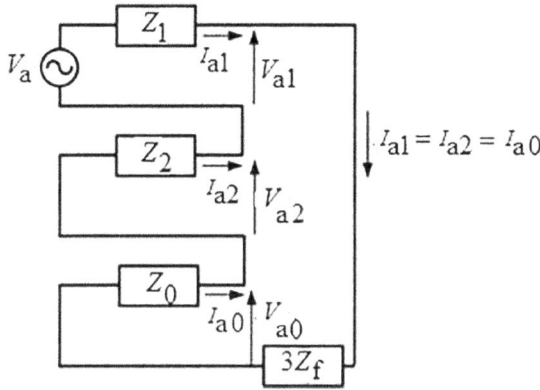

FIGURE 9.9 Single line to ground fault equivalent sequence network.

From the above it can be concluded that (i) for a three-phase fault at generator terminal $I_a = V/jX_1$. But for single LG fault $I_a = 3V /[j(X_1 + X_2 + X_0)]$; considering Z_1, Z_2 and Z_0 are pure reactances X_1, X_2 and X_0 and $Z_f = 0$. With generator neutral solidly earthed and $X_0 \ll X_1$ and also $X_1 = X_2$, it is evident that the single LG fault current is more than the current for a three-phase fault. Thus single LG fault at generator terminal is more severe than three-phase fault with neutral solidly earthed (ii) fault at generator terminals with neutral grounded with a reactance X_n, $I_a = 3V /[j(X_1 + X_2 + X_0 + 3X_n)]$ which implies the severity of three-phase fault and the single LG fault depends on the value of X_n. If X_n is large, the current for a LG fault is lesser than the current for a three-phase fault. If $X_n = (X_1 - X_0)/3$, the line currents for the two types of fault are equal. If $X_n < (X_1 - X_0)/3$, a single LG fault is more severe. If $X_n > (X_1 - X_0)/3$, a single line to three-phase fault is more severe.

9.4.2   LINE TO LINE FAULT

The faulted segment for line to line fault is shown in Figure 9.10. The fault has occurred at node F of the network between phases b and c through the impedance Z_f. Then $I_a = 0$, $I_F = I_b$ and $I_c = -I_b$ also $V_b - V_c = I_b Z_f$; since fault has occurred through an impedance Z_f.

The symmetrical component of fault currents and the symmetrical components of voltages at F under fault are

$$
\begin{bmatrix} I_{a1} \\ I_{a2} \\ I_{a0} \end{bmatrix} = \frac{1}{3} \begin{bmatrix} 1 & \alpha & \alpha^2 \\ 1 & \alpha^2 & \alpha \\ 1 & 1 & 1 \end{bmatrix} \begin{bmatrix} 0 \\ I_b \\ -I_b \end{bmatrix}; \quad \begin{bmatrix} V_{a1} \\ V_{a2} \\ V_{a0} \end{bmatrix} = \frac{1}{3} \begin{bmatrix} 1 & \alpha & \alpha^2 \\ 1 & \alpha^2 & \alpha \\ 1 & 1 & 1 \end{bmatrix} \begin{bmatrix} V_a \\ V_b \\ V_b - Z_f I_b \end{bmatrix}
$$

$$(9.5)$$

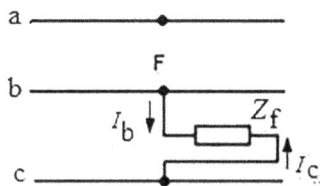

FIGURE 9.10   Line to Line fault.

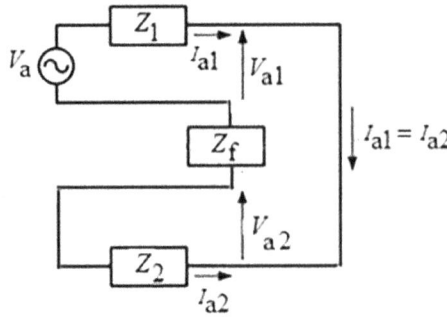

FIGURE 9.11   Line to Line fault equivalent sequence network.

The current relation implies $I_{a2} = -I_{a1}$ and $I_{a0} = 0$. Now, $V_{a1} = V_a + V_b(\alpha^2 + \alpha) - \alpha^2 Z_f I_b$ and $3V_{a2} = V_a + V_b(\alpha^2 + \alpha) - \alpha Z_f I_b$. So, $3(V_{a1} - V_{a2}) = (\alpha - \alpha^2) Z_f I_b = j\sqrt{3}.Z_f I_b$. As $I_{a0} = 0$; $I_b = (\alpha^2 - \alpha) I_{a1}$ as $I_{a2} = -I_{a1}$ and $(V_{a1} - V_{a2}) = Z_f I_{a1}$. The sequence network is then as shown in Figure 9.11.

9.4.3   DOUBLE LINE TO GROUND FAULT

The faulted segment for a double line to ground fault is shown in Figure 9.12. The fault has occurred at node F of the network between phases b and c through the impedance Z_f.

The currents and voltage at the fault can be expressed as $I_a = 0$; i.e. $I_{a1} + I_{a2} + I_{a0} = 0$ and $V_b = V_c = Z_f(I_b + I_c) = 3Z_f I_{a0}$. The symmetrical components of voltages at F under fault is

$$\begin{bmatrix} V_{a1} \\ V_{a2} \\ V_{a0} \end{bmatrix} = \frac{1}{3} \begin{bmatrix} 1 & \alpha & \alpha^2 \\ 1 & \alpha^2 & \alpha \\ 1 & 1 & 1 \end{bmatrix} \begin{bmatrix} V_a \\ V_b \\ V_c \end{bmatrix} \tag{9.6}$$

From which it follows that $V_{a1} = V_{a2} = \left[V_a + V_b(\alpha^2 + \alpha) \right] / 3$ and (9.44) $V_{a0} = \left[V_a + 2V_b \right] / 3$. $V_{a0} - V_{a1} = \left[2 - \alpha - \alpha^2 \right] V_b / 3 = V_b = 3Z_f I_{a0}$ or $V_{a0} = V_{a1} + 3Z_f I_{a0}$. The sequence network is then as shown in Figure 9.13.

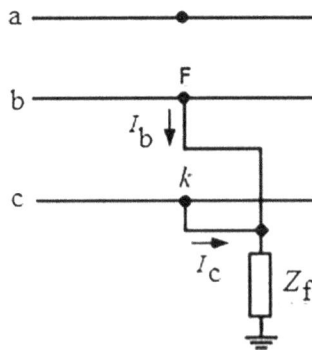

FIGURE 9.12   Line to Line to Ground fault.

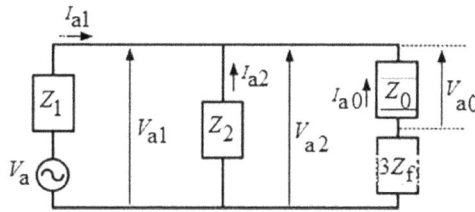

FIGURE 9.13 Line to Line to Ground fault equivalent sequence network.

9.5 EARTHING

Earthing is the first step toward electrical safety. The body of electrical equipment, appliances, etc. is to be connected to earth for safe discharge of electric current to the earth, in the event of any leakage, fault, etc. The earth is considered to be zero potential. The objectives of an earthing system are: (i) to provide safety to personnel during normal and fault conditions by limiting step and touch potential; (ii) to provide means to carry electric currents into the earth under normal and fault conditions without exceeding any operating and equipment limits or adversely affecting continuity of service; (iii) to prevent damage to electrical/electronic apparatus; (iv) to dissipate lightning strokes; and (v) to stabilize voltage during transient conditions and to minimize the probability of flashover during transients. The two main types of earthing are as follows: (i) System earthing/neutral earthing – functions of which are to detect earth faults and to control the fault current, since large fault currents can cause the potential rise of exposed parts of the power system to reach dangerous levels. (ii) Equipment earthing (Safety).

Earthing system is designed to achieve low earth resistance and also to achieve safe "Step Potential" and "Touch Potential". Step Potential is the potential difference between two points on the earth's surface, separated by distance of one pace that will be assumed to be 1 m in the direction of maximum potential gradient. Touch Potential is the potential difference between a grounded metallic structure and a point on the earth's surface separated by a distance equal to the normal maximum horizontal reach, approximately 1 m.

9.5.1 TYPES OF EARTHING

Solid Earthing – In this method neutral is directly connected to the earth without having any intentional resistance between neutral and ground. The single-phase earth fault current in a solidly earthed system may exceed the three-phase fault current. The main advantage of solidly earthed systems is low overvoltages, which makes the earthing design common at low-voltage levels. (ii) Resistance earthing: Resistance is connected between the neutral and the ground to improve the earth fault detection. Resistance Grounding Systems limits the phase-to-ground fault currents, thereby reducing burning and melting effects in faulted electrical equipment like switchgear, transformers, cables, etc. and also reduces the mechanical stresses in circuits/Equipment carrying fault currents. (iii) Reactance earthing: Reactance is connected between the neutral and ground. (iv) Resonant earthing: An adjustable reactor of correctly selected value to compensate the capacitive earth current is connected between the neutral and the earth. The coil is called Arc Suppression Coil or Earth Fault Neutralizer. (v) Effectively earthed system: A system in which the value of the phase to earth voltage of the healthy phases during an earth fault, never exceed 1.39 times the pre-fault phase to ground voltage is effectively earthed system.

Pipe Earthing – In this method of earthing, a galvanized steel pipe of suitable length and diameter is buried vertically in the permanent wet soil under the ground. It is comparatively less costly than plate earthing. The length and diameter of the pipe are determined by the conditions of soil and the current to be carried. Normally minimum diameter and length of the pipe are maintained 40 mm and 2.5 m, respectively, for ordinary condition of soil and greater length is used for rocky and dry soil conditions. The depth under ground level at which the pipe is buried depends upon the moisture condition of soil but it should not be less than 3.75 m under the ground. The earthing pipe is surrounded by alternative layers of charcoal and salt to keep moisture and thereby reduces the earth resistance. Area around the pipe is called pit area. The earth wire is fastened to the top section of the pipe with nut and bolts.

Plate Earthing – In this method a metallic plate of sufficient size is buried in wet soil vertically under the ground. If copper plate is used for this purpose, the minimum dimension of the plates should be 60 cm × 60 cm × 3 mm, and if it is GI plate, then minimum dimension should be 60 cm × 60 cm × 6 mm. The earth plate is embedded in alternative layer of Coke and salts for minimum thickness of about 15 cm.

Strip Earthing – In this type of earthing, a strip electrode of cross section not less than 25 mm × 1.6 mm is buried in a horizontal trench of depth not less than 0.5 m. If copper is used, then the desired cross section is 25 mm × 4 mm, and if galvanized steel/iron is used, then the desired cross section is 3 mm². When using round conductors made of galvanized steel or iron, the cross-sectional area should not be less than 6 mm². The length of the conductor buried should not be less than 15 m. This type of earthing is used in rocky soil earth bed because at such places excavation work for plate earthing is difficult.

9.6   REACTORS

Reactors are inductive coils having large inductive reactance in comparison to their resistance and are used for limiting short circuit currents during fault conditions. The reactor is having low reactance at low currents. In a power system the reactor can be located either (i) in series with generators, (ii) in series with feeders and (iii) in busbars reactors (in ring system or in tie-bar system).

Generator Reactors – When the reactors are inserted between the generator and the generator bus, the reactors are known as generator reactors. Such reactors protect the machines individually. Since modern machines have transient reactance sufficient to protect themselves against dead three-phase short circuit at its terminals, hence separate reactors are not required in the modern installation. However, when new machines are installed in an old power station, generator reactors may be added for the older generators. The magnitude of such reactors is very approximately about 5 per cent or 0.05 per unit.

The main disadvantage of this method is that if a short circuit occurs on one feeder, the voltage at the generator bus may drop to such a low value that the synchronous machines connected to this common busbar may fall out of step. Thus, the whole system will be adversely affected. Moreover, a full-load current is always flowing through these reactors during normal operation resulting in a constant voltage drop and constant power loss.

Feeder Reactors – When the reactors are connected in series with the feeders instead of generator, the reactors are known as feeder reactors. In the event of fault on any one feeder, in this case, the main voltage drop is in its reactor only and the busbar voltage is not affected much, hence other machines continue supplying load. The other advantage is that the fault on a feeder will not affect other feeders and thus the effects of fault are localized.

The disadvantage of locating reactors in this position is that these do not provide any protection to the generators against short-circuit faults across the busbars. This is, however, of no importance, because such faults are rare and also modern alternators have large transient reactance for their protection against short circuits. The other drawback is of constant voltage drop and constant power loss in reactors even during normal operation.

Busbar Reactors – Busbar reactors are used to tie together separate bus sections. In this system sections are made of generators and feeders and these sections are connected to each other at the common busbar. In this system normally one feeder is fed from one generator. Under normal operating conditions small amount of power flows through the reactors, and therefore, voltage drop and power loss in the reactors are low. The reactors can, therefore, be made with a fairly high ohmic resistance and there is not much voltage drop across it. In case of fault on any one feeder, only one generator feeds the fault while the current from other generators is limited because of presence of the busbar reactors. Thus, heavy currents and voltage disturbances caused by a short circuit on a bus section are reduced and confined primarily to that section. Busbar reactors, however, do not protect the generators connected to the faulty sections. They facilitate the parallel operation of large systems and are extensively employed.

Busbar reactor of ring type is a modification of the above system. In general, with this system the voltage regulation between feeder sections is better than the above system. This system is ideally suited to the generating systems where frequently new generators are being added. In this system, the generators are connected to the common busbar through the reactors but the feeders are fed from the generator side of the reactors. The operation of this system is similar to ring system. It has got an additional advantage. If the number of sections is increased, the fault current will not exceed a certain value, which is fixed by the size of individual reactors. Thus, the switchgear designed to operate successfully on this limiting current will continue operating successfully for any number of extensions of the sections and require no modification.

Saturable Reactors – Saturable reactors as shown in Figure 9.14, also known as magnetic amplifiers, have a non-linear voltage-current characteristic so that when the voltage across the reactor rises above a certain threshold, the reactor current increases disproportionately due to core saturation and can be used for power flow control and also as fault current limiters. It has existed since the late 1800s. As a result, its effective impedance reduces below that without saturation. The amount of saturation is controlled by injecting a variable dc current I_{dc} into the dc control winding that provides dc bias flux φ_{dc} of the ferromagnetic core.

With the exception for the core saturation, the line current flowing in the main ac winding is not affected by that in the dc control winding. Because the two windings are orthogonal, the mutual coupling between them is negligible. The dc current generates the bias dc flux and controls the saturation of the core, i.e., the self-inductance of the ac winding. This inductance reaches the maximum when the core is not saturated (at zero dc current) and the minimum when it is fully saturated (at large enough dc current). The overall reactance in the controlled ac circuit changes as a result of the ac winding self-inductance change. The device operates in the saturation region under normal operating condition, when there is no fault in the system by injecting a high DC current in the control winding, thus offering very low reactance in the system and must be quickly taken out of saturation upon the onset of short-circuit current. This is achieved by sensing the fault and turning DC current to zero.

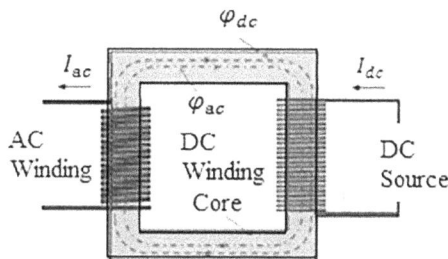

FIGURE 9.14   Saturable reactors.

Ferroresonance – Ferroresonance is a general term applied to a wide variety of interactions between capacitors and iron-core inductors that result in unusual voltages and/or currents. In linear circuits, resonance occurs when the capacitive reactance equals the inductive reactance at the frequency at which the circuit is driven. Iron-core inductors used in power system have a non-linear characteristic and have a range of inductance values. Therefore, there may not be a case where the inductive reactance is equal to the capacitive reactance, but yet very high and damaging over-voltages occur. In power system the ferroresonance occurs when a nonlinear inductor is fed from a series capacitor.

The nonlinear inductor in power system can be due to: (i) The magnetic core of a wound type voltage transformer, (ii) Bank type transformer, (iii) The complex structure of a three-limb three-phase power transformer (core type transformer) and (iv) The complex structure of a five limb three-phase power transformer (shell-type transformer).

The circuit capacitance in power system can be due to a number of elements, such as the following: (i) The circuit-to-circuit capacitance, (ii) Parallel lines capacitance, (iii) Conductor to earth capacitance, (iv) Circuit breaker grading capacitance, (v) Busbar capacitance and (vi) Bushing capacitance.

The phenomena of ferroresonance is a name given to a situation where the non-linear magnetic properties of iron in transformer iron core interact with series capacitance existing in the electrical network to produce a nonlinear tuned circuit with an unexpected resonant frequency. This phenomenon poses a hazard to an electric power system because it generates overvoltages and over-currents.

9.7   LEARNING OUTCOME

The fault analysis of a power system is required in order to provide information for the selection of switchgear, setting of relays, choice of conductors and stability of system operation. Balanced three-phase faults may be analyzed using an equivalent single-phase circuit. With asymmetrical three-phase faults that create severe unbalanced operating conditions, the use of symmetrical components helps to reduce the complexity of the calculations as transmission lines and components are by and large symmetrical, although the fault may be asymmetrical.

In this chapter, readers have been able to learn all the above aspects.

10 Load Flow Analysis

10.1 LOAD FLOW

The study of various methods of solution to power system network is referred to as *load flow study*. The solution provides the voltage various buses, power flowing in various lines and line losses. Purpose of load flow study are as follows: (i) the information obtained from a load flow study are magnitude and phase angles of bus voltages, real and reactive power flowing in each line and line losses which are of utmost importance and frequently provide the starting conditions for other power system analysis such as transient stability, fault analysis and contingency analysis. This is also important as the magnitudes of the bus voltages are required to be held within a specified limit. (ii) The load flow gives us the sinusoidal steady state of the entire system – voltages, real and reactive power generated and absorbed and line losses. (iii) Once the bus voltage magnitudes and their angles are computed using the load flow, the real and reactive power flow through each line can be computed. Also based on the difference between power flow in the sending and receiving ends, the losses in a particular line can also be computed. (iv) Furthermore, from the line flow we can also determine the over- and underload conditions. For load flow studies it is assumed that the loads are constant and they are defined by their real and reactive power consumption. It is further assumed that the generator terminal voltages are tightly regulated and therefore are constant.

Different types of load flow study are as follows: (i) Gauss Siedel Method; invented by two German scientists – Carl Friedrich Gauss and Phillip Ludwig Siedel; (ii) Newton Raphson Method – contains only first-order terms of Taylor Series expansion; (iii) Fast Decoupled Load Flow; (iv) DC Load Flow – only performs active power flow analysis, not a full load analysis; (iv) Second Order Load Flow – consider second-order terms of Taylor Series expansion that requires lesser iteration and better convergence; (v) Multiple Load Flow – consider critical loading condition/voltage collapse; and (vi) Continuation Load Flow – consider high stress contingency situation in bulk power system.

10.2 CLASSIFICATION OF BUSES

Load Buses – $P-Q$ bus [P, Q known, V and δ unknown]	No generators are connected. Hence the generated real power P_{Gi} and reactive power Q_{Gi} are taken as zero. The load drawn by these buses are defined by real power $-P_{Li}$ and reactive power $-PQ_{Li}$ in which the negative sign accommodates for the power flowing out of the bus. The objective of the load flow is to find the bus voltage magnitude $|V_i|$ and its angle δ_i.
Generator or Voltage Controlled Buses P-V bus [P, V known, Q and δ unknown]	Generators are connected. The power generation in such buses is controlled through a prime mover while the terminal voltage is controlled through the generator excitation. Keeping the input power constant through turbine-governor control and keeping the bus voltage constant using automatic voltage regulator, we can specify constant P_{Gi} and $|V_i|$ for these buses. The reactive power supplied by the generator Q_{Gi} depends on the system configuration and cannot be specified in advance. Furthermore, we have to find the unknown angle δ_i of the bus voltage.
Slack or Swing Bus [V, δ known, P and Q unknown]	Usually, this bus is numbered 1 for the load flow studies. This bus sets the angular reference for all the other buses. Since it is the angle difference between two voltage sources that dictate the real and reactive power flow between them, the particular angle of the slack bus is not important. However, it sets the reference against which angles of all the other bus voltages are measured. For this reason, the angle of this bus is usually chosen as 0°. Furthermore, it is assumed that the magnitude of the voltage of this bus is known.

DOI: 10.1201/9781003231240-10

Need for a Slack Bus – The slack bus is needed to account for transmission line losses. In a power system the total power generated will be equal to sum of power consumed by loads and losses. Only the generated power and load power are specified for buses. The slack bus is assumed to generate the power required for losses. Since the losses are unknown, the real and reactive power is not specified for slack bus. They are estimated through the solution of load flow equations.

10.3   BUS ADMITTANCE MATRIX

Load flow calculations can be made using bus admittance matrix or bus impedance matrix. For a network having n nodes or buses excluding ground, a set of following equations, one for each node, can be written as $\bar{I}_i = \sum_{m=1}^{n} \bar{Y}_{im}\bar{V}_m$, $i = 1,2,3,......,n$. $\bar{I}_i$ = complex current entering in the ith bus, $\bar{V}_m$ = complex voltage to ground of the bus m and $\bar{Y}_{im}$ = complex admittance between buses I and m. $Y_{ii} = y_{i0} + y_{i2} + y_{i3} + ... + y_{in}$, where y_{i0} = shunt admittance and $\sum \left[(Y_{ii} + Y_{ik}) = 0 \right]$ for a row, when $y_{i0} = 0$.

The expression $\bar{I}_i = \sum_{m=1}^{n} \bar{Y}_{im}\bar{V}_m$ can be established by using either the bus or loop frame of reference. The coefficients of the equations depend on the selection of the independent variables, i.e., voltage or currents. Thus, either the admittance or impedance network matrix can be used. In the loop frame reference in admittance form, loop admittance matrix is obtained by a matrix inversion and this process is tedious and time consuming and is to be repeated for each subsequent change of network if any. Bus frame reference in admittance form finds widespread application because of the simplicity of data preparation and the case with the bus admittance can be formed and modified for network changes in subsequent cases. Bus admittance matrix is basically nodal admittance matrix. In realistic system this matrix is quite sparse as in real power system each bus is connected to only a few other buses through transmission line. Thus the advantages of having bus admittance matrix are as follows: (i) Y_{BUS} is a Sparse Matrix, that is, most of elements are zero, and hence less memory is required for storage whereas Z_{BUS} is a full matrix and all elements are non-zero so more memory is required; (ii) Y_{BUS} can be easily formed using inspection method whereas Z_{BUS} building algorithm is complicated; (iii) Y_{BUS} can be easily modified in case of any changes in the power system such as addition of transmission line or removal of transmission line; and (iv) Consider an element having admittance y_{ip} connected between buses i and p. Four entries are affected: Y_{ii}, Y_{ip}, Y_{pi}, Y_{pp} and the new entries will be $Y_{ii,\text{new}} = Y_{ii,\text{old}} + y_{ip}$, $Y_{ip,\text{new}} = Y_{ip,\text{old}} - y_{ip}$, $Y_{pi,\text{new}} = Y_{pi,\text{old}} - y_{ip}$ and $Y_{pp,\text{new}} = Y_{pp,\text{old}} + y_{ip}$.

10.4   REAL AND REACTIVE POWER INJECTED IN A BUS

Let the voltage at the ith bus be denoted by $V_i = |V_i| \angle \delta_i = |V_i|(\cos\delta_i + j\sin\delta_i)$. The self-admittance at bus i is $Y_{ii} = |Y_{ii}| \angle \theta_{ii} = |Y_{ii}|(\cos\theta_{ii} + j\sin\theta_{ii}) = G_{ii} + jB_{ii}$ and the mutual admittance between buses i and j can be written as $Y_{ij} = |Y_{ij}| \angle \theta_{ij} = |Y_{ij}|(\cos\theta_{ij} + j\sin\theta_{ij}) = G_{ij} + jB_{ij}$. If the power system contains a total number of n buses, the current injected at bus i is given as $\bar{I}_i = \bar{Y}_{i1}\bar{V}_1 + \bar{Y}_{i2}\bar{V}_2 + ... + \bar{Y}_{in}\bar{V}_n = \sum_{k=1}^{n} \bar{Y}_{ik}\bar{V}_k$. In a power system, complex power is a more important quantity than the current. The complex power input into a bus can be expressed as $P_i + jQ_i = \bar{V}_i\bar{I}_i^*$. Hence, $P_i + jQ_i = \bar{V}_i \sum_{k=1}^{n} \bar{Y}_{ik}^* \bar{V}_k^*$, where $i = 1,2,3,......,n = |V_i| \angle \delta_i \sum_{k=1}^{n} \bar{Y}_{ik} \angle -\theta_{ik} \bar{V}_k \angle -\delta_k = |V_i| \sum_{k=1}^{n} \bar{Y}_{ik}\bar{V}_k e^{j}(\delta_i - \delta_k - \theta_{ik})$, $i = 1,2,3,......,n$.

Remembering that for each bus there are two unknowns and thus for a system of n buses, there are $2n$ unknowns and $2n$ real equations that result from the set of complex equation. Hence the solution is feasible. For systematic analysis, it is convenient to regard loads as negative generators and lump together the generator and load at the buses. Thus at ith bus, the net complex power injected into the bus is given by $S_i = P_i + jQ_i = (P_{Gi} - P_{Di}) + j(Q_{Gi} - Q_{Di})$, where the complex power supplied by the generators is $S_{Gi} = P_{Gi} + jQ_{Gi}$ and the complex power drawn by the loads is $S_{Di} = P_{Di} + jQ_{Di}$. The real and reactive powers injected into ith bus are then $P_i = (P_{Gi} - P_{Di})$ and $Q_i = (Q_{Gi} - Q_{Di})$.

In the load flow problem, it is convenient to use $P_i - jQ_i = \bar{V}_i^* \bar{I}_i$ rather than $P_i + jQ_i = \bar{V}_i \bar{I}_i^*$.

So, $P_i - jQ_i = \bar{V}_i^* \bar{I}_i = \bar{V}_i^* \sum_{k=1}^{n} \bar{Y}_{ik} \bar{V}_k = |V_i|(\cos\delta_i - j\sin\delta_i) \sum_{k=1}^{n} |Y_{ik} V_k|(\cos\theta_{ik} + j\sin\theta_{ik})(\cos\delta_k + j\sin\delta_k) = \sum_{k=1}^{n} |Y_{ik} V_i V_k|(\cos\delta_i - j\sin\delta_i)(\cos\theta_{ik} + j\sin\theta_{ik})(\cos\delta_k + j\sin\delta_k)$. Since, $(\cos\delta_i - j\sin\delta_i)(\cos\theta_{ik} + j\sin\theta_{ik})(\cos\delta_k + j\sin\delta_k) = \cos(\theta_{ik} + \delta_k - \delta_i) + j\sin(\theta_{ik} + \delta_k - \delta_i)$; $P_i = \sum_{k=1}^{n} |Y_{ik} V_i V_k|\cos(\theta_{ik} + \delta_k - \delta_i)$ and $Q_i = -\sum_{k=1}^{n} |Y_{ik} V_i V_k|\sin(\theta_{ik} + \delta_k - \delta_i)$. P_i and Q_i are set of non-linear simultaneous algebraic equations for which there is no general solutions until now. These are to be solved iteratively by some numerical methods. In an iterative method, we progressively compute more accurate estimates of the unknown until results are obtained to a desired degree of accuracy in a finite number of iterations. When this happens the solutions are said to converge.

10.5   LOAD FLOW BY GAUSS-SEIDEL METHOD

To start with, we assume that all the buses except slack bus is load bus, i.e., P and Q are known. The slack bus voltage being specified, there are $(n-1)$ bus voltage starting values of which magnitudes and angles are assumed. These values are updated through an iterative process. Since,

$$P_i - jQ_i = \bar{V}_i^* \bar{I}_i = \bar{V}_i^* \sum_{k=1}^{n} \bar{Y}_{ik} \bar{V}_k = \bar{V}_i^* \sum_{k=1, k \neq i}^{n} \bar{Y}_{ik} \bar{V}_k + \bar{V}_i^* \bar{Y}_{ii} \bar{V}_i ; \; \bar{V}_i = (1/\bar{Y}_{ii})\left[(P_i - jQ_{,i})\left[(P_i - jQ_{,i})/ V_i^* - \sum_{k=1, k \neq i}^{n} \bar{Y}_{ik} \bar{V}_k \right]\right.$$

and if the ith bus is a load bus, then $\bar{V}_i^{(k)} = (1/\bar{Y}_{ii})\left[(P_i - jQ_i)/ V_i^{*(k-1)} - \sum_{m=1}^{i-1} \bar{Y}_{im} \bar{V}_m^{(k)} - \sum_{m=i+1}^{n} \bar{Y}_{im} \bar{V}_m^{(k-1)} \right]$ and $\bar{V}_i = (1/\bar{Y}_{ii})\left[(P_i - jQ_i)/ V_i^* - \sum_{k=1, k \neq i}^{n} \bar{Y}_{ik} \bar{V}_k \right]$

$\bar{V}_i$ can be written as $(A_i / V_i^*) - \sum_{k=1, k \neq i}^{n} \bar{B}_{ik} \bar{V}_k$; where $A_i = (P_i - jQ_i)/ \bar{Y}_{ii}$ and $\bar{B}_{ik} = \bar{Y}_{ik} / \bar{Y}_{ii}$.

δ_i can be calculated from the relation $\delta_i^{(K)} = \angle V_i^{(K)} = $ Angle of $\left[A_i^{(k)} / (V_i^{(k)})^* - \sum_{k=1, k \neq i}^{n-1} \bar{B}_{ik} \bar{V}_k^{(k)} - \sum_{k=n+1, k \neq i}^{n} \bar{B}_{ik} \bar{V}_k^{(k-1)} \right]$.

If the ith bus is a generator bus, then its reactive power loading is not specified but its minimum and maximum limits are mentioned.

Thus, the inequality $Q_{i,min} \leq Q_i \leq Q_{i,max}$ holds and $\bar{Q}_i^{(k)} = -IM\left[V_i^{*(k-1)}\left(\sum_{m=1}^{i-1} \bar{Y}_{im} \bar{V}_m^{(k)} + \sum_{m=i+1}^{n} \bar{Y}_{im} \bar{V}_m^{(k-1)} \right)\right]$ as $P_i - jQ_i = \bar{V}_i^* \bar{I}_i = \bar{V}_i^* \sum_{m=1}^{n} \bar{Y}_{im} \bar{V}_m$. If the value of Q_i, remains within the range indicated by the inequality $Q_{i,min} \leq Q_i \leq Q_{i,max}$, it can be used for solving the aforesaid equation. If, however, the value of Q_i, calculated falls outside the range, the specified magnitude of V_i cannot be maintained. In this case, the limiting value of Q_i, upper or lower, whichever is nearer to the calculated value is to be taken.

Convergence of the Algorithm – The rate of convergence in the Gauss-Siedel iterative method using the nodal admittance is slow because of the sparsity of the admittance matrix. In each iteration, an improvement in each bus affects only the buses directly connected to it. The rate of convergence can be accelerated by multiplying the voltage changes obtained in each iteration by a factor, known as the *acceleration factor*.

Although the Gauss-Siedel method can be easily programmed and does not require a large number of computer storage, it has several limitations. It fails to converge in systems having (i) large number of radial lines, (ii) heavily loaded lines, (iii) negative values of transfer admittances and (iv) long and short lines terminating on the same bus.

10.6   LOAD FLOW BY NEWTON-RAPHSON METHOD

Since $P_i + jQ_i = V_i I_i^* = V \sum_{k=1}^{n} Y_{ik}^* V_k^* {}_i = V_i \sum_{k=1}^{n} Y_{ik} V_k e^{j(\delta_i - \delta_k - \theta_{ik})}$, it can be said $P_i = f(\delta, V)$ and $Q_i = f(\delta, V)$, where δ and V are vectors of phase angles and voltage magnitudes of the system buses. The difference equations can now be written as

$$\Delta P_i = \sum_{k=1}^{n} \frac{\partial P_i}{\partial \delta_k} \Delta \delta_k + \sum_{k=1}^{n} \frac{\partial P_i}{\partial V_k} \Delta V_k \text{ and } \Delta Q_i = \sum_{k=1}^{n} \frac{\partial Q_i}{\partial \delta_k} \Delta \delta_k + \sum_{k=1}^{n} \frac{\partial Q_i}{\partial V_k} \text{ for } i = 1,2,3,\dots,n$$

(10.1)

ΔP_i and ΔQ_i physically signify the differences between the specified values of P and Q and their calculated values.

Hence the above set of equations can be used to determine the complex bus voltages starting from an arbitrary set of these values in an iterative way. For load buses both P and Q are specified but for generator buses, Q remains unspecified. Hence for each load bus, two equations result but for each generator bus, only one involving ΔP. Moreover, if the reactive loadings of the generator buses remain within their limits, ΔV equals to zero. Further for slack bus $\Delta \delta$ is also zero.

Hence, $\Delta P_i = \sum_{\substack{k=1 \\ k \neq s}}^{n} H_{ik} \Delta \delta_k + \sum_{\substack{k=1 \\ k \neq g}}^{n} N_{ik} \Delta V$ for $i = 1,2,3,\dots,n$ but for $i \neq s$ (slack bus); $H_{ik} = \partial P_i / \partial \delta_k$, $N_{ik} = \partial Q_i / \partial \delta_k$. $\Delta Q_i = \sum_{\substack{k=1 \\ k \neq s}}^{n} M_{ik} \Delta \delta_k + \sum_{\substack{k=1 \\ k \neq g}}^{n} L_{ik} \Delta V_k$ for $i = 1,2,3,\dots,n$ but for $i \neq g$ (generator bus); $M_{ik} = \partial P_i / \partial V_k$, $L_{ik} = \partial Q_i / \partial V_k$.

Thus, $\begin{bmatrix} \Delta \mathbf{P} \\ \Delta \mathbf{Q} \end{bmatrix} = \begin{bmatrix} \mathbf{H} & \mathbf{N} \\ \mathbf{M} & \mathbf{L} \end{bmatrix} \begin{bmatrix} \Delta \delta \\ \Delta \mathbf{V} \end{bmatrix}$.

H and **N** can be found out taking partial derivative w.r.t. to δ_k, $k \neq i$ that gives $\partial P_i / \partial \delta_k + j(\partial Q_i / \partial \delta_k) = -jV_i Y_{ik} V_k e^{j\delta_i} e^{-j\delta_k} e^{-j\theta_{ik}} = -jV_i e^{j\delta_i} Y_{ik} e^{-j\theta_{ik}} V_k e^{-j\delta_k} = -j(e_i + jf_i)(G_{ik} - jB_{ik})(e_k - jf_k)$. The product of last two terms on the r.h.s is a current, $I_k = (a_k - jb_k)$. Hence $\partial P_i / \partial \delta_k + j(\partial Q_i / \partial \delta_k) = -j(e_i + jf_i)(a_k - jb_k)$. So, $H_{ik} = \partial P_i / \partial \delta_k = (a_k f_i - b_k e_i)$ and $M_{ik} = \partial Q_i / \partial \delta_k = -(a_k e_i - b_k f_i)$ for $k \neq i$.

For $k = i$, the partial derivatives become $\partial P_i / \partial \delta_i + j(\partial Q_i / \partial \delta_i) = jV_i e^{j\delta_i} \left[\sum_{k=1}^{n} Y_{ik} e^{-j\theta_{ik}} V_k e^{-j\delta_k} \right] - jV_i e^{j\delta_i} Y_{ii} e^{-j\theta_{ii}} V_i e^{-j\delta_i} = j(P_i + jQ_i) - jV_i^2 (G_{ii} - jB_{ii})$. So, $H_{ii} = \partial P_i / \partial \delta_i = -Q_i - B_{ii} V_i^2$ and $M_{ii} = \partial Q_i / \partial \delta_i = P_i - G_{ii} V_i^2$ for $k = i$.

For obtaining N_{ik} and L_{ik}, taking partial derivatives of $P_i + jQ_i = V_i I_i^* = V_i \sum_{k=1}^{n} Y_{ik}^* V_k^* = V_i \sum_{k=1}^{n} Y_{ik} V_k e^{j(\delta_i - \delta_k - \theta_{ik})}$ w.r.t V_k, $k \neq i$ gives $\partial P_i / \partial V_k + j(\partial Q_i / \partial V_k) = V_i e^{j\delta_i} Y_{ik} e^{-j\theta_{ik}} e^{-j\delta_k} = (V_i e^{j\delta_i} Y_{ik} e^{-j\theta_{ik}} V_k e^{-j\delta_k}) / V_k$. The product of last two terms in the numerator is the current $(a_k - jb_k)$.

Expressing voltages in rectangular form and separating real and imaginary parts gives $N_{ik} = \partial P_i / \partial V_k = (a_k e_i + b_k f_i) / V_k$ and $L_{ik} = \partial Q_i / \partial V_k = (a_k f_i - b_k e_i) / V_k$ for $k \neq i$.

If $k = i$, the partial will be $\partial P_i / \partial V_i + j(\partial Q_i / \partial V_i) = e^{j\delta_i} \left[\sum_{k=1}^{n} Y_{ik} e^{-j\theta_{ik}} V_k e^{-j\delta_k} \right] + V_i e^{j\delta_i} Y_{ii} e^{-j\theta_{ii}} e^{-j\delta_i} = [(P_i + jQ_i) / V_i] - V_i(G_{ii} - jB_{ii})$. So, $N_{ii} = \partial P_i / \partial V_i = (P_i / V_i) + G_{ii} V_i$ and $L_{ii} = \partial Q_i / \partial V_i = (Q_i / V_i) - B_{ii} V_i$. If, $N_{ii} = (\partial P_i / \partial V_i) V_i$ and $L_{ii} = (\partial Q_i / \partial V_i) V_i$, then $N_{ii} = P_i + G_{ii} V_i^2$ and $L_{ii} = Q_i - B_{ii} V_i^2$.

Comparison of the G-S and N-R methods of load flow solutions:

In Gauss-Seidel Method (i) the variables are expressed in rectangular coordinates, (ii) computation time per iteration is less, (iii) it has linear convergence characteristics, (iv) the number of iterations required for convergence increases with size of the system and (v) the choice of slack bus is critical.

In Newton-Raphson Method (i) the variables are expressed in polar coordinates, (ii) computation time per iteration is more, (iii) it has quadratic convergence characteristics, (iv) the number of iterations is independent of the size of the system and (v) the choice of slack bus is arbitrary.

10.7   FAST DECOUPLED LOAD FLOW

An important and useful property of power system is that the change in real power is primarily governed by the charges in the voltage angles, but not in voltage magnitudes. On the other hand, the charges in the reactive power are primarily influenced by the charges in voltage magnitudes, but not in the voltage angles. So (i) under normal steady-state operation, the voltage magnitudes are all nearly equal to 1.0; (ii) as the transmission lines are mostly reactive, the conductance are quite small as compared to the susceptance $G_{ij} \ll B_{ij}$; (iii) under normal steady-state operation the angular differences among the bus voltages are very small, i.e., $\delta_i - \delta_j \approx 0$; and (iv) the injected reactive power at any bus is always much less than the reactive power consumed by the elements connected to this bus when these elements are shorted to the ground $Q_j \ll B_{ii}V_i^2$. We know $V_i = |V_i| \angle \delta_i = |V_i|(\cos\delta_i + j\sin\delta_i)$, $Y_{ii} = |Y_{ii}| \angle \theta_{ii} = |Y_{ii}|(\cos\theta_{ii} + j\sin\theta_{ii}) = G_{ii} + jB_{ii}$, where $G_{ii} =$ conductance and $B_{ii} =$ susceptance and $Y_{ik} = |Y_{ik}| \angle \theta_{ik} = |Y_{ik}|(\cos\theta_{ik} + j\sin\theta_{ik}) = G_{ik} + jB_{ik}$.

Since $\quad P_i - jQ_i = \overline{V}_i^* \overline{I}_i = \overline{V}_i^* \sum_{k=1}^{n} \overline{Y}_{ik}\overline{V}_k = |V_i| \sum_{k=1}^{n} |Y_{ik}V_k| e^{j(\theta_{ik} +\delta_k -\delta_i)} = |V_i|^2 \left(+jB_{ii}\right) + \sum_{\substack{k=1 \\ k \neq i}}^{n}$

$|Y_{ik}V_iV_k| e^{j(\theta_{ik} +\delta_k -\delta_i)}$, hence $P_i = |V_i|^2 G_{ii} + \sum_{\substack{k=1 \\ k \neq i}}^{n} |Y_{ik}V_iV_k| \cos(\theta_{ik} +\delta_k -\delta_i)$ and $Q_i = -|V_i|^2 B_{ii} +$

$\sum_{\substack{k=1 \\ k \neq i}}^{n} |Y_{ik}V_iV_k| \sin(\theta_{ik} +\delta_k -\delta_i)$. Hence, $\partial P_i / \partial V_i = 2 V_iG_{ii} + \sum_{\substack{k=1 \\ k \neq i}}^{n} V_kY_{ik} \cos(\delta_i -\delta_k -\theta_{ik}) =$

$2 V_iG_{ii} + \sum_{\substack{k=1 \\ k \neq i}}^{n} V_kY_{ik} \left[\cos(\delta_i -\delta_k)\cos\theta_{ik} + \sin(\delta_i -\delta_k)\sin\theta_{ik} \right] = 2 V_iG_{ii} + \sum_{\substack{k=1 \\ k \neq i}}^{n} V_kY_{ik} [\cos$

$(\delta_i -\delta_k)\cos\theta_{ik} + \sin(\delta_i -\delta_k)\sin\theta_{ik}] = 2 V_iG_{ii} \sum_{\substack{k=1 \\ K \neq i}}^{n} V_k \left[G_{ik}\cos(\delta_i -\delta_k) + + B_{ik}\sin(\delta_i -\delta_k) \right]$ for $k = i$

and $\partial P_i / \partial V_k = V_iY_{ik} \cos(\delta_i -\delta_k -\theta_{ik}) = V_i \left[G_{ik}\cos(\delta_i -\delta_k) + B_{ik}\sin(\delta_i -\delta_k) \right]$, for $k \neq i$.

Now G_{ii} and G_{ik} are very small and negligible and also $\cos(\delta_i -\delta_k) \approx 1$ and $\sin(\delta_i -\delta_k) \approx 0$, hence $\partial P_i / \partial V_i \approx 0$ and $\partial P_i / \partial V_k \approx 0$, so $J_2 = 0$.

$\partial Q_i / \partial \delta_i = \sum_{\substack{k=1 \\ K \neq i}}^{n} V_kV_i \left[G_{ik} \cos(\delta_i -\delta_k) + B_{ik} \sin(\delta_i -\delta_k) \right], k = i$ and $Q_i / \partial \delta_k = - V_iV_k \left[G_{ik}\cos \right.$

$\left. (\delta_i -\delta_k) + B_{ik}\sin(\delta_i -\delta_k) \right]$, $k \neq i$.

G_{ii} and G_{ik} are very small and negligible and also $\cos(\delta_i -\delta_k) \approx 1$ and $\sin(\delta_i -\delta_k) \approx 0$, hence $\partial Q_i / \partial \delta_i \approx 0$ and $\partial Q_i / \partial \delta_k \approx 0$, so $\mathbf{J}_3 = 0$.

So, $\begin{bmatrix} \Delta \mathbf{P} \\ \Delta \mathbf{Q} \end{bmatrix} = \begin{bmatrix} \mathbf{J}_1 & 0 \\ 0 & \mathbf{J}_4 \end{bmatrix} \begin{bmatrix} \Delta \delta \\ \Delta \mathbf{V} \end{bmatrix}$

In other words, $\Delta \mathbf{P}$ depends only on $\Delta \delta$ and $\Delta \mathbf{Q}$ depends only on $\Delta \mathbf{V}$. Thus, there is a decoupling between "$\Delta \mathbf{P} - \Delta \mathbf{\theta}$" and "$\Delta \mathbf{Q} - \Delta \mathbf{V}$" relations. Again,

$$\partial P_i/\partial \delta_i = \partial/\partial \delta_i \left[\sum_{k=1}^{n} V_iY_{ik}V_k \cos(\delta_k -\delta_i +\theta_{ik}) \right]$$

$$= -V_iY_{ii}V_i \sin(\delta_i -\delta_i +\theta_{ii}) - \sum_{k=1}^{n} V_iY_{ik}V_k \sin(\delta_k -\delta_i +\theta_{ik})$$

$$= -B_{ii}V_i^2 - Q_i \approx -B_{ii}V_i^2; \text{ for } k = i$$

$$\partial P_i/\partial \delta_k = -V_iY_{ik}V_k \sin(\delta_k -\delta_i +\theta_{ik}) = V_iY_{ik}V_k \sin(\delta_i -\delta_k -\theta_{ik})$$

$$= V_iY_{ik}V_k \left[\sin(\delta_i -\delta_k)\cos\theta_{ik} - \cos(\delta_i -\delta_k)\sin\theta_{ik} \right] = -V_iV_k$$

$$B_{ik}; \text{ for } k \neq i.$$

This gives $\Rightarrow \Delta P_i = -V_i \sum_{k=1}^{n} B_{ik} V_k \Delta \delta_k$ or, $\Delta P_i / V_i = -\sum_{k=1}^{n} B_{ik} V_k \Delta \delta_k$. Under normal steady-state operating condition, $V_i \approx 1.0$ and $\Delta P_i / V_i = -\sum_{k=1}^{n} B_{ik} \Delta \delta_k$. Hence, $\Delta \mathbf{P} / \mathbf{V} = -[\boldsymbol{B}]\Delta \delta = [\boldsymbol{B'}]\Delta \delta$. Matrix $\boldsymbol{B'}$ is a constant matrix having a dimension of (n-1) X (n-1). Its elements are the negative of the imaginary part of the element (i, k) of the Y_{BUS}, where $i = 1, 2, 3, \ldots, n$ and $k = 1, 2, 3, \ldots, n$.

$\partial Q_i / \partial V_k = -B_{ii} V_i$ for $k = i$ and $Q_i / \partial V_k = -B_{ik} V_i$ for $k \neq i$, so, $\Delta Q_i = -V_i \sum_{k=1}^{n} B_{ik} \Delta V$ and $\Delta Q_i / V_i = -\sum_{k=1}^{n} B_{ik} \Delta V_k$. Since, under normal steady-state operating condition, $V_i \approx 1.0$; so $\Delta P_i / V_i = -\sum_{k=1}^{n} B_{ik} \Delta \delta_k$, i.e., $\Delta \mathbf{P} / \mathbf{V} = -[\boldsymbol{B''}]\Delta \mathbf{V}$. Matrix $\boldsymbol{B''}$ is a constant matrix having a dimension of (n-m) X (n-m).

Its elements are the negative of the imaginary part of the element (i, k) of the $\mathbf{Y_{BUS}}$, where $i = (m + 1), (m + 2), \ldots, n$ and $k = (m + 1), (m + 2), \ldots, n$. As the matrices B' and $\boldsymbol{B''}$ are constant matrices, it is not necessary to invert these matrices in each iteration. Rather, the inverse of these matrices can be stored and used in each iteration, thereby making the algorithm faster. Further simplification in the FDLF algorithm can be made by (i) ignoring the series resistances in calculating the elements $[\boldsymbol{B'}]$. Also, by omitting the elements of $[\boldsymbol{B'}]$ that predominantly affect reactive power flows, i.e., shunt reactance and transformer off-nominal in phase taps. (ii) Omitting from $[\boldsymbol{B'}]$ the angle shifting effect of phase shifter, which predominantly affects real power flow

Alternatively: $\begin{bmatrix} \Delta \mathbf{P} \\ \Delta \mathbf{Q} \end{bmatrix} = \begin{bmatrix} \mathbf{J}_1 & 0 \\ 0 & \mathbf{J}_4 \end{bmatrix}\begin{bmatrix} \Delta \delta \\ \Delta \mathbf{V} \end{bmatrix} = \begin{bmatrix} \boldsymbol{H} & 0 \\ 0 & \boldsymbol{L} \end{bmatrix}\begin{bmatrix} \Delta \delta \\ \Delta \mathbf{V} \end{bmatrix}$ putting N and M equal to zero. This gives $[\Delta \mathbf{P}] = [\boldsymbol{H}][\Delta \delta]$ and $[\Delta \mathbf{Q}] = [\boldsymbol{L}][\Delta \mathbf{V}]$.

Diagonal elements of H, i.e., $H_{ii} = -Q_i - B_{ii} V_i^2$ and off-diagonal elements are $H_{ik} = V_i V_k [G_{ik} \sin(\delta_i - \delta_k) - B_{ik} \cos(\delta_i - \delta_k)]$. Diagonal elements of L, i.e., $L_{ii} (Q_i / V_i) - B_{ii} V_i$ and off-diagonal elements are $L_{ik} = V_i [G_{ik} \sin(\theta_{ik}) - B_{ik} \cos(\theta_{ik})]$. The decoupled algorithm loses quadratic convergence and resorts to linear (geometric convergence).

10.8 DC LOAD FLOW

Full power flow allows for management of both active and reactive power flows. Recently, with the liberalization of electricity markets, active power and reactive power are treated as different products. Active power is a tradable commodity, while reactive power is rather regarded as an ancillary service that has to be provided by the system operator and its costs are socialized among all users of the system. Due to the separation of these products, methods looking only at the active power flow become of increasing interest.

DC power flow is of the variations of the Newton method very similar to fast decoupled method. It is a simplification of a full AC power flow and looks only at active power flows, neglecting voltage support, reactive power management and transmission losses. Thanks to its simplicity, and even more to the fact that DC power flow problem is linear, it is very often used for techno-economic studies of power systems for assessing the influence of commercial energy exchanges on active power flows in the transmission network.

DC power flow can be applied if a number of assumptions are satisfied.

The resistance of transmission circuits is significantly less than the reactance. Usually, it is the case that the x/r ratio is between 2 and 10. So any given transmission circuit with impedance of $z = r - jx$ will have an admittance of

$$y = \frac{1}{(r + jx)} = \frac{r}{(r^2 + x^2)} - \frac{jx}{(r^2 + x^2)} = g + jb; \text{ where } g = \frac{r}{(r^2 + x^2)} \text{ and } b = \frac{-x}{(r^2 + x^2)} \qquad (10.2)$$

If r is very small compared to x, then we observe that g will be very small compared to b, and it is reasonable to approximate $g = 0$, $b = -1/x$. Now, if $g = 0$, then the real part of all of the Y- bus elements will also be zero, that is, $g = 0 \Rightarrow G = 0$.

Since $P_i = \sum_{k=1}^{n} |Y_{ik} V_i V_k| \cos(\theta_{ik} + \delta_k - \delta_i) = \sum_{k=1}^{n} |V_i||V_k||Y_{ik}|[\cos\theta_{ik} \cos(\delta_k - \delta_i) - \sin\theta_{ik} \sin(\delta_k - \delta_i)] = \sum_{k=1}^{n} |V_i||V_k| [G_{ik} \sin(\delta_k - \delta_i) - B_{ik} \cos(\delta_k - \delta_i)]$ and $Q_i = \sum_{k=1}^{n} |V_i||V_k|[G_{ik} \cos(\delta_k - \delta_i) + B_{ik} \sin(\delta_k - \delta_i)]$, P_i can be written as $P_i = \sum_{k=1}^{n} V_k V_i B_{ik} \sin(\delta_i - \delta_k)$ and $Q_i = -\sum_{k=1}^{n} V_k V_i B_{ik} \cos(\delta_i - \delta_k)$.

For most typical operating conditions, the difference in angles of the voltage phasors at two buses k and j connected by a circuit, which is $\delta_k - \delta_i$ for buses k and i, is less than $10°–15°$. It is extremely rare to ever see such angular separation exceed $30°$. Thus, we say that the angular separation across any transmission circuit is "small". This gives $P_i = \sum_{k=1}^{n} V_k V_i B_{ik} \sin(\delta_i - \delta_k) \approx \sum_{k=1}^{n} V_k V_i B_{ik} (\delta_i - \delta_k)$ and $Q_i = -\sum_{k=1}^{n} V_k V_i B_{ik} \cos(\delta_k - \delta_i) \approx -\sum_{k=1}^{n} V_k V_i B_{ik} = -|V_i|^2 B_{ii} - \sum_{k=1, k \neq i}^{n} V_k V_i B_{ik}$.

A significant progress has been achieved at this point, in relation to obtaining linear power flow equations, since the trigonometric terms are eliminated. B_{ik} is not actually a susceptance but rather an element in the Y-bus matrix. If $k \neq i$, then $B_{ik} = -b_{ik}$, i.e., the Y_{BUS} element in row k column i is the negative of the susceptance of the circuit connecting bus k to i. If $k = i$, then $B_{ii} = -b_i + \sum_{k=1, k \neq i}^{n} b_{ik}$.

The classic power flow problem consists of active and reactive power flow and it can be formulated using four variables per each node – voltage angle, voltage magnitude, active and reactive power injections. Active power losses are not known in advance as they depend on active power injection pattern and voltage profile. Other variables are also interdependent, which makes the problem non-linear. This is why it is often made linear and the solution is iterated. The losses are re-estimated at each iteration based on all other variables.

DC power flow, on the contrary, is a linear problem. It neglects active power losses, and assumes that magnitudes of nodal voltages are equal. Furthermore, voltage angle differences are assumed to be small. The only variables are voltage angles and active power injections. Due to the fact that losses are neglected, all active power injections are known in advance. Therefore, the problem becomes linear and there is no need for iterations.

To simplify the power flow problem and make it linear, a number of assumptions are made, which are as follows: (i) voltage angle differences are small, i.e., $\sin\delta = \delta$; (ii) Line resistance is negligible, i.e., $R \ll X$, thus lossless lines; and (iii) Flat voltage profile. However, such assumptions are not always realistic. First, the X/R ratio condition can be difficult to guarantee. The influence of resistance increases with the decrease of voltage, which means that only the high-voltage transport networks can withstand this condition. Moreover, voltages will most likely not be flat but will vary among busses, causing the voltage profile to be different from the assumed one. Each of these assumptions has some influence on the accuracy of the power flow calculations. In the per-unit system, the numerical values of voltage magnitudes $|V_i|$ and $|V_k|$ are very close to 1.0. This makes $P_i = \sum_{k=1, k \neq i}^{n} B_{ik}(\delta_i - \delta_k)$.

Since $\Delta \mathbf{P} = -[B]\Delta\delta = [B']\Delta\delta$, $\mathbf{P}$ is the vector of bus real power injections, B' is the bus susceptance matrix and δ is the vector of bus voltage angles. As to the fact that losses are neglected, all active power injections are known in advance. Since B' is also known, the single solution for this problem can be calculated directly using: $\Delta\delta = [B']^{-1}\Delta\mathbf{P}$.

The DC power flow has three advantages over the standard N-R method, which are as follows: (i) the system matrix B' is about half the size of full size problem, (ii) the problem is noniterative just a single calculation in order to obtain the solution and (iii) the system matrix B' is independent of the system state and therefore to be calculated only once as long as system topology does not change.

The first two advantages make DC power flow seven to ten times faster than AC. The last advantage is of great importance when multiple subsequent solutions are needed since $\left[\boldsymbol{B'}\right]^{-1}$ has to be calculated only once. This significantly reduces calculation time as triangularization is computational heavy.

10.9 LEARNING OUTCOME

Load flow analysis is the most important and essential approach for power system operation and planning. The analysis can provide a balanced steady operation state of the power system, without considering system transient processes. The load flow analysis is an important fundamental tool in planning and designing the future expansion of the system besides economic scheduling of existing systems. Various analysis methods like Gauss-Seidel, Newton-Raphson, Fast Decoupled Load Flow Analysis and DC Load flow analysis are employed. As the system is growing and becoming more complex, faster, simpler and more efficient mathematical formulations with less storage are the requirement ensuring convergence.

In this chapter, readers have been able to learn all the above aspects.

11 Stability Analysis

11.1 REQUIREMENT OF POWER SYSTEM STABILITY

System stability is a major factor that must be maintained for the reliable operation of electric power system to ensure the consumers a continuous source of stable power. A power system never operates at a point of true steady state and thus is always characterized by dynamic behavior. A steady state operating condition may be defined as an operating condition of a power system in which all the operating quantities that characterize it can be considered to be constant for the purpose of analysis.

Whenever a disturbance occurs, the system tries to adjust to new operating condition and the adjustment to the new operating condition is called the transient period. A disturbance is a sudden change or a sequence of changes in one or more parameters of the system or in one or more of the operating quantities, which are physical in nature that can be measured or calculated and that can be used to describe operating conditions. The system behavior under this time is the dynamic system performance. If the oscillatory response of a power system during the transient period following a disturbance is damped and the system settles in a finite time to a new steady operating condition, the system is stable. Thus, stability is a property that enables it to remain in a state of operating equilibrium under normal operating condition and to remain in an acceptable state of equilibrium after being subjected to a disturbance.

The disturbances can be small such as small load changes or random load changes occurring under normal load condition and large such as loss of generator, a fault or loss of line. Small disturbances are those for which the equations that describe the dynamics of the power system may be linearized for the purpose of analysis. Large disturbances are which for those the equations that describe the dynamics of the power system cannot be linearized for the purpose of analysis. Different types of stability are (i) Steady State Stability – studies are restricted to small and gradual changes in the system operating conditions. A power system is steady state stable for a particular steady state operating condition if, following any small disturbance, it reaches a steady state operating condition which is identical/close to the pre-disturbance operating condition. It is often termed as *small signal stability* (ii) Dynamic Stability – The ability of a power system to maintain stability under continuous small disturbances is investigated under the name of *Dynamic Stability or small signal stability*; (iii) Transient Stability – A power system is transiently stable for a particular steady state operating condition and for a particular disturbance, if following the major disturbance, it reaches an acceptable steady state operating condition. It is often termed as large signal stability. Transient Stability limit is lower than the steady state stability limit.

11.2 SWING EQUATION

The differential equation that relates angular momentum, its acceleration and the rotor angle is called swing equation. Solution of swing equation will show how the rotor angle changes with respect to time following a disturbance. Kinetic energy of rotor $\mathrm{KE} = (1/2)I\omega_m^2$, where I = moment of inertia in kg/m^2 and ω_m = angular speed in mechanical rad/s Angular momentum $M = I\omega_m$; so $\mathrm{KE} = (1/2)M\omega_m$. When torque is applied to a body, the body experiences angular acceleration and the torque is equal to the product of angular acceleration and moment of inertia: $T = I\alpha$, where T is the net torque and α is the angular acceleration. The angular acceleration, a, may be expressed in terms of θ_m, a mechanical angle which is measured from a non-rotating reference frame. So, $\alpha = d^2\theta_m/dt^2$. Hence, $T = I\alpha = I\left(d^2\theta_m/dt^2\right)$.

DOI: 10.1201/9781003231240-11

The net torque, which produces acceleration, is the algebraic difference of mechanical torque applied and electrical torque output $= T_m - T_e = T_a$. Thus, the equation of motion of the machine is governed by $I\left(d^2\theta_m/dt^2\right) = T_m - T_e = T_a$; T_m is the mechanical torque supplied by the prime mover in N-m, T_e is the electrical torque output of the alternator in N-m and θ is the angular position of the rotor in rad. Multiplying both sides with ω, we get

$$I\omega\frac{d^2\theta_m}{dt^2} = T_m\omega - T_e\omega = T_a\omega, \text{ or } M\frac{d^2\theta_m}{dt^2} = P_m - P_e = P_a \qquad (11.1)$$

where P_m, P_e and P_a respectively are the mechanical, electrical and accelerating power in MW. Neglecting the losses, the difference between the mechanical and electrical torque gives the net accelerating torque T_a. In the steady state, the electrical torque is equal to the mechanical torque, and hence the accelerating power will be zero. During this period the rotor will move at synchronous speed ω_s in rad/s.

As the rotor is continuously rotating at synchronous speed in steady state θ_m will also be continuously varying with respect to time. To make the angle θ_m constant in steady state we can measure this angle with respect to a synchronously rotating reference instead of a stationary reference. Hence, we can write the angular position θ_m with respect to the synchronously rotating frame, and we define $\theta_m = \omega_{ms}t + \delta_m$. So,

$$\frac{d\theta_m}{dt} = \omega_{ms} + \frac{d\delta_m}{dt} \text{ and } \frac{d^2\theta_m}{dt^2} = \frac{d^2\delta_m}{dt^2} \qquad (11.2)$$

To convert them into electrical radians and electrical radians per second respectively we have to take the number of poles (M) of the synchronous machine rotor into consideration. Hence, the electrical angle and electrical speed can be represented as $\delta = (P/2)\delta_m$ elect. rad and $\omega = (P/2)\omega_m$ elect. rad per second. Rather than giving the moment of inertia of a synchronous machine for dynamic studies, it is more convenient to use per unitized quantity called inertia constant H defined as the stored kinetic energy per unit KVA of machine rating; the swing equation can be represented as in pu system

$$\frac{2H}{\omega_s}\frac{d^2\delta}{dt^2} = P_m - P_e = P_a; \text{ where } H = \frac{1}{2}M\omega_s; \text{so, } \frac{d^2\delta}{dt^2} = \frac{\pi f_s}{H}\left[P_m - P_e\right] = \frac{\pi f_s}{H}\left[P_m - P_{\max}\sin\delta\right] \quad (11.3)$$

H is the inertia constant and is equal to the stored kinetic energy per unit MVA and is expressed as MJ/MVA or MW-s/MVA. If MVA rating $= G$ then $GH = KE = (1/2)I\omega^2 = (1/2)M\omega$. Now, $\omega = 360f$ elect. deg./s, so $GH = (1/2)M(360f)$ or $M = GH/180f$. The equation that describes the behavior of the rotor dynamics is the swing equation, which is as follows:

$$M\frac{d^2\delta}{dt^2} = P_m - P_e = P_a \qquad (11.4)$$

The equation describes the behavior of the rotor dynamics and hence is known as the *swing equation*.

11.3   STEADY STATE STABILITY OR SMALL SIGNAL STABILITY

The swing equation is described as

$$\frac{2H}{\omega_s}\frac{d^2\delta}{dt^2} = P_m - P_e = P_a; \text{or } \frac{d^2\delta}{dt^2} = \frac{\omega_s}{2H}\left(P_m - P_e\right) \Rightarrow \frac{d^2\delta}{dt^2} = \frac{\pi f_s}{H}\left(P_m - P_e\right) = \frac{\pi f_s}{H}\left(P_m - P_{\max}\sin\delta\right)$$

$$(11.5)$$

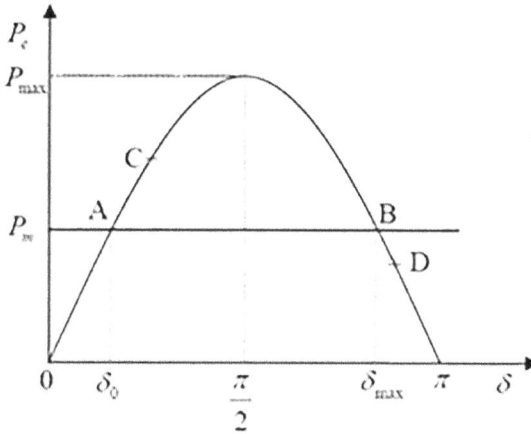

FIGURE 11.1 $P - \delta$ curve of synchronous machine.

The synchronous generator real power output can be represented as $P_e = P_{e,\max} \sin \delta = P_{e,\max} \sin \delta$. If $P_m = P_{\max} \sin \delta$, then there will be no speed change and there will be no angle change. But, if $P_m \neq P_{\max} \sin \delta$ due to disturbance in the system, then the speed either increases or decreases with respect to time. Let us take the case of $P_m > P_{\max} \sin \delta$, there is more input mechanical power than the electrical power output. In this case, as the energy has to be conserved, difference between the input and output powers will lead to increase in the kinetic energy of the rotor and speed increases.

Similarly, if $P_m < P_{\max} \sin \delta$, then the input power is less than the required electrical power output. Again the balance power, to meet the load requirement, is drawn from the kinetic energy stored in the rotor due to which the rotor speed decreases. Since the mechanical power input P_m and the maximum power output of the generator max $P_{\max}$ are known for a given system topology and load, we can find the rotor angle δ from $\delta = \sin^{-1}[P_m/P_{\max}]$ or $\pi - \sin^{-1}[P_m/P_{\max}]$. The rotor angle δ has two solutions at which $P_m = P_{\max} \sin \delta$. If $\delta_0 = \sin^{-1}[P_m/P_{\max}]$ and $\delta_{\max} = \pi - \sin^{-1}[P_m/P_{\max}] = \pi - \delta_0$, then there is a very important implication of the two solutions $(\delta_0, \delta_{\max})$ of swing equation on the stability of the system that can be written as

$$\frac{H}{\pi f_s} \frac{d^2 \delta}{dt^2} = P_m - P_{\max} \sin \delta \tag{11.6}$$

This can be understood from what is called as swing curve or $P - \delta$ curve or the swing curve as shown in Figure 11.1.

The $P - \delta$ curve shows the plot of electrical power output P_e with respect to the rotor angle δ. As can be observed, the curve is a sine curve with δ varying from 0 to π. The maximum power output of the generator $P_{\max}$ occurs at an angle $\delta = \pi/2$. On the same curve the input mechanical power P_m can also be represented. Since P_m is assumed to be constant and does not vary with respect to δ, a straight line is drawn which cuts the $P - \delta$ curve at points A and B. It can be seen that at points A and B the mechanical power input P_m is equal to P_e. The rotor angles at point A and B are δ_0 and $\delta_{\max}$, the solutions of equation $P_m = P_{\max} \sin \delta$.

11.4 TRANSIENT STABILITY OR LARGE SIGNAL STABILITY

The main difference between small-disturbance and large-disturbance stability analysis is that in small-disturbance, the system that can be represented by differential equation can be linearized, but for transient stability, the disturbance is large and the aforementioned linearization of the differential equation cannot be implemented. Consider the swing equation,

$$\frac{d^2\delta}{dt^2} = \frac{\pi f_s}{H}\left(P_m - P_{\max}\sin\delta\right) \tag{11.7}$$

Multiplying both sides by $2\dfrac{d\delta}{dt}$

$$2\frac{d\delta}{dt}\frac{d^2\delta}{dt^2} = \frac{\pi f_s}{H}\left(P_m - P_{\max}\sin\delta\right)2\frac{d\delta}{dt} \tag{11.8}$$

$$\text{or, } \frac{d}{dt}\left[\frac{d\delta}{dt}\right]^2 = \frac{2\pi f_s}{H}\left(P_m - P_{\max}\sin\delta\right)\frac{d\delta}{dt}; \text{ or, } d\left[\frac{d\delta}{dt}\right]^2 = \frac{2\pi f_s}{H}\left(P_m - P_{\max}\sin\delta\right) \tag{11.9}$$

Integrating on both sides and taking square root we get

$$\frac{d\delta}{dt} = \left[\int_{\delta_0}^{\delta}\frac{2\pi f_s}{H}\left(P_m - P_{\max}\sin\delta\right)d\delta\right]^{1/2} \tag{11.10}$$

For the system to be stable the rotor angle δ should settle down to its steady state value and hence $d\delta/dt$ should be zero (rate of change of angle with respect to time should become zero in steady state)

$$\frac{d\delta}{dt} = 0 \Rightarrow \int_{\delta_0}^{\delta}\left(P_m - P_{\max}\sin\delta\right)d\delta = 0; \text{ or, } \int_{\delta_0}^{\delta}P_a\,d\delta = 0 \tag{11.11}$$

The condition of stability can therefore be stated as: the system is stable if the area under P_a (accelerating power) $-\delta$ curve reduces to zero at some value of δ. In other words, the positive (accelerating) area under the $P_a - \delta$ curve must be equal to negative (decelerating) area and hence the name equal area criterion of stability.

11.5   EQUAL AREA CRITERION

Considering a salient pole machine which is delivering power at a terminal voltage V_t, the real power delivered is $P_e = P_m = \left(\left[|E_0||V_t|\right]/|X_s|\right)\sin\delta$. When the synchronous machine is operating in steady state delivering an electrical power P_e which is equal to mechanical power P_m. Whenever a disturbance occurs, the system tries to adjust to new operating condition and the adjustment to the new operating condition is called the transient period. The system behavior under this time is the dynamic system performance. Since a system with energy storage cannot respond instantaneously, when subjected to a disturbance, the output cannot follow the input disturbance immediately and exhibits an oscillatory behavior, before a steady state can be reached. If the oscillatory response of a system during the transient period following a disturbance is damped and the system settles in a finite time to a new steady operating condition, the system is stable. Thus, stability is a property that enables it to remain in a state of operating equilibrium under normal operating condition and to remain in an acceptable state of equilibrium after being subjected to a disturbance.

If the output of a system at steady state does not exactly agree with the input, the system is said to have a steady state error. This error is indicative of the accuracy of the system. Thus, time response of a control system consists of two parts: the transient and the steady state response. Transient response refers to that portion of the response which goes from the initial state to the final state. Steady state response refers to that portion of the response as time t approaches to infinity.

Stability is a condition of equilibrium between opposing forces. A system is said to be in equilibrium if, in the absence of any disturbance or input, the output remains in the same state.

Absolute stability is whether the system is stable or unstable. A linear time-invariant system is critically stable if the oscillations of the output continue forever. It is unstable if the output diverges without bound from its equilibrium. The system is said to be locally stable about an equilibrium point if, when subjected to small disturbances, it remains within a small region surrounding the equilibrium point. If, t increases, the system returns to the original state, it is said to be asymptotically stable in the small. Steady state operation is characterized by the fact that mechanical torque is essentially in equilibrium with the electrical torque developed, so that rotor acceleration is characterized by the small deviations from zero and the speed varies around the synchronous speed corresponding to the common electrical frequency of the power system.

Before any disturbance or load changes occur, generator delivers power $P_{e(1)}$ at a load angle δ_1. Then the power is increased suddenly with a sudden increase in shaft input power in such a way that the power output changes to $P_{e(2)}$ at a load angle δ_2 as shown in Figure 11.2. The sudden increase is then $(P_{e(1)} - P_{e(2)})$. The difference in the power gives rise to the rate of change of stored kinetic energy in the rotor masses. Thus, the rotor will accelerate under the constant influence of non-zero accelerating power and hence the load angle will increase and the rotor tends to accelerate towards the new equilibrium at the angle to new angle δ_2. By the time the rotor reaches δ_2, the speed is in excess to synchronous speed because of the stored kinetic energy in the rotor masses, which, provided there is no damping, is proportional to the area shaded by full lines. In consequence, the rotor will overshoot, but immediately after passing the position δ_2, a retarding torque builds up which tends to slow down the rotor. When the rotor reaches a maximum value δ_3, the rotor swings back following the operating curve towards δ_2. The increase in the angle to δ_3 if eventually does not stop and the rotor does not start decelerating, the generator will lose synchronism. Ultimately the rotor settles back to point δ_2, which is ultimate steady state operating point.

In Figure 11.2a, rotor overshoots to a position δ_3 which is left to the steady state maximum power limit, i.e. $\delta = 90°$. Neglecting damping, this position is such that the shaded area which is proportional to the decrease in kinetic energy due to deceleration to the area representing gain in kinetic energy. Since the torque at the load angle δ_3 is a decelerating torque, the rotor will swing back and tend to overshoot to a load angle of δ_1. Oscillations will thus be set up within δ_1–δ_3. However, as there must be some damping, the swing ultimately settles at angle δ_2. Thus δ_2 must be below the transient stability limit for the sudden increase in load. For stability, the area under the graph of accelerating power versus δ must be zero for some value of δ; i.e., the positive (accelerating) area under the graph must be equal to the negative (decelerating) area. This criterion is therefore known as the equal area criterion for stability.

In Figure 11.2b, the load increase is greater. Rotor overshoots to a position δ_3 which is just after right to the steady state maximum power limit, i.e. $\delta = 90°$. Oscillations will be set up within δ_1–δ_3.

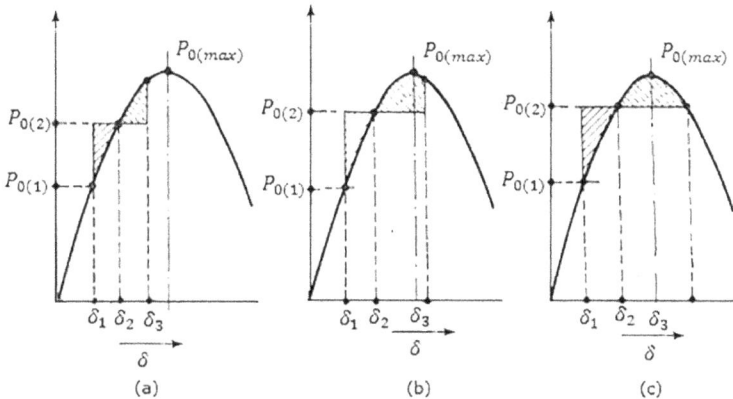

FIGURE 11.2 Transient stability of synchronous generator (equal area criterion).

However, as there must be some damping, the swing ultimately settles at angle δ_2 without losing synchronism. Thus δ_2 must be below the transient stability limit for the sudden increase in load.

With a sufficient large change in load, the limit of stability is reached as shown in Figure 11.2c. If the change in load is still further, the initial accumulation of kinetic energy because of such load change would carry the rotor beyond the position of δ_3 that results the angle δ continue to increase and the generator will be out of step. So, it can be concluded for any increase in load or the input, the steady increase in rotor angle δ due to insufficient synchronizing torque or damping torque rotor oscillation of increasing magnitude causes instability. If the oscillations damp out, the machine will be stable. This is shown in Figure 11.3.

The maximum value of δ_2 beyond which the increase in load angle δ causes an instability is called critical clearing angle. If the system is to remain stable and the equal-area criterion is to be satisfied, value of δ should be less than critical clearing angle.

Equal area criterion is a graphical method to assess the transient stability of power system. This criterion is applicable to all two-machine systems, whether they actually have only two machines or whether they are simplified representations of systems with more than two machines. Two-machine systems may be divided into two types: (i) those having one finite machine swinging with respect to an infinite bus and (ii) those having two finite machines swinging with respect to each other.

The basic equation for equal area criterion as per Figure 11.2a is $\int_{\delta_1}^{\delta_3} P_a \, d\delta = 0$. This equation applies to any two points say δ_1 and δ_1 on the power-angle diagram, provided they are points at which the rotor speed is synchronous. If we perform the integration of the aforesaid equation in two steps, we can write $\int_{\delta_1}^{\delta_2} (P_m - P_e) \, d\delta + \int_{\delta_2}^{\delta_3} (P_m - P_e) \, d\delta = 0$. The same can be written as

$$\int_{\delta_1}^{\delta_2} (P_m - P_e) \, d\delta = \int_{\delta_2}^{\delta_3} (P_e - P_m) \, d\delta \tag{11.12}$$

The left integral applies to the fault period, whereas the right integral corresponds to the immediate post-fault period up to the point of maximum swing. P_e is zero during the fault. The shaded area A1 is given by the left-hand side of equation and the shaded area A2 is given by the right-hand side. So, the two areas A1 and A2 are equal.

The shaded area A1 is dependent on the time taken to clear the fault. If there is delay in clearing, the angle δ_2 is increased; likewise, the area A1 increases and the equal-area criterion requires that area A2 also increase to restore the rotor to synchronous speed at a larger angle of maximum swing δ_3.

If the delay in clearing is prolonged so that the rotor angle δ swings beyond the angle δ_3, then the rotor speed at that point on the power-angle curve is above synchronous speed when positive

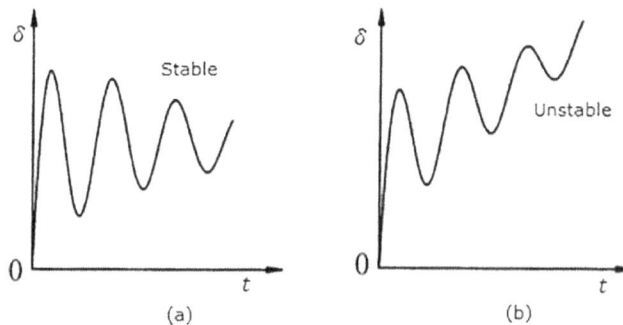

FIGURE 11.3   Nature of rotor angle oscillation.

accelerating power is again encountered. Under the influence of this positive accelerating power, the angle δ will increase without limit and instability results. Therefore, there is a critical angle for clearing the fault in order to satisfy the requirements of the equal-area criterion for stability. This angle is called the critical clearing angle δ_{cr}.

The corresponding critical time for removing the fault is called the critical clearing time t_{cr}. Thus, the critical clearing time is the maximum elapsed time from the initiation of the fault until its isolation such that the power system is transiently stable.

Let us now consider a synchronous machine is connected to infinite busbar as discussed in previous paragraphs through a transformer and double circuit line as shown in Figure 11.4a. Curve 1 in Figure 11.4b is the power output curve for normal steady state operation. In determining the curve the reactance to be considered is the sum of synchronous reactance X_s, transformer reactance X_t and the transmission line reactance X_l. Line reactance X_l is parallel combination of line reactance X_{l1} and line reactance X_{l2}. The load angle at which the machine operates is δ_1. Let the power delivered normal steady state operation is P_0. The load angle at which power is delivered is δ_1.

Let a fault in transmission line occur and immediately the power transmitted will fall. The removal of any transmission line will increase line reactance X_l. Thus, at the occurrence of the fault, the operating point "a" changes to the corresponding point "b". Because of the reduction of power output, the rotor will accelerate and the operating point will move from "b" to point "c". If at point "c" fault is cleared, so that alternator now supplies the power through the transformer and one transmission line, the operating point will change suddenly from point "c" to a corresponding to a point "d" on a much higher curve 3. Deceleration then commences but increase in load angle will continue until point "e" is reached. At this point, which is determined according to the equal area criterion, the speed of rotation is synchronous. Due to the decelerating torque, the speed will begin to fall and consequently the rotor drops back towards the load angle given by point "f", which is equilibrium position for the post fault condition. Due to the decrease in speed the load angle overshoots and oscillates about the equilibrium position until the same is damped out. If there is

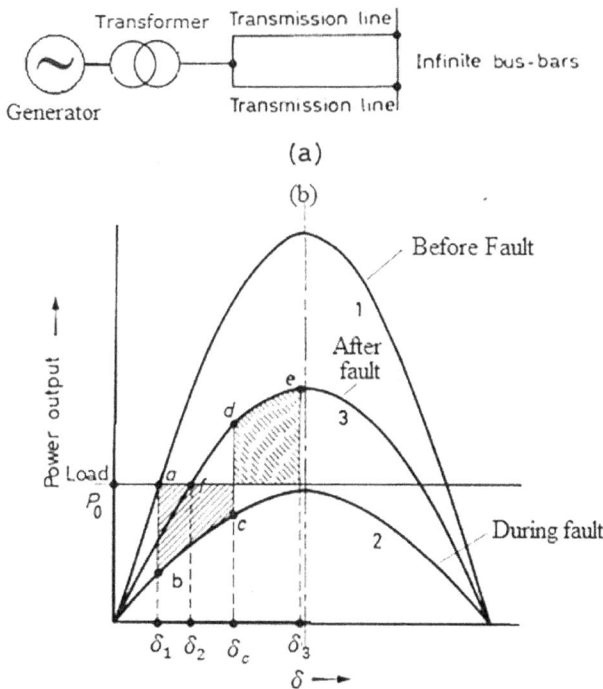

FIGURE 11.4 Equal are criteria applied to an alternator connected to infinite bus.

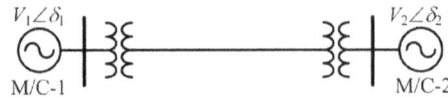

FIGURE 11.5 Multimachine stability.

a considerable delay in clearing the fault, the area representing the gain in kinetic energy may be too large to make an equal loss area possible. In such case the machine will lose synchronism. Similarly, the criterion can be applied to find the critical switching angle, i.e. the largest value that δ_c can be permitted to be without the machine losing stability.

11.6 MULTIMACHINE STABILITY

Consider the simple power system shown in Figure 11.5 in which two machines are operating. For multimachine stability analysis each synchronous machine is represented by a constant voltage source behind the transient reactance with saturation and network transients being neglected. Let us assume that starting with the initial angles δ_1 and δ_2 with respect to some reference at nominal frequency, machine 1 accelerates while machine 2 decelerates from this nominal frequency.

When a disturbance occurs, accelerating power P_{ai} will not be zero like stable operation. For machine i to be stable, δ_i must settle to a constant value so that P_{ai} becomes zero. We then have $d\delta_i/dt = \omega_i$; $(2H_1/\omega_s)\ddot{\delta}_1 = P_{m1} - P_{e1}$ and $(2H_2/\omega_s)\ddot{\delta}_2 = P_{m2} - P_{e2}$, where the subscripts 1 and 2 refer to machines 1 and 2, respectively. Let us assume that the transmission line is lossless.

Then in the simple case where the power from machine 1 flows to machine 2, we get $P_{e1} = -P_{e2} = \left[(|V_1||V_2|)/|X|\right]\sin(\delta_1 - \delta_2) = \left[(|V_1||V_2|)/|X|\right]\sin\delta_{12}$ where $\delta_{12} = \delta_1 - \delta_2$. Since the system is lossless, we may write $P_{m1} = -P_{m2}$. This means that in the steady state, the power generated at machine 1 is absorbed through machine 2. So,

$$\frac{2H_1}{\omega_s}\ddot{\delta}_1 - \frac{2H_2}{\omega_s}\ddot{\delta}_2 = 2P_{m1} - P_{e1} + P_{e2} = 2P_{m1} - \frac{|V_1||V_2|}{|X|}\sin\delta_{12} \qquad (11.13)$$

Let us assume $H_1 = H_2 = H$, $V_1 = V_2 = 1.0$ per unit and $P_{m1} = 0$, we get $\ddot{\delta}_{12} = -\omega^2\sin\delta_{12}$, where the oscillation frequency ω is given by $\omega = \sqrt{\omega_s/HX}$. Thus, the weighted difference of angles will approximate simple harmonic motion for small changes in δ_{12} and the frequency will decrease for an increase in inertia H or impedance. Another aspect can be seen by adding the system to give $H_1\ddot{\delta}_1 + H_2\ddot{\delta}_2 = P_{m1} + P_{m2} = 0$. Thus, the overall acceleration of the machine group will depend on the overall balance between power generated and consumed. Usually there are governors on the generators to reduce generated power if the system frequency increases.

11.7 LEARNING OUTCOME

Power system stability analysis namely steady state stability, Dynamic stability and Transient stability involves the study of the dynamics of the power system under disturbances. Evaluating power system stability is equivalent to evaluating the dynamic performance of the power system. To achieve operational efficiency of power system including quality power, analysis of stability of the system is an essential requirement as instability hinders the secure operation of the system.

In this chapter, readers have been able to learn all the above aspects.

12 Fuses and Circuit Breakers

12.1 FUSES

A fuse is a circuit element designed to melt when the current exceeds some limit, thereby opening the circuit. The fuse is the current interrupting devices which break or open the circuit by fusing the element and thus remove the faulty device from the main supply circuit. The basic design equation for fuses is the Preece equation, for wires in free air, i.e. $I = A^* D^{3/2}$: where I is the fusing current in Amps, A is a constant depending on the metal and D is the diameter of the wire (in mm or inches). "*" Exponent in the equation should be adjusted to 1.287 for silver and 1.32 for tungsten. I is the fusing current.

12.1.1 TYPES OF FUSES

DC Fuse – The DC fuse opens or breaks the circuit when the excessive current flows through it. The difficulty with the DC fuse is that the arc produced by the direct current is very difficult to extinct because there are no zero current flows in the circuit. For reducing the DC fuse arcing, the electrodes are placed more distance apart due to which the size of the fuse increases as compared to AC fuse.

AC Low-Voltage Rewirable Fuses – Rewirable Fuse of semi-enclosed type has two essential parts: a fuse base, which has the in and out terminal, and the other is a fuse carrier, which holds the fuse element. Both the fuse base and carrier are made of porcelain. The fuse carrier can be plugged in and removed from the base of the fuse. The fuse carrier can be safely withdrawn from the and inserted into the fuse base without touching live parts. The carrier is shown Figure 12.1.

AC Low-Voltage Cartridge Type Fuse – This type of fuse has a completely closed structure with the Fuse Links enclosed in the container. The cartridge may vary in length to match the fuse rating of the circuit to be protected. This is shown in Figure 12.2.

AC Low-Voltage D Type Cartridge Fuse – This type of fuse has a fuse base, adapter ring, cartridge and a fuse cap as the main elements. The cartridge is moved in the fuse cap. The cap is fixed on the fuse base.

AC Low-Voltage Link Type Cartridge Fuse – This type of fuse is also called High Rupturing Capacity (HRC) fuse. The enclosure of the HRC fuse is filled with powdered pure quartz, which acts as an arc extinction medium. The silver and copper wires are used for making the fuse wire.

Dropout Fuse – The melting of fuse causes the fuse element to drop out under gravity about its lower support.

Striker Fuse – It is a mechanical device having enough force and displacement which can be used for closing tripping/indicator circuits. This not only provides the user with a visual indication that the fuse link has operated but can also be used to operate other switching out circuit.

Switch Fuse – which is a combined unit known as iron clad switch. It may be double pole for controlling single-phase two-wire circuit, triple pole for controlling three-phase three-wire circuit or triple pole with neutral link for controlling three-phase four wire circuit.

High-Voltage HRC Fuse – The main problem of the high-voltage fuses is that of the corona. Therefore, the high-voltage fuses have special design. These types of fuses are: (i) Cartridge

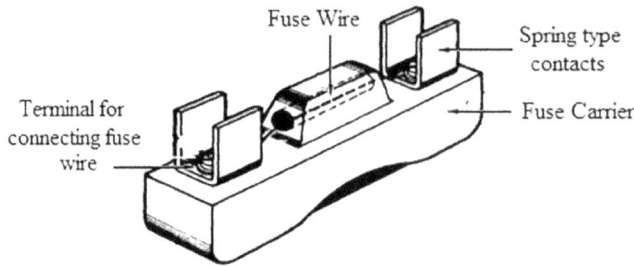

FIGURE 12.1 Rewirable fuse carrier.

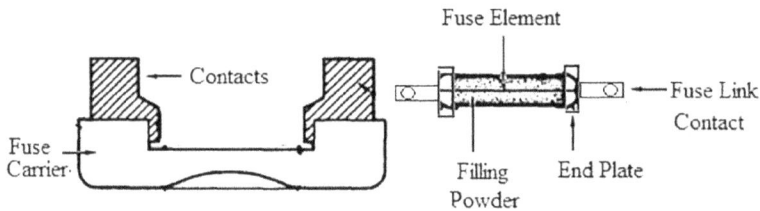

FIGURE 12.2 Cartridge type fuse and carrier.

Type Fuse – The fuse element of the HRC fuse is wound in the shape of the helix which avoids the corona effect at the higher voltages. It has two fused elements placed parallel with each other; one is of low resistance and the other is of high resistance. The low resistance wire carries the normal current which is blown out, reducing the short-circuit current during the fault condition. (ii) Liquid Type HRC Fuse – Such type of fuses is filled with carbon tetrachloride and sealed at both the ends of the caps. When the fault occurs, then the current exceeds beyond the permissible limit, and the fuse element is blown out. The liquid of the fuse acts as an arc extinguishing medium for the HRC fuses; (iii) Expulsion Type HV Fuse.

The fuses are characterized by the (i) Rated current – it is the maximum current that the fuse can continuously conduct without interrupting the circuit; (ii) Speed – the speed at which a fuse blows to isolate the circuit. The operating time is not a fixed interval but decreases as the current increases. Fuses have different characteristics of operating time compared to current. A standard fuse may require twice its rated current to open in 1 second, a fast-blow fuse may require twice its rated current to blow in 0.1 seconds, and a slow-blow fuse may require twice its rated current for tens of seconds to blow. (iii) The I^2t rating – It is related to the amount of energy let through by the fuse element when it clears the electrical fault. (iv) Breaking Capacity – It is the maximum current that can safely be interrupted by the fuse. (v) Rated Voltage – The voltage rating of the fuse must be equal to or, greater than, what would become the open-circuit voltage.

12.2 SWITCHGEAR

Switching devices can be classified as (i) Circuit breakers – Circuit breakers must make and break all currents within the scope of their ratings, from small inductive and capacitive load currents up to the short-circuit current. (ii) Switches – Switches must make and break normal currents up to their rated normal current and be able to make on existing short circuits (up to their rated short-circuit making current). However, they cannot break any short-circuit currents.

(iii) Contactors – Contactors are load-breaking devices with a limited making and breaking capacity. They are used for high switching rates, but can neither make nor break short-circuit currents. (iv) Disconnectors – Disconnectors are used for no-load closing and opening operations. Their function is to "isolate" downstream equipment so they can be worked on.

A *switchgear* or *electrical switchgear* is a generic term which includes all the switching devices associated with mainly power system protection. It also includes all devices associated with control, metering and regulating of electrical power system. Assembly of such devices in a logical manner forms switchgear. Switchgear has to perform the function of carrying, making and breaking the normal load current like a switch and it has to perform the function of clearing the fault in addition to that it also has provision of metering and regulating the various parameters of electrical power system. Thus, the switchgear includes circuit breaker, current transformer, voltage transformer, protection relay, measuring instrument, electrical switch, electrical fuse, miniature circuit breaker, lightening arrestor or surge arrestor, isolator and other associated equipment. *Switchgear protection* plays a vital role in modern power system network, right from generation through transmission to distribution end. The current interruption device or switching device is called circuit breaker in *Switchgear protection* system.

12.2.1 Circuit Breaker

Electrical Circuit Breaker is a switching device which can be operated manually as well as automatically for controlling and protecting electrical power system respectively. As the modern power system deals with huge currents, special attention is given during designing of *circuit breaker* to safe interruption of arc produced during the *operation of circuit breaker*. For saving electrical equipment that are used in electrical power system and the power networks, the fault current should be cleared from the system as quickly as possible. Again after the fault is cleared, the system must come to its normal working condition as soon as possible for supplying reliable quality power to the receiving ends. In addition to that for proper controlling of power system, different switching operations are required to be performed. So for timely disconnecting and reconnecting different parts of power system network for protection and control, there must be some special type of switching devices which can be operated safely under huge current carrying condition. During interruption of huge current, there occurs arcing in between switching contacts, and the objective is to quench these arcs in a safe manner.

12.2.2 Different Types of Circuit Breaker

According to different criteria there are different types of circuit breakers.

- According to their arc quenching media, the circuit breaker can be divided as (i) Oil Circuit Breaker – bulk oil and minimum oil, (ii) Air Circuit Breaker, (iii) SF_6 Circuit Breaker and (iv) Vacuum Circuit Breaker.
- According to their services, the circuit breaker can be divided as (i) Outdoor Circuit Breaker and (ii) Indoor Breaker.
- According to the operating mechanism of circuit breaker, they can be divided as (i) Manually Spring-operated Circuit Breaker, where operating spring is charged manually; (ii) Electrically Spring-operated Circuit Breaker, where operating spring is charged electrically through a motor; and (iii) Solenoid operated Breaker.
- According to the voltage level of installation; types of circuit breaker are referred as (i) High-Voltage Circuit Breaker, (ii) Medium-Voltage Circuit Breaker and (iii) Low-Voltage Circuit Breaker.
- According to tripping mechanism, circuit breakers are (i) series trip and (ii) shunt trip.

- In Metal-Clad Switchgear, the main circuit breaker is draw-out type (removable) and arranged with a mechanism for moving it physically between connected and disconnected positions. It is equipped with self-aligning and self-coupling primary and secondary disconnecting devices.

12.2.3 Rating of Circuit Breaker

- **Rated Voltage** – The rated voltage is the upper limit of the highest system voltage the device is designed for.
- **Rated Normal Current** – The rated normal current is the current the main circuit of a device can continuously carry under defined conditions. The heating of components – especially of contacts – must not exceed defined values. Permissible temperature rises always refer to the ambient air temperature.
- **Rated Insulation Level** – The rated insulation level is the dielectric strength from phase to earth, between phases and across the open contact gap, or across the isolating distance. The dielectric strength is verified by a lightning impulse withstand voltage test with the standard impulse wave of 1.2/50 μs and a power-frequency withstand voltage test (50 Hz/1 minute).
- **Rated Peak Withstand Current** – The rated peak withstand current is the peak value of the first major loop of the short-circuit current.
- **Rated Breaking Current** – The rated breaking current is the load breaking current in normal operation.
- **Rated Short-Circuit Breaking Current** – The rated short-circuit breaking current is the root mean square value of the breaking current in case of short circuit at the terminals of the switching device.
- **Rated Short-Circuit Making Current** – The rated short-circuit making current is the peak value of the making current in case of short circuit at the terminals of the switching device. This stress is greater as that of the rated peak withstand current, as dynamic forces may work against the contact movement. Making capacity = 2.55 × symmetrical breaking capacity.
- **Switching Duties** – (i) O – 3 minutes – CO – 3 minutes – CO at full rated short-circuit breaking current; (ii) O – 0.3 seconds – CO – 3 minutes – CO up to a rated short-circuit breaking current for rapid load transfer; and (iii) O – 0.3 seconds – CO for auto reclosing.
- **Short Time Rating** – It is the period for which the CB is able to carry fault current while remaining closed.
- **First Pole to Clear Factor** – ratio of RMS voltage between healthy phase and faulty phase and phase to neutral voltage when fault is removed.
- **Protection Classes for Switchgear** – identified by the two letters "IP" (stands for ingress protection) followed by two digits denoting the degree of protection. The first digit refers to degree of protection against contact and ingress of foreign objects/particles solid in nature, and the second digit refers to degree of protection against ingress of water. First digit varies from 1 (implies protection against ingress of large foreign bodies) to 6 (implies protection against ingress of dust). Second digit varies from 1 (implies protection against ingress of vertically falling water drops) to 8 (implies protection against continuous submersion in water). If the digits starts with 0 (zero), it implies no protection.

12.2.4 Formation of Arc

During opening of current carrying contacts in a circuit breaker, the contacts become current carrying state to voltage withstand state. When the contacts begin to separate, the contact area decreases rapidly and large fault current causes increased current density and hence rise in temperature.

The heat produced in the medium between contacts is sufficient to ionize in case of air or vaporize and ionize the medium. The ionized air or vapor acts as conductor, and an arc is struck between the contacts. The potential difference between the contacts is quite small and is just sufficient to maintain the arc. The arc provides a low resistance path and consequently the current in the circuit remains uninterrupted so long as the arc persists. For total interruption of current the circuit breaker is essential to quench the arc as quick as possible. As soon as the contacts separate out, an arc is formed. The voltage across the contacts during the arcing period is known as the arc voltage and is relatively low with heavy current arcs of short length. At current zero it rises rapidly to the peak value since a short-circuit current is almost 90° lagging.

12.2.5   ARC INTERRUPTION OR ARC QUENCHING OR ARC EXTINCTION IN CIRCUIT BREAKER

Heat Loss from Arc – Heat loss from arc in circuit breaker takes place through conduction, convection as well as radiation. In circuit breaker with plain break arc in oil, arc in chutes or narrow slots, nearly all the heat loss is due to conduction. In air blast circuit breaker or in breaker where a gas flow is present between the electrical contacts, the heat loss of arc plasma occurs due to convection process. During opening of electrical contacts, the arc in circuit breaker is produced, and it is extinguished at every zero crossing of the current and then it is again reestablished during next cycle. The final arc extinction or arc quenching in circuit breaker is achieved by rapid increase of the dielectric strength in the medium between the contacts so that reestablishment of arc after zero crossing cannot be possible. This rapid increase of dielectric strength in between circuit breaker contacts is achieved either by deionization of gas in the arc media or by replacing ionized gas by cool and fresh gas. There are various deionization processes applied for arc extinction in circuit breaker, let us discuss in brief.

There are two methods of arc extinction: (i) High resistance method, in which arc resistance is made to increase with time by lengthening of arc or cooling of arc or reducing cross section of arc or splitting of arc (by letting the arc pass through a narrow opening or by having smaller area of contacts) so that current is reduced to a value insufficient to maintain the arc. Consequently, the current is interrupted or the arc is extinguished. It is employed only in DC circuit breakers and low-capacity AC circuit breakers. (ii) Low resistance method or zero current method, in which arc resistance is kept low until current is zero where the arc extinguishes naturally and is prevented from restriking in spite of the rising voltage across the contacts. All modern high power AC circuit breakers employ this method for arc extinction. In an AC system, current drops to zero after every half-cycle. At every current zero, the arc extinguishes for a brief moment.

The medium between the contacts contains ions and electrons so that it has small dielectric strength and can be easily broken down by the rising contact voltage known as restriking voltage. If such a breakdown does occur, the arc will persist for another half cycle. If immediately after current zero, the dielectric strength of the medium between contacts is built up more rapidly than the voltage across the contacts, the arc fails to restrike and the current will be interrupted. The transient voltage appearing across the contacts at current zero during arc period is called the *re-striking* voltage. This voltage will probably re-strike the arc so that it persists for another half cycle. This voltage is given by the expression $V_{\text{Restrike}} = V_m \left[1 - \cos\left(t/\sqrt{LC}\right) \right]$. The normal frequency rms voltage that appears across the breaker contacts after final arc extinction has occurred is called the recovery voltage. Figure 12.3 shows short-circuit current waveform along-with restriking voltage transient, recovery voltage waveform.

If a three-phase fault occurs not involving ground, the voltage across the circuit breaker pole, first to clear, is 1.5 times the phase voltage. In a three-phase circuit breaker, the arc extinction in the three poles is not simultaneous as the currents in three phases are mutually 120° apart. Hence the

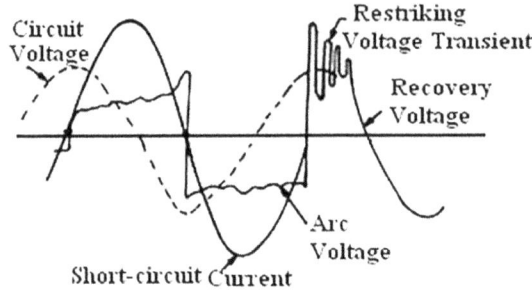

FIGURE 12.3 Short-circuit current, restriking voltage transient and recovery voltage waveform.

power frequency recovery voltage of the phase in which the arc gets extinguished first is about 1.5 times the phase voltage. To consider the effect of the first pole to clear on the power frequency component of the recovery voltage, the first pole to clear factor is considered, which is the ratio of rms voltage between healthy and faulty line to the phase to neutral voltage with fault removed at the location of the circuit breaker.

The rate of rise of restriking voltage or Rate of Rising of Restriking Voltage (RRRV) is the rate of increase in restriking voltage and expressed in volts per micro-second. If v is the restriking voltage in volts, then RRRV is dv/dt in V/µs. Since $V_{\text{Restrike}} = V_m\left(1 - \cos\left(t/\sqrt{LC}\right)\right)$, maximum value of RRRV is $V_m/\sqrt{LC}$ at $t = \sqrt{LC}\left(\pi/2\right)$.

Resistance switching in circuit breaker refers to a method adopted for reducing the rate of rise of restriking voltage and the overvoltage transients. In this method, a shunt resistance is connected across the contacts of circuit breaker. Figure 12.4 shows the resistance switching circuit.

When the fault occurs, the contacts of the circuit breaker are open, and an arc is struck between the contacts. With the arc shunted by the resistance R, a part of arc current is diverted through the resistance. This results in the decrease of arc current and an increase in the rate of deionization of the arc path. Thus, the arc resistance is increased, leading to further increase in current through the shunt resistance R. This build-up process continues until the current becomes so small that it fails to maintain the arc; eventually the arc is extinguished, and the circuit gets interrupted. The shunt resistance also ensures the damping of the high frequency restriking voltage transient. Frequency is

$$f = \left(1/2\pi\right)\sqrt{\left(1/LC\right) - 1/\left(4R^2C^2\right)}.$$

Thus, when the value of the resistance is equal to the $0.5\sqrt{L/C}$, the oscillatory nature of the transient will be zero. Thus, putting resistor across the circuit breaker contacts may be used to perform any one or more of the functions: (i) It reduces the RRRV burden on the circuit breaker.

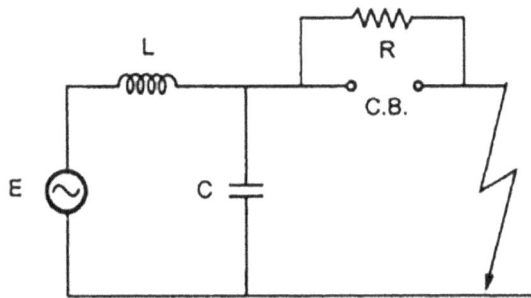

FIGURE 12.4 Resistance switching circuit current.

(ii) It reduces the high-frequency restriking voltage transients during switching out inductive or capacitive loads. (iii) In a multi-break circuit breaker, it helps in distributing the transient recovery voltage more uniformly across the contact gaps.

Current Chopping in circuit breaker is defined as a phenomenon in which current is forcibly interrupted before the natural current zero. When interrupting low inductive currents such as magnetizing currents of the transformer shunt reactor, the rapid deionization of the contact space and blast effect may cause the current to be interrupted before the natural current zero. Current Chopping is mainly observed in Vacuum Circuit Breaker, Air Blast Circuit Breaker and SF_6 circuit breaker and is not the phenomena in Oil Circuit Breaker. The current chopping is considered a serious drawback because it sets up high-voltage transient across the breaker contacts, as when it occurs, the energy stored in the load side inductances oscillates through the system line to earth capacitances (winding and cable capacitances) and causes an increase in the voltage. The voltage developed across the breaker contacts v, when current chopping occurs, is $v = i\sqrt{LC}$, where i is the arc current, L is the inductance of the circuit and C is the capacitance between the breaker contacts.

12.2.6 AIR BLAST CIRCUIT BREAKER

The working principle of *air circuit breaker* is rather different from those in any other types of circuit breakers. The main aim of all kinds of circuit breaker is to prevent the reestablishment of arcing after current zero by creating a situation where in the contact gap will withstand the system recovery voltage. The *air circuit breaker* does the same but in different manner. For interrupting arc it creates an arc voltage in excess of the supply voltage. Arc voltage is defined as the minimum voltage required for maintaining the arc. This circuit breaker increases the arc voltage by mainly three different ways: (i) It may increase the arc voltage by cooling the arc plasma. As the temperature of arc plasma is decreased, the mobility of the particle in arc plasma is reduced; hence more voltage gradient is required to maintain the arc. (ii) It may increase the arc voltage by lengthening the arc path. As the length of arc path is increased, the resistance of the path is increased, and hence to maintain the same arc current, more voltage is required to be applied across the arc path. That means arc voltage is increased; (iii) Splitting up the arc into a number of series arcs also increases the arc voltage.

There are mainly two types of ACB available: (i) Plain air circuit breaker and (ii) Air blast circuit breaker; can be either Axial Blast ACB or Axial Blast ACB with side moving contact or Cross Blast ACB. Air blast circuit breakers are used for the system voltage of 245, 420 kV and even more, especially where faster breaker operation was required.

Advantages of Air Blast Circuit breakers are the following: (i) there is no chance of fire hazard as in case of oil; (ii) the breaking speed of circuit breaker is much higher; (iii) arc quenching is much faster; (iv) the duration of arc is same for all values of small as well as high current interruptions; (v) as the duration of arc is smaller, lesser amount of heat is realized from arc to current-carrying contacts, hence the service life of the contacts becomes longer; (vi) the stability of the system can be well maintained as it depends on the speed of operation of circuit breaker; and (vii) requires much less maintenance compared to oil circuit breaker. The disadvantages are: (i) In order to have frequent operations, it is necessary to have sufficiently high-capacity air compressor. (ii) Frequent maintenance of compressor, associated air pipes and automatic control equipment is also required; (iii) due to high-speed current interruption, there is always a chance of high rate of rise of re-striking voltage and current chopping; (iv) there is also a chance of air pressure leakage from air pipe junctions.

12.2.7 OIL CIRCUIT BREAKER

Bulk Oil Circuit Breaker or *BOCB* is such *types of circuit breakers* where oil is used as arc quenching media as well as insulating media between current carrying contacts and earthed parts of the breaker. The oil used here is same as transformer insulating oil. The basic construction of BOCB is quite simple. Here all moving contacts and fixed contacts are immersed in oil inside closed iron vessel or iron tank. Whenever the current carrying contacts are being open within the oil, the arc is produced in between the separated contacts. The large energy will be dissipated from the arc in oil which vaporizes the oil as well as decomposes it. Because of that a large gaseous pressure is developed inside the oil which tries to displace the liquid oil from the surrounding of the contacts. The inner wall of the oil tank has to withstand this large pressure of the displaced oil. Thus, the oil tank of BOCB has to be sufficiently strong in construction.

Minimum Oil Circuit Breaker (MOCB) utilizes oil as the interrupting media. However, unlike *BOCB*, a MOCB places the interrupting unit in insulating chamber at live potential. The insulating oil is available only in interrupting chamber. The features of designing *MOCB* are to reduce requirement of oil, and hence these breakers are called *MOCB*. As the volume of the oil in BOCB is huge, the chances of fire hazard in bulk oil system are more. For avoiding unwanted fire hazard in the system, one important development in the design of oil circuit breaker has been introduced where use of oil in the circuit breaker is much less than that of BOCB. It has been decided that the oil in the circuit breaker should be used only as arc quenching media not as an insulating media. Then the concept of *MOCB* comes.

There are a number of disadvantages of using oil as quenching media in circuit breakers. Flammability and high maintenance cost are two such disadvantages, although cost is less for such type of breakers. Users were forced to search for different medium of quenching.

12.2.8 Vacuum Circuit Breaker

A *vacuum circuit breaker* is such kind of circuit breaker where the arc quenching takes place in vacuum. The technology is suitable for mainly medium voltage application. For higher voltage vacuum technology has been developed but not commercially viable. The operation of opening and closing of current carrying contacts and associated arc interruption takes place in a vacuum chamber in the breaker which is called vacuum interrupter. The vacuum interrupter consists of a steel arc chamber in the center symmetrically arranged ceramic insulators. The vacuum pressure inside a vacuum interrupter is normally maintained at 10^{-5} to 10^{-7} tor. The material used for current carrying contacts plays an important role in the performance of the vacuum circuit breaker. CuCr is the most ideal material to make VCB contacts. The contacts have to open and close within the vacuum interrupter; it follows that the mechanical drive to the moving contact has to be able to conduct movement into the vacuum interrupter through air-tight seal. The moving contact is attached to the end plate of the vacuum interrupter by metal bellows.

Arc Quenching – During the galvanic separation of the contacts, the current to break produces a metal-vapor arc discharge. The current flows through this metal-vapor plasma until the next current zero. The density distribution of metal vapor plasma is non-homogenous. Near the current zero, contact vapor production ceases. The dielectric strength of the interrupter also increases, and the arc extinguishes interrupting the circuit. The metal vapor loses its conductivity after few microseconds already – the insulating capability of the contact gap recovers quickly.

The vacuum interrupter contacts have to satisfy a number of design criteria. In the closed position the contacts must carry the steady-state current without excessive overheating. When the contacts open, they must provide sufficient metal vapor to permit the arc to carry the circuit current smoothly to current zero. At current zero, however, there must be insufficient metal vapor for the arc to reignite when the restored voltage appears across the contacts. The opening contacts must be capable of interrupting a wide range of currents from the very low currents involved with switching cables and capacitor banks, to the normal continuous duty current, to overload currents and finally to high short-circuit currents. The contacts must not weld and must have strength to survive many thousands of operations.

Service life of Vacuum Circuit Breaker is much longer than other types of circuit breakers. There is no chance of fire hazard as oil circuit breaker. It is much environment friendly than SF_6 Circuit breaker. VCB is much user friendly. Replacement of Vacuum Interrupter is much convenient.

12.2.9 SF$_6$ CIRCUIT BREAKER

SF$_6$ has excellent insulating property. SF$_6$ has high electro-negativity. That means it has high affinity of absorbing free electron. Whenever a free electron collides with the SF$_6$ gas molecule, it is absorbed by that gas molecule and forms a negative ion. Not only the gas has a good dielectric strength, but it also has the unique property of fast recombination after the source energizing the spark is removed. The gas has also very good heat transfer property. Due to its low gaseous viscosity (because of less molecular mobility), SF$_6$ gas can efficiently transfer heat by convection. So due to its high dielectric strength and high cooling effect SF$_6$ gas is approximately 100 times more effective arc quenching media than air.

Arc Quenching – In the closed position of the breaker, the contacts remain surrounded by sulfur hexafluoride gas (SF$_6$) gas at a pressure of about 2.8 kg/cm^2. When the breaker operates, the moving contact is pulled apart and an arc is struck between the contacts. The movement of the moving contact is synchronized with the opening of a valve which permits sulfur hexafluoride gas at 14 kg/cm^2 pressure from the reservoir to the arc interruption chamber. The high-pressure flow of sulfur hexafluoride gas rapidly absorbs the free electrons in the arc path to form immobile negative ions which are ineffective as charge carriers. The result is that the medium between the contacts quickly builds up high dielectric strength and causes the extinction of the arc. After the breaker operation (i.e. after arc extinction), the valve is closed by the action of a set of springs.

The sulfur hexafluoride circuit breakers have certain advantages, which are: (i) due to the superior arc quenching property of sulfur hexafluoride gas, such circuit breakers have very short arcing time; (ii) since the dielectric strength of sulfur hexafluoride gas is 2–3 times that of air, such breakers can interrupt much larger currents; (iii) sulfur hexafluoride gas circuit breaker gives noiseless operation due its closed gas circuit and no exhaust to atmosphere unlike the air blast circuit breaker.

12.3 LEARNING OUTCOME

Fuses and circuit breakers are essential components for the protection of power system. The importance and need of circuit protection are continuously increasing besides its complexity. The traditional circuit breaker technologies are changing to accommodate the present-day requirement.

In this chapter, readers have been able to learn all the above aspects.

13 Power System Protection

13.1 PROTECTION

The objective of *power system protection* is to isolate a faulty section of electrical power system from rest of the live system so that the rest portion can function satisfactorily without any severe damage due to fault current. Protection can be classified as (i) Apparatus protection that deals with detection of a fault in the apparatus and consequent protection. Transmission Line Protection, Transformer Protection, Generator Protection, Motor Protection, Busbar Protection, etc. fall under this category; (ii) System protection that deals with detection of proximity of system to unstable operating region and consequent control actions to restore stable operating point and/or prevent damage to equipment. Out-of-Step Protection, Under-frequency/Over frequency protection, etc. are such kinds of protection.

13.2 RELAY

A relay is a logical element which processes the inputs (mostly voltages and currents) from the system/apparatus and issues a trip decision if a fault within the relay's jurisdiction is detected. To monitor the health of the apparatus, relay senses current through a current transformer (CT), voltage through Potential Transformer (PT). The relay element analyzes these inputs and decides whether (i) there is an abnormality or a fault and (ii) if yes, whether it is within jurisdiction of the relay. If the fault is in its jurisdiction, relay sends a tripping signal to circuit breaker (CB) which opens the circuit. Functions of Relay are (i) to sound an alarm or to close the trip circuit, (ii) to disconnect the abnormally operating part so as to prevent subsequent fault, (iii) to disconnect the faulty part so as to minimize the damage of faulty part, (iv) to localize the fault and (v) to disconnect the faulty part quickly so as to improve the stability.

Types of Relay – (i) Electromechanical Relays – are first generation of relays that uses the principle of electromechanical energy conversion. They are immune to electromagnetic interference and rugged. (ii) Solid State Relays – More flexible; Less power consumption and low burden; Improved dynamic performance characteristics; Reduced panel space. (iii) Numerical Relays – Maximum flexibility; Provides multiple functionality; Self checking and communication facility; can be made adaptive, i.e. settings can be changed automatically thus attuned to the prevailing power system condition

Desirable Attributes of Relay – (i) Reliability – when a fault occurs, the relays must respond instantly and correctly. A quantitative measure for reliability is defined as the ratio of number of correct trips to the number of desired trips plus number of incorrect trips. (ii) Dependability – A relay is said to be dependable if it trips only when it is expected to trip. This happens when the fault is in its primary jurisdiction. However, false tripping of relays or tripping for faults that is either not within its jurisdiction, or within its purview, compromises system operation. Dependability is the degree of certainty that the relay will operate correctly and can be expressed as the ratio of number of correct trips to the number of desired trips. (iii) Selectivity – The relay must be operated in only those conditions for which relays are commissioned in the electrical power system. There may be some typical condition during fault for which some relays should not be operated or operated after some definite time delay; hence, protection relay must be sufficiently capable to select appropriate condition for which it would be operated. (iv) Sensitivity – Sensitivity may be defined as the smallest value of operating/actuating quantity at which it starts operating in relation with the minimum value of fault current in the protected zone. Sensitivity factor is the ratio of minimum value of fault current

DOI: 10.1201/9781003231240-13

in zone to the minimum operating current. (v) Speed – The protective relays must operate at the required speed. This maximizes safety and minimizes equipment damage/system instability. (vi) Discrimination – There must be a correct coordination provided in various power system protection relays in such a way that for fault at one portion of the system should not disturb other healthy portion. Fault current may flow through a part of healthy portion since they are electrically connected, but relays associated with that healthy portion should not be operated faster than the relays of faulty portion otherwise undesired interruption of healthy system may occur.

Time Grade Protection – A straightforward way of obtaining selective protection is to use time grading. The principle is to grade the operating times of the relays in such a way that the relay closest to the fault spot operates first. Time-graded protection is implemented using over current relays with either definite time characteristic or inverse time characteristic. The operating time of definite time relays does not depend on the magnitude of the fault current, while the operating time of inverse time relays is shorter the higher the fault current magnitude is. The time-graded protection is best suited for radial networks.

Current Grade Protection – In certain cases, protection principle based on current grading can be used to essentially accelerate the operation of the protection in faults arising close to the relaying point. The protection is implemented by using one directional or non-directional stage of the over current relay.

Zones of Protection – The protected zone is that part of the power system, guarded by certain protection and usually contains one or at the most two elements of the power system. The zones are arranged to overlap so that no part of the system remains unprotected. A relay's zone of protection is a region which defines its jurisdiction. It is shown by demarcating the boundary. It is essential that primary zones of protection should always overlap so that no portion of the system ever remains unprotected. Depending upon the rating of the equipment, its location, relative importance and probability of faults, each equipment is covered by a protective zone. The complete power system is covered by several protective zones as explained in Figure 13.1. While creating a zone, neighboring zones overlap so that no "dead spot" are left.

Unit and Non-Unit Protection – A unit protective system is one in which only faults occurring within its protected zone are isolated. Faults occurring elsewhere in the system have no influence on the operation of a unit system. A non-unit system is a protective system which is activated even when the faults are external to its protected zone. Unit system has absolute discrimination and its zone of protection is absolutely defined. Examples are differential protection, pilot wire and carrier current protection. A non-unit system does not possess absolute discrimination. The discrimination is obtained by time grading, current grading or a combination of time and current grading. Example – distance protection.

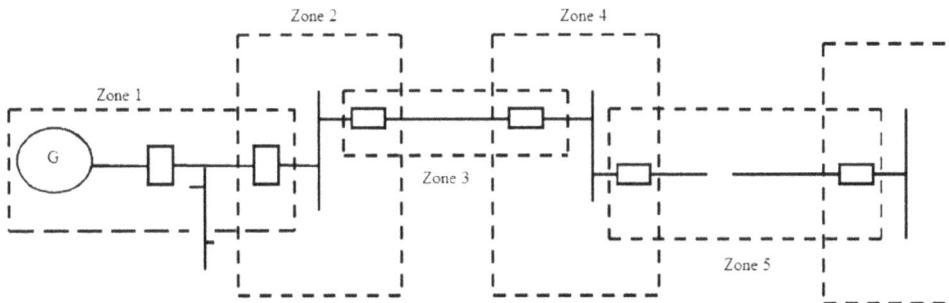

FIGURE 13.1   Protective zones.

Pick-up level of actuating signal is the value of actuating quantity (voltage or current) which is on threshold above which the relay initiates to be operated. If the value of actuating quantity is increased, the electromagnetic effect of the relay coil is increased, and above a certain level of actuating quantity, the moving mechanism of the relay just starts to move.

Reset level is the value of current or voltage below which a relay opens its contacts and comes in original position.

Operating Time of Relay is just after exceeding pick-up level of actuating quantity, the moving mechanism (e.g. rotating disc) of relay starts moving and it ultimately closes the relay contacts at the end of its journey. The time which elapses between the instant when actuating quantity exceeds the pick-up value and the instant when the relay contacts close is the operating time of the relay.

Reset time of Relay is the time which elapses between the instant when the actuating quantity becomes less than the reset value and the instant when the relay contacts return to its normal position.

Reach of relay can be viewed from the operation of distance relay. A distance relay operates whenever the distance seen by the relay is less than the pre-specified impedance. The actuating impedance in the relay is the function of distance in a distance protection relay. This impedance or corresponding distance is called reach of the relay.

Based on Characteristic the protection relay can be categorized as (i) Instantaneous Relays in which no intentional time delay is there. The relay has only pick setting and no time setting. (ii) Definite time Relays in which setting can be adjusted to issue trip output at a definite time, which is adjustable. (iii) Inverse time Relays with definite minimum time (IDMT) whose characteristic is inverse in the initial part which tends to a definite minimum operating time as the current becomes very high. Characteristics are of three types, namely, normal inverse where time of operation $t_{op} = 0.14 \, (TMS)/\left[(PSM)^{0.02} - 1\right]$; very inverse where time of operation $t_{op} = 13.5 \, (TMS)/\left[(PSM) - 1\right]$; and extremely inverse where time of operation $t_{op} = 0.14 \, (TMS)/\left[(PSM)^{0.02} - 1\right]$.

Based on application the protection relay can be categorized as Primary Relay or Backup Relay. Primary relay or primary protection relay is the first line of power system protection whereas Backup relay is operated only when primary relay fails to be operated during fault. Hence backup relay is slower in action than primary relay.

Any relay may fail to be operated due to any of the reasons: (i) the protective relay itself is defective; (ii) auxiliary trip voltage supply to the relay is unavailable; (iii) trip lead from relay panel to circuit breaker is disconnected; (iv) trip coil in the circuit breaker is disconnected or defective; (v) current or voltage signals from CT or PT respectively are unavailable.

13.2.1 Types of Electromagnetic Relay

Attracted Armature type relay – the relay can be either in AC or DC. The minimum current at which the armature gets attracted to close the trip circuit is called pick up current. The relay does not have directional feature unless they are provided with additional polarized coil. It has high operating speed and the operating time as small as 0.5 seconds can be used termed as instantaneous. VA burden of such relay is less in the range of 0.1–0.6.

For DC operation, the force exerted on an armature is constant and directly proportional to the square of the magnetic flux in the air gap or the square of the current passes through the coil; $F = KI^2$, where F is the net force, K is constant and I is the current of armature coil. When the force exceeds restraining force, the relay operates. The threshold condition for relay operation would therefore be reached when $KI^2 = K'$, where K' is the restraining force. Thus the relay operation is dependent on the constants K' and K for a particular value of the coil current and is influenced by: (i) Ampere – turns developed by the relay operating coil; (ii) the size of air gap between the relay core and the armature; and (iii) restraining force on the armature.

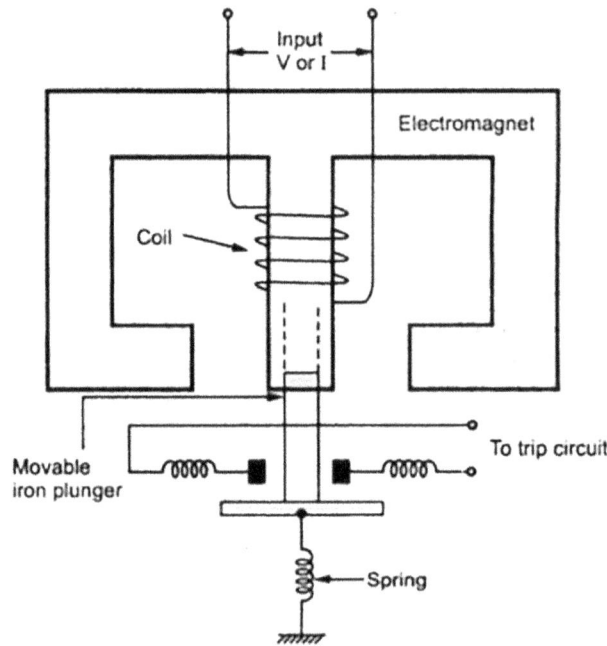

FIGURE 13.2   Attractive armature type relay.

AC operation can be characterized by $F = KI^2 = K(I_m \sin \omega t)^2 = 0.5\ KI_m^2 - 0.5\ KI_m^2 \cos 2\omega t$. The force is not constant, as it contains two components: one is constant and is equal to $0.5\ KI_m^2$ and independent of time, and the other pulsating at double frequency of the applied ac quantity equal to $0.5\ KI_m^2 \cos 2\omega t$. Figure 13.2 shows the attractive armature type relay.

Induction Disc type relay – every induction disc type relay works on the Ferraris principle. This principle says, a torque is produced by two-phase displaced fluxes, which is proportional to the product of their magnitude and phase displacement between them. Mathematically it can be expressed as $T = \varphi_1 \varphi_2 \sin \theta$. The deflecting torque is produced by the eddy currents in an aluminum or copper disc by the flux. The relays are of two types: (i) shaded pole induction disc type and (ii) watt-hour meter type induction disc relay.

The shaded-pole structure shown in Figure 13.3 is usually actuated by current flowing in a single coil wound on a magnetic structure containing an air gap. The air-gap flux produced by the actuating current is split into two fluxes displaced in time and space by a so-called shading ring, generally of copper, that encircles part of half of the pole face of each pole, as shown.

Let $\varphi_1 = \varphi_{m1} \sin \omega t$ is the flux of un-shaded portion of the pole and $\varphi_2 = \varphi_{m2} \sin(\omega t + \theta)$ is the flux of shaded portion of the pole. The disc is normally made of aluminum so as to have low

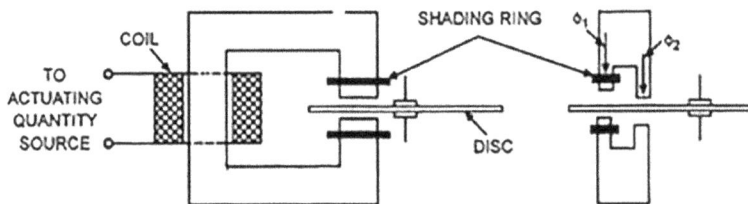

FIGURE 13.3   Shaded pole induction disc type relay.

inertia and, therefore, needs less deflecting torque for its movement. The two rings have currents induced in them by the alternating flux of the electromagnet, and the magnetic fields developed by these induced currents cause the flux, in the portions of the iron surrounded by the rings to lag in phase by 40°–50° behind the flux in the un-shaded portions of the pole. θ is the phase difference between the two fluxes. As these two fluxes link with the disc, there must be an induced emf e_1 and e_2 in the disc. As the disc is purely resistive, the induced current in the disc will be i_1 and i_2. Voltage induced e_1 is proportional to $d\varphi_1/dt \propto \omega\varphi_{m1}\cos\omega t$ and e_2 is proportional to $d\varphi_2/dt \propto \omega\varphi_{m2}\cos\cos(\omega t + \theta)$, and as the paths of eddy currents in the rotor have negligible resistance, it can be assumed $i_1 \propto e_1 \propto \omega\varphi_{m1}\cos\omega t$ and $i_2 \propto e_2 \propto \omega\varphi_{m2}\cos(\omega t + \theta)$. The currents produced by the flux interact with the other flux and vice versa. Thus, the forces produced are $F_1 \propto \varphi_1 i_2 \propto \varphi_{m1}\varphi_{m2}\sin\omega t\cos(\omega t + \theta)$ and $F_2 \propto \varphi_2 i_1 \propto \varphi_{m1}\varphi_{m2}\sin(\omega t + \theta)\cos\omega t$.

Since the forces are in opposition, net force $F = F_2 - F_2 \propto \varphi_{m1}\varphi_{m2}[\sin\omega t\cos(\omega t + \theta) - \sin(\omega t + \theta)\cos\omega t] = \varphi_{m1}\varphi_{m2}\sin\theta$. Hence the net force or torque acting on the disc is same at every instant. When $\theta = 0$, $F = 0$ and relay will not operate. When $\theta = 90°$, torque will be maximum. The torque production is shown in Figure 13.4.

In the watt-hour type construction shown in Figure 13.5 an E-shaped electromagnet and a U-shaped electromagnet is provided with a disc placed in between that is free to rotate. The E-shaped magnet carries both primary coil which is energized from CT and produces a flux φ_1 and secondary coil which is a closed coil in series with the U-shaped magnet energizing coil that produces a flux φ_2. As the alternating flux φ_1 induces a voltage in the secondary coil, which is closed by itself creates a current that in turn creates the φ_2. The phase angle between them is θ.

The primary coil can be tapped at intervals through a plug setting bridge. There are mainly three types of shape of rotating disc available for induction disc type relay. They are spiral, round and vane shaped. Directional feature can be provided in the watt-hour type construction with two

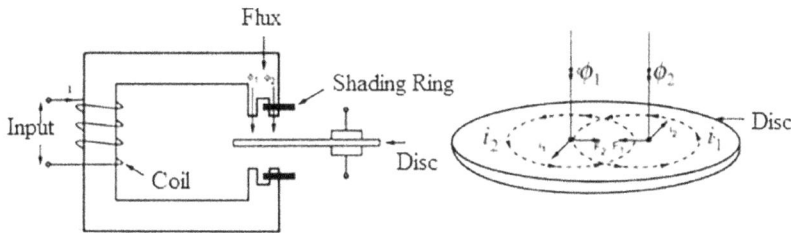

FIGURE 13.4 Torque production of induction disc type relay.

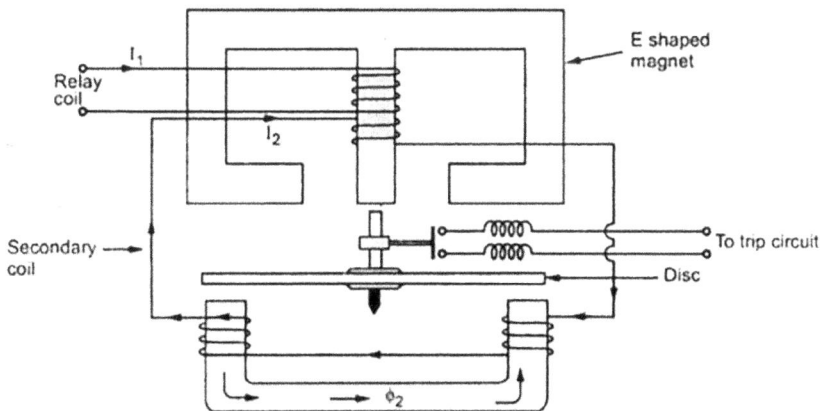

FIGURE 13.5 Watt-hour type induction disc type relay.

actuating coil, one being current coil and other voltage coil, which are energized from CT and PT, respectively.

Operating time/current characteristic is inverse; the time reduces as the current increases. However, the operating time is adjusted with the help of a time setting multiplier. TSM is generally in the form of a back stop, which decides the arc length through which the disc travels. If a relay takes S seconds with TSM 1, the same relay will take a time equal to $T \times S$ seconds for TSM T. A 1.3 seconds normal inverse IDMT relay implies operation time 1.3 seconds at TMS = 1.0 and PSM = 10. VA burden is less in the range of 2.5.

Induction Cup Type Relay – the relay can be considered as a different version of induction disc type relay, actually disc is replaced by aluminum cup, reducing inertia. Induction cup type relay is used where very high-speed operation along with polarizing and/or differential winding is requested. Generally, four pole and eight pole designs are available. The number of poles depends upon the number of windings to be accommodated. Induction cup relay shown in Figure 13.6 consists of (i) Stationary iron core (stator), (ii) Hollow cylindrical cup (rotor), (iii) Electromagnet and (iv) Coil. Stationary iron core acts as stator. Hollow cylindrical cup acts as rotor similar to a disc in other two relays. It is free to rotate in air gap between electromagnet and stationary iron core. Electromagnet creates magnetic field. Induction cup relay consists of two pairs of coil; when these coils are energized, flux is produced in the coils. These fluxes will have same frequency but will have phase delay. These coils will then energize iron core thus producing a rotating flux in the air gap. Due to these fluxes, eddy currents are made to circulate in rotor.

The inertia of cup type design is much lower than that of disc type design. Hence very high-speed operation is possible in induction cup type relay. Further, the pole system is designed to give maximum torque per KVA input. In a four-pole unit almost all the eddy currents induced in the cup by one pair of poles appear directly under the other pair of poles – so that torque/VA is about three times that of an induction disc with a c-shaped electromagnet. Induction cup type relay is practically suited as directional or phase comparison units. This is because, besides their sensitivity, induction cup relay has steady non-vibrating torque and their parasitic torque due to current or voltage alone is small.

In Induction Cup-type-Directional/Power Relay, a four-pole induction cup type relay, one pair of poles produces flux proportional to voltage and the other pair of poles produces flux proportional

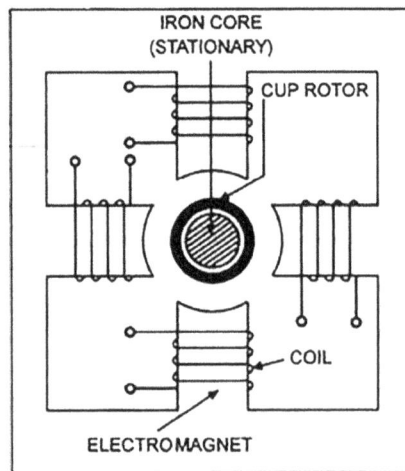

Induction Cup Relay

FIGURE 13.6 Induction cup type relay.

to current. The angle between system voltage e and current i is θ. The current produces a flux φ_1 which is in phase with i, and the voltage produces a flux φ_2 which is in quadrature with e. Hence, angle between φ_1 and φ_2 is $(90° - \theta)$. The torque $T \infty \varphi_1\varphi_2 \sin(90° - \theta) = k\ \varphi_1\varphi_2 \sin(90° - \theta)$. By design, the angle can be made to approach any value. Accordingly, induction-cup type relay can be designed to produce maximum torque when system angle $\theta = 0°$ or 30° or 45° or 60°. The former is known as power relays as they produce maximum torque when $\theta = 0°$, and the latter is known as directional relays – they are used for directional discrimination in protective schemes under fault conditions, as they are designed to produce maximum torque at faulty conditions. By manipulating the current or voltage coil arrangements and the relative phase displacement angle between various fluxes, induction cup type relay can be made to measure pure reactance of a power circuit.

Balanced Beam Type Relay – The relay can be said a variant of attraction armature type relay, but still these are treated as different types of relay as they are employed in different field of application. Balanced beam type relays were used in differential and distance protection schemes. The use of these relay becomes absolute as sophisticated induction disc type relay and induction cup type relays supersede them. There are two coils, on each side of a beam, which is centrally pivoted. The beam remains in horizontal position till operating force becomes more than restraining force. Neglecting the spring effect, the net torque is given by $T = K_1[I_1]^2 - K_2[I_2]^2$ where I_1 is the current in the operating coil and I_2 is the current in the restraining coil. At the instant of relay operation, $K_1[I_1]^2 = K_2[I_2]^2$. The relay is fast operating and has high ratio of resetting quantity to operating quantity. The relay has lower VA burden.

Moving Coil Type Relay – The moving coil relay or polarized DC moving coil relay (polarized relays are one where permanent magnet is used in magnetic structure) is most sensitive electromagnetic relay. This is shown in Figure 13.7. Because of its high sensitivity, this relay is used widely for sensitive and accurate measurement for distance and differential protection. This type of relays is inherently suitable for DC system. Although this type of relay can be used for AC system, necessary rectifier circuit should be provided in current transformer.

In a moving coil relay the movement of the coil may be rotary or axial. Both of them have been perfected to a large extent by the various manufactures, but the inherent limitation of a moving coil relay remains, i.e. to lead the current in and out of the moving coil system which, for reasons of sensitivity, has to be designed to be very delicate. Between these two types of moving coil relay an axial moving type has twice sensitivity than that of rotary type. With moving coil relay, sensitivities

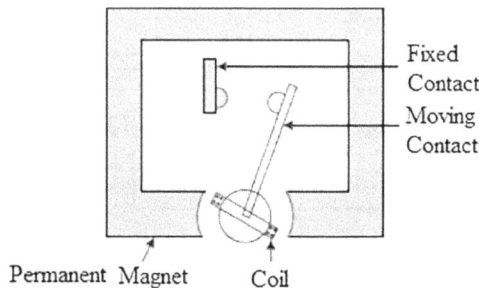

Fixed
Contact
Moving
Contact

Permanent Magnet   Coil

FIGURE 13.7   Moving coil type relay.

of the order of 0.2–0.5 mW are typical. Speed of operation depends upon damping provided in the relay. The operating torque is proportional to current in the coil and is given by $F \propto NHIL$, where F =Force, H =magnetic field in the air gap, I =current in the coil and L =length of the coil. The torque is given by $T = 2\pi rF$, where r =radius of the coil. The relay of this kind has uniform torque for various position of the coil.

The other version of moving coil type relay is auxiliary moving coil type relay that has one air gap and is more sensitive than previous one. The coil is supported axially and moves horizontally when current is passed. The relay is faster than the previous one. The polarized moving coil type relay is basically moving iron type with an additional polarizing coil that gives polarizing feature. Polarizing quantity is one that produces flux in addition to the main flux, which increases sensitivity.

Directional Relay – A directional relay can be compared to a contact making wattmeter. A wattmeter develops maximum positive torque when the current and voltage supplied to the current coil and the pressure coil are in phase. If we define the maximum torque angle (MTA) as the angle between the voltage and current at which the relay develops maximum torque, then a wattmeter can be called a directional relay with MTA of zero degree. There are of two types.

Directional Power Relay – The relay operates when the power in the circuit flows in a particular direction. The construction and operating principle of this relay is similar to watthour type induction relay. This is shown in Figure 13.8. Here the primary coil of E-shaped magnet is energized from potential transformer. There is no secondary coil in the E-shaped magnet. The U-shaped magnet is energized from the current transformer.

Hence the torque production in this type of relay is due to interaction of the flux produced from both voltage and current of the circuit. The current coil winding can be tapped through a plug setting bridge. This type of relay is not suitable for use under short circuit condition as under short circuit condition the voltage falls drastically and may not produce driving torque for operation. Hence directional over current relay is used.

Directional Over Current Relay – which has directional element which is of directional power type and non-directional element which is non-directional over current relay type. The current coil of the non-directional element is connected in series with the primary coil of the non-directional element. The plug setting bridge is connected in this element to adjust current setting as per requirement. The trip contact is in the non-directional element. This is shown in Figure 13.9.

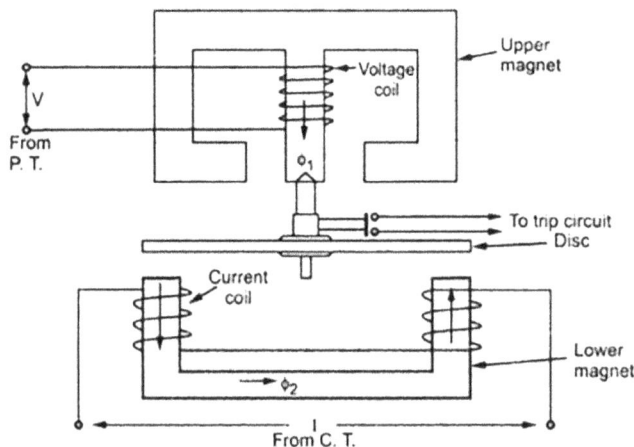

FIGURE 13.8 Directional power relay.

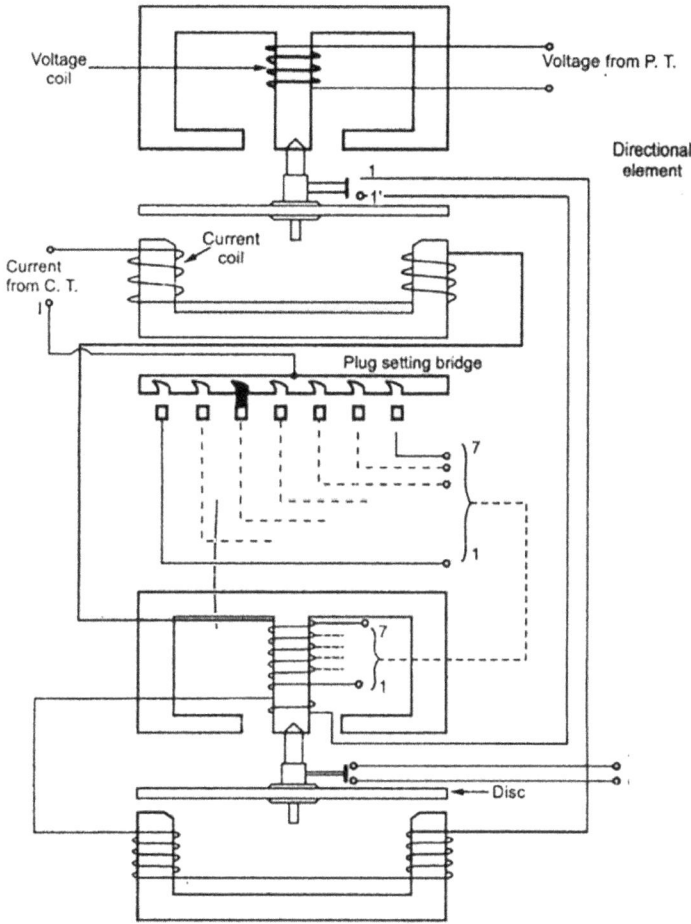

FIGURE 13.9   Directional overcurrent relay.

13.2.2   STATIC RELAY

Electromagnetic relays use moving parts such as armature in their control circuitry. The relays which do not use moving parts and use the solid-state electronic components such as diodes, transistors, etc. are called static relays. The circuits such as comparators, level detectors, zero crossing detectors, etc. are designed using electronic components in the static relays for measurement and comparison of electrical quantities. The static relay is designed in such a way that whenever a quantity under consideration exceeds a particular level, the static circuit produces a response, which is manipulated and given to a tripping circuit. Figure 13.10 shows the block diagram of static relay.

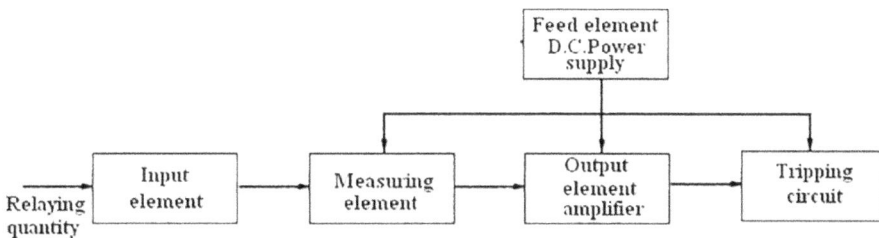

FIGURE 13.10   Block diagram of static relay.

Static relays have some advantages like (i) moving parts are absent in the relay; (ii) burden gets considerably reduced, thus CTs of smaller burden can be used; (iii) power consumption is less; (iv) response is quick; (v) the resetting time can be reduced and overshoots can be reduced due to absence of mechanical inertia and thermal storage; (vi) sensitivity is high as signal amplification can be achieved very easily; and (vii) the relay is smaller and compact. The limitations are (i) the characteristics of electronic components are temperature sensitive; (ii) reliability is unpredictable; (iii) relay has low short time overload capacity; (iv) additional DC supply is required; (v) susceptible to voltage fluctuations and transients; and (vi) less robust.

13.2.3 DIGITAL RELAY

Any digital relay can be thought of as comprising three fundamental subsystems: (i) a signal conditioning subsystem, (ii) a conversion subsystem, (iii) a digital processing relay subsystem. The first two subsystems are generally common to all digital protective schemes, while the third varies according to the application of a particular scheme. Each of the three subsystems is built up of a number of components and circuits.

The construction of the relay, the block diagram of which is shown in Figure 13.11, can be broadly classified into (i) Analog Input Subsystem, (ii) Discrete Input Subsystem, (iii) A/D Converter, (iv) Microprocessor, (v) Discrete Output Subsystem and (vi) Operating Signaling and Communication Subsystems.

Advantages of Digital Relay: (i) High level of functionality integration, (ii) Additional monitoring functions, (iii) Functional flexibility, (iv) Capable of working under a wide range of temperatures, (v) Self-checking and self-adaptability, (vi) Able to communicate with other digital equipment (pear to pear), (vii) Less sensitive to temperature, aging, (viii) More Accurate, (ix) Signal storage is possible.

Limitations of Digital Relay: (i) Short lifetime due to the continuous development of new technologies, (ii) Susceptibility to power system transients, (iii) As digital systems become increasingly more complex, they require specially trained staff for Operation.

The components are: (i) Transducers – In digital relays, magnitudes of current from CT (5/1 A) and voltage from PT (110 V) are both further reduced using auxiliary transducers and/or mimic impedances that remove exponentially decaying DC offset present in the signal within the relays to suit the requirements of the components used. (ii) Surge protection circuit – The current and voltage from the secondary of the CT and PT is connected to surge protective circuits, which typically consist of capacitors and isolating transformers. Zener diodes are also commonly used to protect electronic circuits against surges. (iii) Analogue filtering – Such filtering is usually performed using low-pass filters to remove unwanted high frequencies before sampling; (iv) Analogue-to-Digital conversion; and (v) Digital-to-analogue conversion

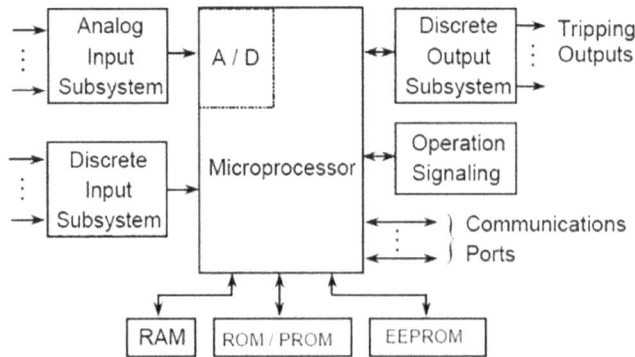

FIGURE 13.11 Block diagram of digital relay.

13.3   DISTANCE RELAY

Distance protection in its basic is a non-unit protection and provides (i) more accurate as more information is used for taking decision; (ii) directional, i.e. it responds to the phase angle of current with respect to voltage phasor; (iii) fast and accurate having usually induction cup design; (iv) primarily used in transmission line protection. Also it can be applied to generator backup, loss of field (LOF) and transformer backup protection. Since the impedance of a transmission line is proportional to its length, for distance measurement it is appropriate to use a relay capable of measuring the impedance of a line up to a predetermined point (the reach point). All distance elements are based on measuring an apparent impedance between the line terminal and the fault location. The basic principle of distance protection involves the division of the voltage at the relaying point by the measured current. The apparent impedance so calculated is compared with the reach point impedance. If the measured impedance is less than the reach point impedance (Z_R), it is assumed that a fault exists on the line between the relay and the reach point. The reach point of a relay is the point along the line impedance locus that is intersected by the boundary characteristic of the relay. Consider the single-phase circuit shown in Figure 13.12.

A distance relay measures the voltage (V) and current (I) at one end of the line. We want a distance element to respond to faults short of a predetermined reach point and restrain for faults beyond that reach point. Universal torque equation of relay is $T = K_1 I^2 + K_2 V^2 + K_3\ VI\cos(\theta - \tau) + K_4$; where T is the net torque acting of the relay, which implies algebraic sum of operating torque, that tends to close the trip contact and the restraining torque that tends to oppose the closing of trip contact. $K_1 I^2$ is torque due to current coil. $K_2 V^2$ is torque due to voltage coil. $K_3\ VI\cos(\theta - \tau)$ is torque due to directional unit. θ is the angle between voltage and current fed to the relay. τ is the maximum torque angle. K_4 is the torque due to control spring and can be neglected. Types of Distance relays

Impedance Relay – The relays that respond to the ratio of the rms voltage at the line terminal and rms current flowing in the line are classified as impedance relays. The magnitude of the ratio of the voltage and current phasors is the magnitude of the measured impedance. In these relays, the magnitude of the measured impedance is compared with a specified magnitude (usually 80%–90% of the line impedance). If the magnitude of the measured impedance is less than the specified magnitude, the relay indicates that the fault is in the protected zone. The relays are double actuating quantity relays with one coil energized by voltage and the other coil energized by current. The torque produced by a current element is balanced against the torque of a voltage element. The current element produces positive (pick-up) torque, whereas the voltage element produces negative (reset) torque. In other words, an impedance relay is a voltage-restrained overcurrent relay having no directional feature. If we let the control-spring effect be $-K_4$ and K_3 equal to zero, the torque equation is: $T = K_1 I^2 - K_2 V^2 - K_4$, where I and V are the rms value of current and voltage. At the balance point, when the relay is on the verge of operating, the net torque is zero, and $K_2 V^2 = K_1 I^2 - K_4$. So, $V/I = Z = \sqrt{(K_1/K_2) - (K_4/K_2 I^2)}$. It is customary to neglect the effect of the control spring (K_4), since its effect is noticeable only at current magnitudes

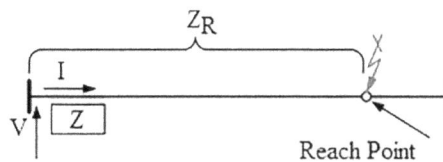

FIGURE 13.12   Distance element reach.

well below those normally encountered. Consequently, if we let K_4 be zero, the preceding equation becomes: $Z = V/I = \sqrt{K_1/K_2}$. In other words, an impedance relay is on the verge of operating at a given constant value of the ratio of V to I, which may be expressed as an impedance. The operating characteristic of an impedance relay in terms of voltage and current on V-I plane is straight line.

R-X Diagram – It is more convenient way of describing the operating characteristic of a distance relay is by means of "Impedance Diagram" or R-X diagram. The relay is represented in a plane, called R-X plane, having X-axis as R (Resistance) and Y-axis as Reactance (X). $|Z| = \sqrt{R^2 + X^2}$, or $Z^2 = R^2 + X^2$. This is the equation of a circle, and magnitude of impedance Z represents the radius of the circle, with center at the origin. $\tan\varphi = X/R$. The numerical values of V and I determine the length of the radius vector Z while the phase angle φ represents between V and I determines the exact position Z. The operation of the relay is independent of φ. Since the relay operates for certain value, less than the set value of, the Z operating characteristic is a circle of radius Z. The numerical value of the ratio of V to I is the length of a radius vector Z, and the phase angle φ between V and I determines the position of the vector. If I is in phase with V, the vector lies along the $+R$ axis; but, if I is 180 degrees out of phase with V, the vector lies along the $-R$ axis. If I lags V, the vector has a $+X$ component; and, if I leads V, the vector has a $-X$ component. Any value of Z less than the radius of the circle will result in the production of positive torque, and any value of Z greater than this radius will result in negative torque, regardless of the phase angle between V and I.

Plane impedance relay has the following disadvantages like (i) it is non-directional. The reach of the simple impedance relay is independent of the phase angle between voltage and current at the relay location. It responds to the faults on both sides of CT and PT location. Hence it cannot discriminate between internal and external faults; (ii) it is affected by arc resistance of line fault and result in under-reach. (iii) It is sensitive to power swings as a large area is covered by the circle on each side on R-X plane; (iv) at some point during the power swing, the apparent impedance enters the trip region of the relay operating characteristic causes the relay to trip, putting the line out of service. Various types of actuating structure are used in the construction of impedance relays. Inverse-time relays use the shaded-pole or the watt-metric structures. High-speed relays may use a balance-beam magnetic-attraction structure or an induction-cup or double-loop structure.

To keep the relay from operating during faults on an adjacent line, a directional relay is used in conjunction with impedance relays. The directional relay determines if the fault is on the line side of the relay or on the bus side of the relay. If the fault is on the line side, the impedance relay is allowed to open the line circuit breaker. If the fault is on the bus side of the relay, the trip signal from the impedance relay is blocked. This is achieved by connecting the trip contacts of the directional relay and the impedance relay in series. If the relay has directional feature, then the characteristic will be a semi-circle.

Reactance Relay – The reactance-relay unit of a reactance-type distance relay has, in effect, an over current element developing positive torque, and a current-voltage directional element that either opposes or aids the over current element, depending on the phase angle between the current and the voltage. In other words, a reactance relay is an over current relay with no directional property or directional restraint. If we let the control-spring effect be $-K_4$ and K_2 equal to zero, the torque equation is: $T = K_1 I^2 - K_3 VI \sin\theta - K_4$, where θ is defined as positive when I lags V. At the balance point, the net torque is zero, and hence $K_1 I^2 = K_3 VI \sin\theta + K_4$. This gives $(V/I)\sin\theta = Z \sin\theta = (K_1/K_3) - (K_4/K_3 I^2)$. If we neglect the effect of the control spring (K_4); $X = K_1 / K_3 = $ constant. In other words, this relay has an operating characteristic such that all impedance radius vectors whose heads lie on this characteristic have a constant X component. This describes the straight line parallel to X axis. The operating characteristic of the relay on the R-X plane is a straight line with intercept of X, on the reactance axis. The entire area below this straight line represents the trip region. The significant thing about this characteristic is that the resistance component of the impedance has no effect on the operation of the relay; the relay responds solely to the reactance component. The relay responds for the fault in the forward direction and also in the reverse direction for an unlimited distance.

MHO Relay – A reactance-type distance relay for transmission-line protection could not use a simple directional unit as in the impedance-type relay, because the reactance relay would trip under normal load conditions at or near unity power factor, as will be seen later when we consider what different system-operating conditions look like on the R-X diagram. The reactance-type distance relay requires a directional unit that is inoperative under normal load conditions. The type of unit used for this purpose has a voltage-restraining element that opposes a directional element, and it is called an admittance or mho relay. In other words, this is a voltage-restrained directional relay. When used with a reactance type distance relay, this unit has also been called a "starting unit". If we let the control-spring effect be $-K_4$ and K_1 equal to zero, the torque of such a unit is: $T = K_3 VI \cos(\theta - \tau) - K_2 V^2 - K_4$; where θ and τ are defined as positive when I lags V. θ is the phase angle between the voltage and current fed to the relay. At the balance point, the net torque is $K_2 V^2 = K_3 VI \cos(\theta - \tau) - K_4$. This gives $V/I = Z = (K_3/K_2)\cos(\theta - \tau) - [K_4/(K_2 VI)]$. If we neglect the control-spring effect, $Z = (K_3/K_2)\cos(\theta - \tau)$. This equation is like that of the directional relay when the control spring effect is included, but here there is no voltage term, and hence the relay has one circular characteristic. A mho relay is a directional relay with voltage restraint.

The operating characteristic of a mho distance relay, also known as an admittance relay, is a circle that passes through the origin of the R-X plane. Since the third quadrant of the R-X plane is outside the operating characteristic of the relay, the faults on the bus side are not seen by this relay. Another advantage of using mho relays for transmission line protection is that, when protecting the same line, their reach along the R axis is substantially less than that of impedance relays. Because of these advantages, the use of mho relays is always preferred over the use of impedance relays. Mho relay is also known as admittance relay and measures a component of admittance $Y < \varphi$. The impedance relay trips if the magnitude of the impedance is within the circular region. Since the circle spans all the quadrants, it leads to non-directional protection scheme. In contrast, the mho relay which covers primarily the first quadrant is directional in nature.

Quadrilateral Relay – The circular and straight-line characteristics of distance relays were originally developed using electromechanical technology. The circular shape of the relays was a natural outcome of that technology. The straight line is a special type of circle; it is a circle of infinite radius. When analog electronics technology became acceptable, it became possible to develop relay shape characteristics other than circles. The most important development in this area was the introduction of the quadrilateral characteristic, which is discontinuous. Two approaches are used in developing these relays. The first approach is to develop special equations that describe the quadrilateral and implement them. The second approach is to include the design of four blocking characteristics using operational amplifiers based on analog electronic circuits. An advantage of this characteristic is that the "reach" of the relay in the R and X directions can be controlled somewhat independently.

13.4   DIFFERENTIAL RELAY

Differential protection is based on the principle that any fault within electrical equipment would cause the current entering it, to be different from that leaving it. Thus, comparing the two currents either in magnitude or in phase or both, issue a trip output if the difference exceeds a predetermined set value. Usually, the currents entering and leaving the equipment to be protected are stepped down with the help of CTs on either side. The currents transformed by the two CTs placed at the two ends of the equipment to be protected, being equal in magnitude as well as in phase, there will be no current in the spill path, and the relay connected to spill path will not operate. The differential relaying scheme should also remain stable for any fault which is outside its protective zone. Such faults are called external faults or through faults. During external faults too, the current leaving the protected zone is the same as that entering it. Types of Differential Relay: (i) Current Differential Relay – where two current transformers are fitted on the either side of the equipment to be protected. The secondary circuits of CTs are connected in series in such a way that they carry secondary CT current in same direction. The operating coil of the relaying element is connected across the CT's secondary circuit. (ii) Voltage Differential Relay – In this arrangement the current transformer is connected to either side of the equipment in such a manner that EMF induced in the secondary of both current transformers will oppose each other. That means the secondary of the current transformers from both sides of the equipment are connected in series with opposite polarity. The differential relay coil is inserted somewhere in the loop created by series connection of secondary of current transformers.

Drawbacks of Simple Differential Protection: (i) In the differential schemes as discussed above, it is tacitly assumed that the CTs are ideal and identical, which in practical very difficult to get. CTs are subject to ratio and phase angle errors. Both these errors depend upon the burden on the CTs, which in turn depend on the lead lengths and the impedance of the relay coil. The errors, in general, increase as the primary current increases, as in the case of external faults. (ii) Ideally, for "through faults", secondary currents of both the CTs would be equal in magnitude and in phase with each other, and thus the spill current would be zero. However, let CT on the incoming side has an actual ratio of n_i, and phase-angle error of θ_i; while CT, on the outgoing side has an actual ratio of n_o and phase angle error of θ_o. The difference between these two currents, therefore, ends up as spill current. Since both the ratio and phase angle errors aggravate as primary current increases, the spill current builds up as the "through fault" current goes on increasing. As the "through fault" current goes on increasing, various imperfections of the CTs get magnified. This causes the spill current to build up. Therefore, as the "through fault" current goes on increasing, there comes a stage when the spill current, due to the difference between the secondary currents of the two CTs, exceeds the pick-up value of the over current relay in the spill path. This causes the relay to operate, disconnecting the equipment under protection from rest of the system. Since the magnetizing currents of the two CTs will generally vary widely, there is a substantial spill current during "through fault" conditions. This results in loss of stability and mal-operation of the simple differential scheme. Thus, the simple differential scheme, which looks attractively simple, cannot be used in practice without further modifications. This is especially true in case of transformer protection. The CTs on the two sides of the transformer have to work at different primary system voltage. Because the currents on the two sides of the transformer are, in general, different, the ratios of transformation of the CTs are also different. Their designs are therefore different, making it impossible to get a close match between their characteristics. This explains why the spill current goes on increasing as the "through fault" current increases.

The simple differential relay can be made more stable, if somehow, a restraining torque proportional to the "through fault" current could be developed, the operating torque still being proportional to the spill current. This idea has been implemented in the percentage differential relay. The relay is designed to operate to the differential current in the term of its fractional relation to the current flowing through the protected section. In this type of relay, there are restraining coils in addition to the operating coil of the relay as shown in Figure 13.13. The operating coil is connected to the mid-point of the restraining coil. The restraining coils produce torque opposite to the operating torque. Under normal and through fault conditions, restraining torque is greater than operating torque. Thereby relay remains inactive.

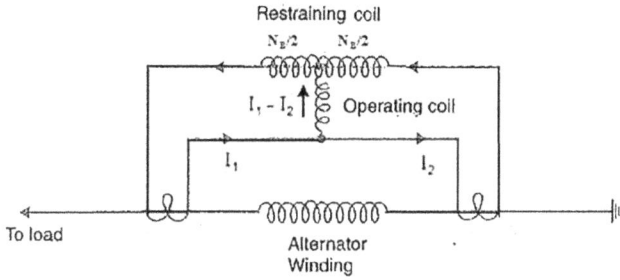

FIGURE 13.13   Bias differential relay.

Let us take the operating coil is having turns N_O and the restraining coil or bias coil is having turns N_B. For operation, $(I_1 - I_2)N_O > I_1 N_B/2 + I_2 N_B/2;$ or $(I_1 - I_2) > (N_B/N_O)(I_1 + I_2)/2.$ $(I_1 - I_2)$ is the operating current, say equal to I_O and $(I_1 + I_2)/2$ is the bias current say equal to I_B. $[N_B/N_O]$ is the bias quantity equal to B. This gives $I_O > BI_B$. The percentage differential relay does not have a fixed pick-up value. The relay automatically adapts its pick-up value to the "through fault" current. Percentage-differential characteristics are available as fixed-percentage or variable percentage. The difference is that fixed-percentage relays exhibit a constant percentage restraint, and for a variable-percentage relay the percentage restraint increases as the restraint current increases.

In case of fault outside the protected zone, as the "through fault" current goes on increasing, various imperfections of the CTs at the two ends get magnified. This causes the spill current to build up. Therefore, as the "through fault" current goes on increasing, there comes a stage when the spill current, due to the difference between the secondary currents of the two CTs, exceeds the pick-up value of the relay in the spill path. This causes the relay to operate, disconnecting the equipment under protection from rest of the system. This is a case of mal-operation, since the relay has tripped on external fault. In such instances, the differential scheme is said to have lost stability and crossed through fault stability limit, which is defined as the maximum "through fault" current beyond which the scheme loses stability.

$$\text{Stability ratio} = \frac{\text{Minimum through fault current beyond which the scheme maloperates}}{\text{Minimum internal fault current required for tripping}}$$

Percentage Differential relay for transformer protection can be implemented through some modifications as simple percentage differential scheme tends to maloperate due to magnetizing inrush. One way to combat this problem is to desensitize the relay for a brief period of time, just after switching on. However, this is not desirable, since the probability of insulation failure just after switching on is quite high, and a desensitized relay would be blind to faults taking place at that crucial time. If the waveforms are compared due to internal fault current with that of the inrush current and a corrective action is taken, the problem can be minimized. The inrush waveform is rich in harmonics whereas the internal fault current consists only of the fundamental. Thus, we can develop additional restraint based on harmonic content of the inrush current. This additional restraint comes into picture only during the inrush condition and is ineffective during faults. A harmonic restraint percentage differential relay is ideal for transformer protection.

Zone of Protection of the Differential Relay: The differential scheme generates a well-defined and closed zone of protection. This zone encompasses everything between the two CTs. Thus, we talk of any fault between the two CTs as an "internal fault". To the differential scheme, all other faults are "external faults" or "through faults". Ideally, therefore, a differential scheme is supposed to respond only to internal faults and restrain from through faults.

To signify the spread between the minimum internal fault current at which the scheme operates and the maximum "through fault" current beyond which the scheme (mal)operates, we define a term called stability ratio as: *Stability ratio*=maximum "through fault" current beyond which the scheme (ma1)operates/minimum internal fault current required for tripping.

The higher the stability ratio, the better is the ability of the system to discriminate between external and internal faults. The stability ratio can be improved by improving the match between the two CTs.

Differential protection of alternator or Merz-Price circulating current protection system is an only protection system that protects alternator stator from burning through at least 80% of it. Makers of protective gear speak of "protecting 80% of the winding" which means that faults in the 20% of the winding near the neutral point cannot cause tripping, i.e. this portion is unprotected. It is a usual practice to protect only 85% of the winding because the chances of an earth fault occurring near the neutral point are very rare due to the uniform insulation of the winding throughout.

13.5   OVER CURRENT PROTECTION

An over current (OC) relay has a single input in the form of ac current. The output of the relay is a normally open contact, which changes over to closed state when the relay trips. Over current includes short-circuit protection. Short-circuit currents are generally several times (5–20) full load current. Hence fast fault clearance is always desirable on short circuits. The protection should not operate for starting currents, permissible over current, current surges. To achieve this, the time delay is provided (in case of inverse relays). The operating time of an instantaneous relay is of the order of a few milliseconds. Such a relay has only the pick-up setting and does not have any time setting. The over current relay may be

i. **Instantaneous Over Current Relay** – Instantaneous actually means no intentional time delay. The operating time of an instantaneous relay is of the order of a few milliseconds. Such a relay has only the pick-up setting and does not have any time setting.

ii. **Definite Time Over Current Relay** – A definite time over current relay can be adjusted to issue a trip output at a definite (and adjustable – Desired definite operating time can be selected with the help of an intentional built in time delay unit) amount of time, after it picks up. Thus, it has a time-setting adjustment and a pick-up adjustment. The relay operates after a predetermined time when the current exceeds its pick-up value. The operating time is constant, irrespective of the magnitude of current above pick-up value. To ensure selectivity of operation under all circumstances in a radial feeder, the operating time of the protection is increased from the far end of protected circuit towards the generating source as shown in Figure 13.14.

iii. **Inverse Time Over Current Relay** – Inverse time characteristic fits in very well, with the requirement that the more severe a fault is, the faster it should be cleared to avoid damage to the apparatus.

Inverse Definite Minimum Time (IDMT Relay) – This is possibly the most widely used characteristic. The characteristic is inverse in the initial part, which tends to a definite

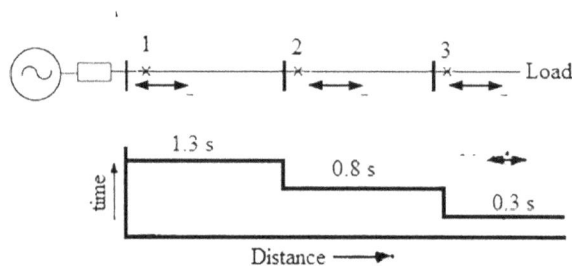

FIGURE 13.14   Time grading overcurrent protection for radial feeders.

minimum operating time as the current becomes very high. There are standard inverse, very inverse and extremely inverse types. Discrimination is included by both "Time" and "Current". There are two basic adjustable settings on all inverse time relays; one is the *time multiplier setting* (TMS), and the other is the current setting usually known as the *plug setting multiplier* (PSM). The relay operation time is inversely proportional to the fault current. The operating time of an over current relay can be moved up (made slower) by adjusting the "time dial setting". The lowest time dial setting (fastest operating time) is generally 0.5 and the slowest is 10.

The time multiplier setting for an inverse time relay is defined as TMS = T/T_m, where T = the required time of operation and T_m = the time obtained from the relay characteristic curve at TMS = 1.0, and using the PSM equivalent to maximum fault current.

Current setting is adjusted by means of a tapped plug bridge hence known as PSM. PSM = Primary Current/(Relay current setting × CT Ratio).

Over current relays generally have 50%–200% setting. Let us consider a 1.0 A relay (i.e. a relay with current coil designed to carry 1.0 A on a continuous basis) whose plug has been set at 0.5 A, i.e. at 50%. Assume that, for a certain fault, the relay current is 5.0 A. The relay, therefore, is said to be operating at a PSM of (5.0/0.5) = 10.

iv. **Directional Over Current Relay** – The relays are combinations of directional and over current relay units. Directional over current protection responds to over currents for a particular direction flow. If power flow is in the opposite direction, the directional over current protection remains un-operative.

If there is a radial feeder, then a non-directional overcurrent relay can be used. However, for a double-end-fed power system, the zones to be generated by the relays use directional overcurrent relay. There are other situations where it becomes necessary to use directional relays to supervise OC relays.

One such situation is a single-end-fed system of parallel feeders, where a fault on any of the parallel lines is fed not only from the faulted line but from the healthy line as well. If directional relays are not provided, in conjunction with OC relays, then the desired zones will not be generated. This will result in both lines being tripped out for any fault on any one of the lines. The scheme is shown in Figure 13.15.

At the sending end 1 and 4 of the feeders, non-directional relays are required. The symbol ↔ indicates non-directional relay. At the other end of the feeders, i.e. at 2 and 3, directional overcurrent relays are required. The arrow mark for directional relays placed at 2 and 3 indicate that the relay will operate if current flow in the direction of arrow. If a fault occurs at F, in between feeder 4 and 3, the directional relay at 3 will trip. The directional relay at 2 will not trip. The relay at 4 will trip on fault. Thus the feeder is isolated during a fault. The directional relays with tripping direction away from the bus will be required at locations "2" and "3" in Figure. However, at locations "1" and "4", non-directional over current relays will suffice. Since directional relay units cost more and also need the provision of PTs, they should be used only when absolutely necessary. In the case of a ring

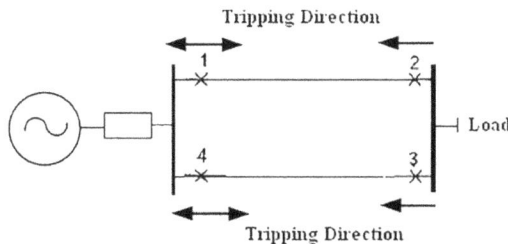

FIGURE 13.15   Parallel feeders single end feed system protection.

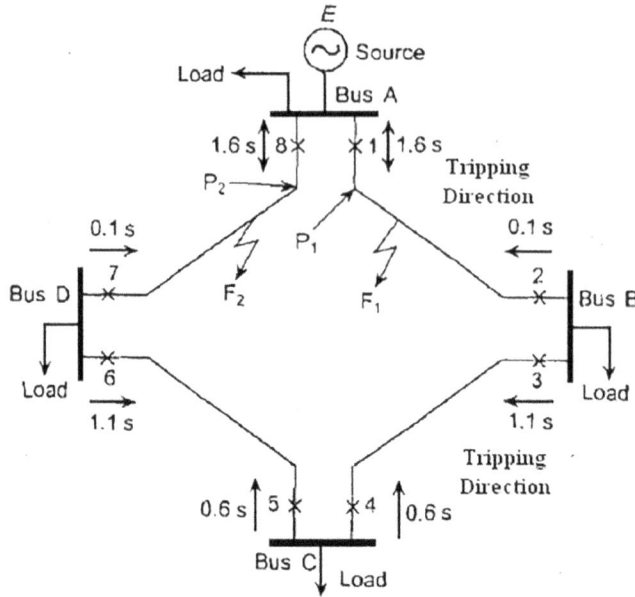

FIGURE 13.16   Ring main feeder protection using directional overcurrent protection.

main feeder system as shown in the Figure 13.16, where also directional supervision of OC relays is called for. It is well known that the ring main feeder allows supply to be maintained to all the loads in spite of fault on any section of the feeder. A fault in any section causes only the CBs associated with that section to trip out, and because of the ring topology, power flows from the alternate path.

A directional relay can be compared to a contact making wattmeter. A wattmeter develops maximum positive torque when the current and voltage supplied to the current coil and the pressure coil are in phase. The maximum torque angle (MTA) as the angle between the voltage and current at which the relay develops maximum. In case of application of directional relays to a three-phase feeder, phase faults need to be considered separately from ground faults. There are various possibilities of energizing these relays; hence the various alternatives need to be carefully considered. The directional relay must meet the requirements: (i) the relay must operate for forward faults; (ii) the relay must restrain during reverse faults; (iii) the relay must not operate during faults other than for which it has been provided, i.e. the relay must not maloperate.

Earth-fault relay is used to protect feeder against faults involving ground protection which can be provided with normal overcurrent relays. The magnitude of earth-fault current is usually low compared to the phase-fault currents because the fault impedance is much higher for earth-faults than for phase-faults. Hence earth-fault relays are set at low settings between 30% and 70% but low values of current settings. Relays used for earth fault protection generally have settings either 10%–40% or 20%–80% setting. The combined overcurrent and earth-fault relay intended to be used for the selective short-circuit and earth-fault protection of three-phase radial feeders in solidly earthed, resistance earthed or impedance earthed power systems. The integrated protection relay includes two overcurrent unit connected on two outer phases and an earth-fault unit connected to central phase with flexible tripping and signaling facilities from economic considerations. Under normal operating conditions and inter-phase faults, not involving ground, no current would flow through the earth-fault relay. The overcurrent relays trip. When a single or double earth-fault occurs, the zero sequence current flows through the earth fault relay and the relay trips. The scheme is shown in Figure 13.17.

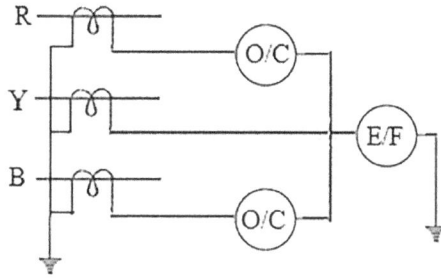

FIGURE 13.17 Combined overcurrent and earth fault protection.

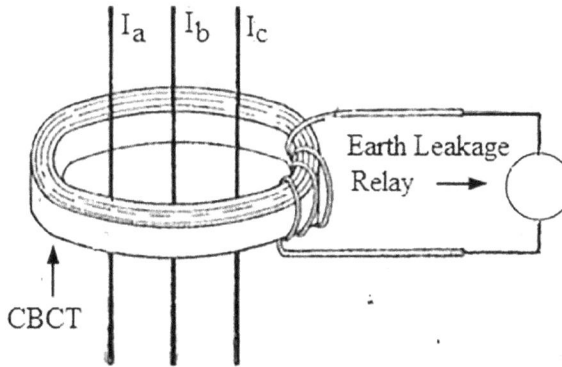

FIGURE 13.18 Earth leakage protection.

Many times because of insulation failure the body of the equipment becomes live. This causes a leakage of current to earth from the body of the equipment as the body is always connected to earth. However, the leakage current may be too small for an over current relay to operate. This poses danger to the personnel who come in contact with the body of the equipment. A special type of differential relay known as the earth leakage relay or current balance relay can easily detect such faults. The scheme is shown in Figure 13.18. Secondary of Core Balance Current Transformer (CBCT) is connected to Earth Fault Relay. During normal operating condition as the vector sum of three-phase current, i.e. $(\bar{I}_a + \bar{I}_b + \bar{I}_c = 0)$ is zero, no residual current in the primary will be present, and hence no current in the secondary circuit of CBCT. During any fault or leakage involving earth, three-phase current passing through the CBCT will not be balanced rather a zero-sequence current will flow, and the secondary reflection of the current will be able to operate the relay.

13.6 PILOT PROTECTION

Pilot protection schemes generally use communication channels to send information from the local relay terminal to the remote relay terminal, for a trip decision. The protection of transmission lines is divided into the following two main categories: (i) Non-pilot schemes are the schemes in which relays installed in each terminal of the transmission line do not have any communications with each other and the relaying decision is only made by analysis of local measurements at location of the relay. Non-pilot protection schemes are usually applicable to short or medium transmission lines. Directional over current and step distance protection schemes belong to this group of transmission protection scheme. (ii) Pilot schemes are the schemes in which relays installed in both end of the transmission line utilize a communication link in order to make a relaying decision.

Pilot relaying is an adaptation of the principles of differential relaying for the protection of transmission-line sections. The term "pilot" means that between the ends of the transmission line there is an interconnecting channel of some sort over which information can be conveyed. Three different types of such a channel are presently in use, and they are called (i) wire pilot – consists generally of a two-wire circuit of the telephone-line type, either open wire or cable; (ii) carrier-current pilot – in which low-voltage, high-frequency (30–200 kz) currents are transmitted along a conductor of a power line to a receiver at the other end, the earth and ground wire generally acting as the return conductor; (iii) microwave pilot is an ultra-high-frequency radio system operating above 900 megacycle.

The distance relays provide fast protection up to 80% of the primary line length. However, primary protection for remaining 20% is deliberately slowed down by coordination time interval. Pilot protection is used for lines to provide the high-speed simultaneous detection of phase and ground faults for 100% of the primary line. Since distance relays are directional relays, the corresponding schemes are known as directional comparison schemes.

Different forms of pilot schemes are: (i) Directional comparison blocking – A form of pilot protection in which the relative operating conditions of the directional units at the line terminals are compared to determine whether a fault is in the protected line section. It uses directional fault detectors to detect faults in the direction of primary line and blocking signal from the remote end in case the fault is not on the primary line. (ii) Directional comparison unblocking – After detecting a fault in the right direction, put the relays in "block mode" and use unblock signals from the remote if the fault is on the primary line. Trip when blocking signal is not received and with supervision from a local terminal forward overreaching element. (iii) Overreaching transfer trip – If fault is detected from both ends of the line, initiate trip. Else, initiate back up protection. (iv) Under reaching transfer trip – Non-permissive and Permissive – The under reaching terminology implies that the functional distances are to be set so as always to overlap but not over reach any remote terminal under all operating condition. (v) Phase-comparison protection – a form of pilot protection that compares the relative phase-angle position of specified currents at the terminals of a circuit.

13.7   TRANSFORMER PROTECTION

There are a number of protections that can be provided in the transformer like

Incipient Faults Protection – Faults which are not significant in the beginning but which slowly develop into serious faults are known as incipient faults. Buchholz relay provides protection against such incipient faults. Buchholz devices provide protection against all kinds of incipient faults, i.e. slow-developing faults such as insulation failure of windings, core heating, fall of oil level due to leaky joints, etc.

Buchholz relay is a gas-actuated relay installed in oil immersed transformers for protection against all kinds of internal faults. Named after its inventor, Buchholz, it is used to give an alarm in case of incipient (i.e. slow-developing) faults in the transformer and to disconnect the transformer from the supply in the event of severe internal faults. It is the simplest form of transformer protection. It detects the incipient faults at a stage much earlier than is possible with other forms of protection. It can only be used with oil immersed transformers equipped with conservator tanks.

When an incipient fault such as a winding-to-core fault or an inter-turn fault occurs on the transformer winding, there is severe heating of the oil. This causes gases to be liberated from the oil around 350°C. There is a build-up of oil pressure causing oil to rush into the conservator. A vane is placed in the path of surge of oil between the transformer and the conservator. A set of contacts, operated by this vane, is used as trip contacts of the Buchholz relay. This output of Buchholz relay may be used to trip the transformer.

Overload Protection – generally overloads can be sustained for long periods, being linked only by the permitted temperature rise in the winding and cooling medium. Excessive overload will result in deterioration of insulation and subsequent failure. It is usual to monitor the winding and oil temperature conditions and an alarm is initiated when the permitted temperature limits are exceeded.

Over Current and Earth Fault Protection – generally called backup protection connected to in-feed side of the transformer to protect from external short circuits and overloads. The current setting must be above sustained overload allowance and below short circuit. The pick-up value of the phase-fault over current units is set such that they do not pick up on maximum permissible overload but are sensitive enough to pick up on the smallest phase fault. The pick-up of the earth fault relay, on the other hand, is independent of the loading of the transformer. The neutral current under load conditions is quite small.

Differential Protection – The differential relay actually compares between primary current and secondary current of power transformer, if any unbalance found in between primary and secondary currents, the relay will actuate and inter trip both the primary and secondary circuit breaker of the transformer.

The power transformer is star connected on one side and delta connected on the other side. The CTs on the star connected side are delta-connected and those on delta-connected side are star-connected. The neutral of the current transformer star connection and power transformer star connections are grounded.

The major operating challenge to transformer differential protection is maintaining security during CT saturation for external faults while maintaining sensitivity to detect low magnitude internal faults. CT saturation reduces the secondary output current from the CT and causes a false differential current to appear to the relay. Biased current differential protection is most commonly applied for transformer protection.

Inrush Phenomenon – The flux in the transformer be written as $\varphi = \varphi_m \sin \omega t$. The induced voltage can then be written as $e = N(d\varphi/dt) = N\varphi_m \cos \omega t = N\varphi_m \sin(\omega t + 90°)$. The applied voltage is exactly equal to the induced voltage and the flux in a transformer lags the applied voltage by 90° in the steady state. Therefore, when voltage is passing through zero and becoming positive the flux should be at its negative maxima and increasing. In a time equal to $T/2$ (half cycle), the flux changes from $-\varphi_m$, to $+\varphi_m$ as shown in Figure 13.19. The change in flux is therefore $2\varphi_m$, in $T/2$ seconds. This is the steady-state picture. If the transformer is switched on at positive zero crossing of the voltage waveform and if we assume that the residual flux is zero, then the initial value of flux is zero, but subsequently the flux must have the same rate of change and same waveform as it had in the steady-state. Thus, the flux must reach a peak value of $2\varphi_m$, in half a cycle. Since power transformers

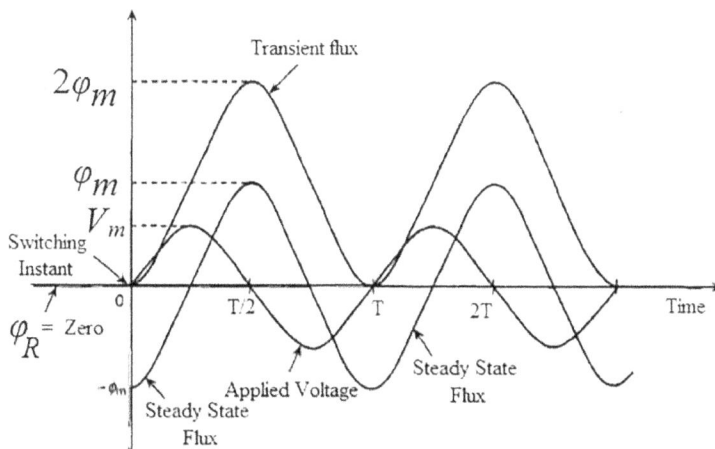

FIGURE 13.19 Inrush phenomenon.

operate near the knee of the saturation curve, a flux demand of $2\varphi_m$ drives the transformer core deep into saturation, causing it to draw a very large magnetizing current with a peaky non-sinusoidal waveform. The magnetizing current is, therefore, very high, of the order of 8–30 times the full-load current. This current is known as inrush current.

Inrush is also experienced whenever there are sudden changes in the system voltage such as sudden recovery of system voltage on clearing of a fault, somewhere in the system. This is called recovery inrush. If an unloaded transformer is being switched on, experiences an inrush, an adjacent transformer, which is in service, may also experience a smaller degree of inrush. This is known as sympathetic inrush. Further, as such a high current flows only on one side of the transformer (on the side which is being connected to the supply), it looks like an internal fault to the differential scheme and ends up as spill current. One way to combat this problem is to desensitize the relay for a brief period of time, just after switching on. However, this is not desirable, since the probability of insulation failure just after switching on is quite high, and a desensitized relay would be blind to faults taking place at that crucial time. A solution to this problem suggests itself, when we compare the waveforms of internal fault current with that of the inrush current. The inrush waveform is non-sinusoidal as rich in harmonics whereas the internal fault current consists of only of the fundamental. Any current of non-sinusoidal wave form may be considered as being composed of a direct-current component plus a number of sine-wave components of different frequencies. The transformer inrush current is rich in second harmonic component. Thus, additional restraint based on harmonic content of the inrush current can be developed. The additional restraint comes into picture only during the inrush condition and is ineffective during faults. Thus, the conceptual scheme of a *harmonic restraint differential relay* comes into effect.

Restricted Earth Fault Protection – Differential including percentage differential and over current protection do not provide adequate protection for star connected windings with grounded neutrals. Faults close to the neutral produce lesser fault current. A percentage differential relay has a certain minimum value of pick-up for internal faults. Faults with fault current below this value are not detected by the percentage differential relay. Winding-to-core faults, which are of the single phase-to-ground type, involving high resistance, fall in this category. Therefore, there must be a more sensitive relaying scheme to cater for high resistance ground faults. Further, the reach of such a protection must be restricted to the winding of the transformer; otherwise it may operate for any ground fault, anywhere in the system, beyond the transformer. When the fault occurs very near the neutral point of the transformer, the voltage available for providing earth fault current is small. Hence the fault current would be low. If the relay is to sense such small signals, it has to be too sensitive and would therefore operate for signals like external faults or switching surges. Hence the practice is to set the relay such that it operates for earth fault current of the order of 15% of rated current; such setting protects restricted portion of the winding, hence the name restricted earth fault protection.

Over Fluxing Protection – Whenever there is an over-voltage (frequency remaining constant), the transformer core is subjected to a higher value of flux in order to be able to support the higher applied voltage ($v = 4.44 \ f\varphi_m N$). By design, power transformers operate at the knee of the saturation curve at normal voltage. Hence, any increase in applied voltage, and the consequent increase in flux density, drives the transformer deeper into saturation. The transformer, therefore, draws an excessive magnetization current. Hence, this condition is described as over-excitation. This, considerably, increases the core losses giving rise to overheating of the transformer. Further, saturation of the core causes the flux to flow into adjacent structures, causing high eddy current losses in the core and adjacent conducting materials. Such an operating condition cannot be allowed to continue for long, and the transformer should be tripped if there is a prolonged over-excitation. It can be easily seen that over-excitation can also occur in case of low-frequency operation of the transformer at rated voltage. Over-excitation can be detected by measuring the V/f ratio by a so-called volts/hertz relay. During over fluxing of a transformer, the transformer current is rich in fifth harmonic current.

13.8   FEEDER PROTECTION

Combined over current and earth fault protection is used for the selective short-circuit and earth-fault protection of radial feeders. Protection of parallel feeders can be provided by the time-graded overload relay with inverse time characteristic at the sending end and instantaneous reverse power or directional relays at the receiving end. Protection of ring main feeders can be provided with a non-directional relay at in-feed end and a directional relay at the load end. The operating times of the relays are determined by considering the grading. Protection of a radial feeder, with an in-feed from one side can be provided with distance relay.

Differential protection of Merz-Price voltage balance scheme or Translay scheme can be used for protection of feeders. In Merz-Price voltage balance scheme, the pair of CTs in each line placed at the two ends is connected through a pilot wire and in series with a relay in such a way that under normal conditions, their secondary voltages are equal and in opposition, i.e. they balance each other. The advantages of Merz-Price system are as follows: (i) can be used for parallel as well as ring main system and (ii) provides instantaneous protection to the ground faults. The limitations of this method are the following: (i) the CTs used must match accurately; (ii) the pilot wires must be healthy without discontinuity; (iii) economically not suitable as the cost is high due to long pilot wires; (iv) due to long pilot wires, capacitive effects may affect the operation of the relays.

In Translay scheme, which is similar to Merz-Price voltage balance system, balance or opposition is between the voltages induced in the secondary windings wound on the relay magnets and not between the secondary voltages of the line current transformers. This permits to use current transformers of normal design and eliminates one of the most serious limitations of original voltage balance system. The three line CTs connected at each end are connected to the tapped primary of the upper element of the relay much like summation transformer. The upper element of the relay also carries a secondary winding that is connected in series with the operating winding on the lower magnet and is closed through pilot wire with the lower magnet of the relay placed at the other end.

13.9   ALTERNATOR PROTECTION

Failure of Prime-Mover – When input to the prime-mover fails; the alternator runs as a synchronous motor and draws some current from the supply system. This motoring condition is known as "inverted running", driving the prime mover at synchronous speed. In steam turbines, it may lead to overheating of blades. Normally, steam flow removes the heat from blades and other affected parts. The heat causes blade distortion and softening, while in hydro turbine it would cause cavitation of the turbine blades. The motoring of generator can be detected by reverse power flow relays having sensitivity of 0.5% of rated power output with time delay of approximate 2 seconds.

Loss of Field – LOF or excitation happens in the generated due to excitation failure. Failure of field system in the generator makes the generator run at a speed above the synchronous speed. In that situation the generator or alternator becomes an induction generator which draws magnetizing current from the system. Slip-induced eddy currents heat rotor surface. If the system cannot provide adequate reactive power support for induction generator mode of operation, then synchronism is lost which can trigger system/area voltage collapse. High reactive current drawn by generator overloads stator also. LOF protection needs coordination of loss of excitation relay with generator capability curve (GCC), steady-state stability limits (SSSL) and excitation controls in order to have an optimum protection scheme.

An undercurrent relay connected in shunt with main field winding circuit is used to detect the fault. This relay will operate if the excitation current comes below its predetermined value. If the relay is to operate for complete loss of field, it must have a setting below the minimum excitation current value which can be 8% of the rated full load current. Again when loss of field occurs due to failure of exciter but not due to problem in the field circuit (field circuit remains intact), there will be an induced current at slip frequency in the field circuit. This situation makes the relay to pick up and drop off as per slip frequency of the induced current in the field. This problem can be overcome by providing a setting of 5% of normal of full load current. Alternatively, impedance-based protection can be provided using directional offset mho distance relay as the direct axis reactance of the machine (X_d) affects the SSSL, although the same has no direct effect on GCC.

Over Current Protection – Modern tendency is to design alternators with very high values of internal impedance so that the machine will stand a complete short-circuit at their terminals for sufficient time without serious overheating. Over current protection for alternator is unnecessary. Standard overcurrent relays are not recommended for backup protection of a generator. The backup relay must be capable of detecting the minimum generator fault current. This minimum current is the sustained current following a three-phase fault assuming no initial load on the generator and assuming the manual voltage regulator in service. Industry standards specify the time duration at which generators must sustain overcurrents without exceeding safe temperature limits, and thus determines the maximum allowable time for backup relay operation. Salient-pole synchronous generator shall be capable of withstanding three-phase short-circuit at its terminals when operating at rated MVA and power factor for 30 seconds, at 5% overvoltage, with fixed excitation. Round rotor generators shall be capable of operating at 130% of rated armature current for at least 60 seconds, starting from stabilized temperatures at rated conditions. The overcurrent relays which are used for the backup function are specially constructed to make their operating characteristics a function of voltage as well as current. There are two types of these relays which are customarily used: the voltage restraint consisting of a conventional induction disk overcurrent unit with a voltage restrained element that applies a torque which opposes the operating torque produced by the current coil controlling the pick-up current of the relay over a 4–1 range; and the voltage controlled overcurrent relay of low burden induction disk type overcurrent relay whose torque is controlled by a high-speed voltage relay which has a dropout level continuously adjustable over a range of 65%–83% of rated voltage.

Overvoltage Protection – It implies internal over voltage. This fault is very rare, as the excitation current that controls the voltage has closed supervision with automatic voltage regulator system. Protection can be given through overvoltage relay. Load rejection in hydro generators may cause voltage rise due to speed rise. In units connected to extra high voltage (EHV) lines a further proportional rise in voltage may occur if receiving end breaker of long interconnecting transmission line is tripped. The uncompensated capacitance of the line will further increase the voltage. Protection for generator overvoltage is provided by frequency insensitive overvoltage relay. It has both instantaneous and time delay units with inverse time characteristics. Instantaneous inverse time is set up to about 150%, while inverse time is set to pick up 110% of normal voltages.

Differential Protection – Differential protection provides fast protection to the stator winding against phase to phase faults and phase to ground faults. It functions on the concept of comparing the two currents in and out of stator coil. In normal condition the two current will be same, if fault occurs there will be some difference, and Merz price circulating current scheme works by detecting this difference or differential current.

Merz-price is the most common type of protection used for stator windings against phase-to-phase or phase-to-ground faults. In each phase there are two identical Current transformers: one is at the machine side and the other on the line side. So, for three phases there are six numbers of CTs. The secondaries of all CTs are connected in star, and the end terminals of each set star are connected with the corresponding end terminals of other set through cable called pilot cable for a star connected alternator stator. This type of scheme is also known as longitudinal differential scheme in order to differentiate it from another differential scheme known as transverse differential scheme, which is used to detect inter-turn faults. Three relay coils are connected in pilot cable. They are connected in equi-potential point of pilot cable. The neutral of the current transformer and the relay are connected to the common terminal. If alternator neutral is not grounded or is grounded through impedance, additional sensitive ground faults protection relays are to be provided.

Phase and/or Ground Faults in the Stator and Associated Protection Zone – Most faults in a generator are a consequence of insulation failure. They may lead to turn-to-turn faults and ground faults. Hence ground fault protection is very essential for generators. The method of grounding the generator neutral affects the protection afforded by differential relays. A resistor is installed between the generator neutral and grounding rod. This resis-tor is called a neutral grounding resistor. The grounding resistor limits the fault current when one phase of circuit shorts or arcs to ground. For example, if sufficient grounding impedance is used so that a ground fault at the generator terminal draws full load current, then for a fault at the midpoint of the winding, the fault current will be approximately one half of the full load current as only half the phase voltage is available to cause a flow of current. Three types of grounding schemes are used in practice: (i) Fault Protection with high impedance grounding is used to limit the maximum fault current due to fault in winding near generator terminals to 1–10 A. This reduces iron burning in the generator, and it helps in avoiding costly repairs. There is an inverse time overvoltage unit connected across the resistor to trip the breaker on overvoltage, which is a consequence of large zero sequence currents flowing through the high resistance due to the fault. High impedance grounding reduces sensitivity for both feeder ground protection and differential protection in the stators of the generators. Alternative to high impedance grounding is low imped-ance grounding. (ii) Fault protection with low impedance grounding – The advantage of low impedance grounding is improved sensitivity of the protection. However, if the fault is not cleared quickly, the damage to equipment can be much higher. It is possible to engineer ground (zero sequence) differential protection using a directional ground over current relaying. (iii) Hybrid Grounding – Hybrid grounding systems are designed to allow for the benefits of both the high impedance and low impedance grounding systems. This system offers advantages of a low impedance grounded system during normal generator operation, i.e. when fault is not detected, and high impedance grounding when a fault is detected within the generator to minimize damage to the stator. The grounding impedance is switched depending on the presence of a ground fault within the generator.

When the neutral is solidly grounded then the generator is completely protected against the earth faults. But when neutral is grounded through resistance, then the stator winding gets partly protected against earth faults. The percent of winding protected is dependent on the value of the earthing resistance and the relay setting. The earth fault near to the neutral point is rare as the voltage of neutral point with respect to earth is less. But when earth fault occurs near the neutral point, then the insufficient voltage across the fault drives very low fault current than the pick-up current of relay coil; hence, the relay remains inoperative for the Merz-price circulation current differential protection. For the alternator with stator having high impedance neutral grounding, a sensitive ground/earth fault relay is used in addition to the differential protection. This is called restricted earth fault protection. The protection function will work only within a specified zone that is pre-determined, i.e. the REF protection is a unit protection. To avoid the magnetizing inrush current, the stabiliz-ing resistor in conjunction with the relay in series is connected.

Standby Earth Fault Protection – is a backup protection of restricted earth protection as it is capable of operating when REF is failed to trip the circuit, due to heavy earth fault outside of the REF protective zone, and all other earth faults. Thus, standby earth fault protection is non-unit protection. One CT is installed at a neutral side of the alternator across which the standby earth fault relay is connected. Hundred percent stator earth fault protection using a low frequency injection technique detects earth faults in the entire winding, including the generator neutral point where conventional earth fault protection cannot detect faults. If an earth fault in the generator star point or close to the star point is not detected, the generator is effectively running with a low impedance earth bypassing the high impedance earth typically used on large machines. A second earth fault can then cause a very high current to flow which can cause a lot of damage to the machine. This is why 100% stator earth fault protection is a common requirement for large machines. Low frequency injection technique can be used to provide protection for 100% of the stator winding in case of a ground fault. The low frequency injection technique provides protection when the machine is stopped running and also when the machine is running up and down. A subharmonic voltage (usually one fourth of the system frequency) (approximately 25 V, 12.5 Hz) is injected through the secondary of the neutral grounding transformer. Such a frequency is used in order to easily suppress interfering signals with power system frequency and/or its harmonics (e.g. third). It is connected in parallel with grounding resistor.

Stator Inter-turn Fault Protection – The Merz-Price protection system gives protection against phase-to-phase faults and earth faults. It does not give protection against inter-turn faults. The inter-turn fault is a short circuit between the turns of the same phase winding. Thus, the current produced due to such fault is local circuit current, and it does not affect the currents entering and leaving the winding at the two ends, where CTs are located. Hence Merz-Price protection cannot give protection against stator inter-turn faults. The protection is given to the generators by splitting the stator winding in two halves and by cross differential connection. Each phase of the generator is doubly wound and split into two parts. The current transformers are connected in the two parallel paths of the each phase winding. The secondaries of the current transformers are cross connected. The current transformers work on circulating current principle. The relay is connected across the cross connected secondaries of the current transformers.

Rotor Earth Fault Protection – The rotor circuit of the generator is not earthed, and hence single ground fault in rotor does not cause circulating current to flow through the rotor circuit. Hence single ground fault does not cause any damage in the rotor circuit. But this can cause an increase in stress to ground at other points in the field winding when voltage is induced in the rotor during transients. Thus, probability of second ground fault may increase. If the second ground fault occurs, then part of the rotor winding is bypassed and the current in the remaining portion increases abruptly, which causes an unbalance in the rotor circuit and hence mechanical and thermal stresses on the rotor for which rotor may get damaged. A high resistance is connected across the field winding of the rotor. The midpoint of the resistor is grounded through a sensitive earth fault relay. When the fault occurs, the relay detects the fault and sends the tripping command to the breaker. The relay detects the earth fault for most of the rotor circuit except the rotor center point. Alternatively, earth fault protection of rotor can be given through AC/DC injection. A small DC power is connected to the field circuit. A fault detecting sensitive relay and the resistance are connected in series with the circuit. The high resistance limits the current through the circuit. An earth fault in the field circuit will cause a flow of current of sufficient magnitude through the relay to cause its operation. In case of AC injection, high resistance is replaced with a capacitor.

Unbalance Protection – When a generator is subjected to unbalance short-circuit currents, a negative-sequence current is caused to flow in the stator windings, establishing a counter rotating flux that induces double-frequency rotor currents. The rotor currents flow in the surface parts of the rotor and cause detrimental heating. The generator is susceptible to varying degrees of damage depending on the magnitude of the negative-sequence current. Positive sequence currents cannot discriminate between balanced and unbalanced operating conditions. On the other hand, negative sequence currents clearly indicate the abnormality. Hence, it can be used as an effective discriminate for unbalanced system operation. Negative sequence currents create an mmf wave in opposite direction to the direction of rotation of rotor. Hence, it sweeps across the rotor that induces second harmonic currents in rotor, which can cause severe overheating and, ultimately, the melting of the wedges in the air gap. An inverse-time overcurrent relay excited by negative sequence current can be used for this protection.

13.10 MOTOR PROTECTION

	Cause
Thermal Overload • Extreme Starting Condition • Locked Rotor • High Overload • Under Voltage • Intermittent Operation	A motor can run overloaded without a fault in motor or supply. A primary motor protective element of the motor protection relay is the thermal overload element, and this is accomplished through motor thermal image modeling. The overload protection is provided in such a fashion that it matches the heating curve of the motor. When a motor stalls, a current equal to the starting current flows and causes a severe damage.
Electrical Overvoltage	Overall result of an overvoltage condition is a decrease in load current and poor power factor. Although old motors had robust design, new motors are designed close to saturation point for better utilization of core materials, and increasing the V/Hz ratio causes saturation of air gap flux leading to motor heating. The overvoltage element should be set to 110% of the motors nameplate unless otherwise started in the data sheets.
Electrical Under Voltage	Overall result of an under-voltage condition is an increase in current and motor heating and a reduction in overall motor performance. Under voltage protection element can be thought of as backup protection for the thermal overload element. In some cases, if an under-voltage condition exists, it may be desirable to trip the motor faster than thermal overload element. Under voltage trip should be set to 80%–90% of nameplate unless otherwise stated on the motor data sheets.
	Electrical Unbalance
Electrical • Ground Fault	Protection is given through Zero Sequence CT Connection. Under normal circumstances, the three-phase currents will sum to zero, resulting in an output of zero from the Zero Sequence CT's secondary. If one of the motors phases were too shorted to ground, the sum of the phase currents would no longer equal zero causing a current to flow in the secondary of the zero sequence. This current would be detected by the motor relay as a ground fault.
Electrical • Differential	Differential protection may be considered the first line of protection for internal phase-to-phase or phase-to-ground faults. In the event of such faults, the quick response of the differential element may limit the damage that may have otherwise occurred to the motor.
Electrical • Short Circuit	The short circuit element provides protection for excessively high over current faults. Phase-to-phase and phase-to-ground faults are common types of short circuits. When a motor starts, the starting current (which is typically six times the Full Load Current) has asymmetrical components. These asymmetrical currents may cause one phase to see as much as 1.7 times the RMS starting current. To avoid nuisance tripping during starting, set the short circuit protection pick up to a value at least 1.7 times the maximum expected symmetrical starting current of motor.

Types of Faults

Stator Faults – Stator short circuit protection for small motors is provided with the help of thermal or dash pot type over current tripping devices giving inverse time characteristics and instantaneous feature at high current. Combined over current and earth-fault relay with highest instantaneous feature can be used for large motors. Differential protection is also used for very large motors.

Rotor Faults – Any form of unbalance either in supply voltage or in the loading pattern will cause sequence currents to flow in the stator which will induce high frequency currents in the rotor. The frequency of these currents in the rotor is (2 – s) times the nominal frequency of supply. The rotor heating due to positive sequence component of the stator current is proportional to the DC resistance value, while the heating effect on the rotor windings of the negative sequence component is proportional to (2 – s) f AC resistance value. The protection can be provided by an instantaneous over current relay.

Overloads – The overload protection is provided in such a fashion that it matches the heating curve of the motor. When a motor stalls, a current equal to the starting current flows and causes a severe damage.

Unbalanced Supply Voltages Including Single Phasing – Single phasing causes negative sequence current to flow. The motor has a limited ability to carry negative sequence current, because of thermal limitations. Single phasing causes the motor to develop insufficient torque, leading to stalling, making the motor to draw excessive current.

Under Voltage
Reverse or open phase starting and Loss of synchronism in case of synchronous motor.

13.11 POWER SWING BLOCKING

Power Swing is a variation in three-phase power flow which occurs when the generator rotor angles are advancing or retarding relative to each other in response to changes in load magnitude and direction, line switching, loss of generation, faults and other system disturbances. Power Swing caused by the large disturbances in the power system which if not blocked could cause wrong operation of the distance relay and can generate wrong or undesired tripping of the transmission line circuit breaker. Particularly, distance relays should not trip unexpectedly during dynamic system conditions such as stable or unstable power swings, and allow the power system to return to a stable operating condition. Thereby, a Power Swing Block (PSB) function is adopted in modern relays to prevent unwanted distance relay element operation during power swing. The main purpose of the PSB function is to differentiate between power faults and power swings, and block distance or other relay elements from operations during a power swing. However, faults that occur during a power swing must be detected and cleared with a high degree of selectivity and dependability. Power swings can be classified as either stable or unstable. Conventional PSB schemes are based mostly on measuring the positive-sequence impedance at a relay location. During normal system operating conditions, the measured impedance is the load impedance, and its locus is away from the distance relay protection characteristics. When a fault occurs, the measured impedance moves immediately from the load impedance location to the location that represents the fault on the impedance plane. During a system fault, the rate of impedance change seen by the relay is determined primarily by the amount of signal filtering in the relay. During a system swing, the measured impedance moves slowly on the impedance plane, and the rate of impedance change is determined by the slip frequency of an equivalent two-source system. Conventional PSB schemes use the difference between impedance rate of change during a fault and during a power swing to differentiate between a fault and a swing. Another method to determine power swing condition is based on a continuous impedance calculation.

13.12 AUTO-RECLOSING

Once a circuit component or equipment is automatically disconnected by protective functions, its reconnection is mostly a control function. Reconnection of a circuit component or equipment (after its trip due to protective device action) is performed either manually or automatically. Overhead Transmission and distribution circuits are the only circuits which are often reconnected automatically without any testing or maintenance performed after its trip due to protection action. This is because 80%–90% of the faults that occur in transmission lines are transient in nature. Isolation during a fault followed by re-energization process is exactly what an auto-reclosing does. In high-speed reclosing, reclosure of the breaker is not delayed, intentionally. The only existing delay is the necessary time needed for deionization of the arc path. In a time-delayed reclosure, the breaker is being closed after an intentional time-delay which is longer than what is considered for deionization of the arc path. The length of delay is mostly between 1 second and 1 minute depending on the situation. The main object of auto-reclosing system is the segregation of permanent faults from transient faults. The auto-reclosing relay recloses the circuit breaker after tripping due to any types of transient faults and blocks the operation if the fault is permanent in nature. There are two types of auto-reclosing; single phase and three phase. Single-phase auto-reclosure is one in which only the faulted phase is opened in the presence of a single-phase fault and reclosed after a controlled delay period. For multiphase faults, all three phases are opened and reclosure is not attempted. In case of single-phase faults which are in majority, synchronizing power can still be interchanged through the healthy phases. In the case of single-phase auto-reclosing each phase of the circuit breaker is segregated and provided with its own closing and tripping mechanism. Also, there is a phase selecting relay that will detect and select the faulty phase. Thus, single-phase auto-reclosing is more complex and expensive as compared to three-phase auto-reclosing. When single-phase auto-reclose is used, the faulty phase be de-energized for a longer interval of time, than in the case of three-phase auto-reclose, owing to the capacitive coupling between the faulty phase and the healthy conductors which tends to increase the duration of the arc. Single-phase auto-reclosing improves the transient state stability, system reliability and availability and reduces the switching over voltages. Three-phase auto-reclosure is one in which the three phases of the transmission line are opened after fault incidence, independent of the fault type, and are reclosed after a predetermined time period following the initial circuit breaker opening. For a single circuit interconnector between two power systems, the opening of all the three phases of the circuit breaker makes the generators in each group start to drift apart in relation to each other, since no interchange of synchronizing power can take place.

The most important parameters of an auto-reclose scheme are: (i) dead time – the time between the auto-reclose scheme being energized and the completion of the circuit through the circuit breaker closing. The dead time consists of opening time, i.e. the time needed for the contacts of faulted phase to separate; arcing time in which the secondary arc fed by the charge of the line and also by the non-faulted phases becomes extinguished and waiting time for reclosing; (ii) reclaim time – time from the making of the closing contacts on the auto-reclose relay to the completion of another circuit within the auto-reclose scheme which will reset the scheme or lock out the scheme or circuit breaker as required; (iii) single or multi-shot – The number of attempts at reclosing which an auto-reclose scheme will make before locking out on a permanent fault. The number of shots may be fixed or adjustable.

13.13 LEARNING OUTCOME

System and equipment protection increases operational efficiency, productivity and safety. Containment of fault in a localizing zone and isolation of the faulty part prevents unnecessary outage. The economic impact of a power interruption is high. Equipment/system protection varies with the nature of the equipment/system. The type of protection varies with the rating.

In this chapter, readers have been able to learn all the above aspects.

14 DC Transmission

14.1 EVOLUTION OF DC TRANSMISSION SYSTEM

The transmission and distribution of electrical energy started with direct current. In 1882, a 50-km-long 2-kV direct current (DC) transmission line was built between Miesbach and Munich in Germany. At that time, conversion between reasonable consumer voltages and higher DC transmission voltages could only be realized by means of rotating DC machines. The reasons for alternating current (AC) technology to be introduced at a very early stage in the development of electrical power systems are: (i) in an AC system, voltage conversion is simple; (ii) an AC transformer allows high power levels and high insulation levels within one unit and has low losses. It is a relatively simple device, which requires little maintenance. (iii) A three-phase synchronous generator, which is universally used for AC power generation, is superior to a DC generator in every respect, and it was soon accepted as the only feasible technology for generation, transmission and distribution of electrical energy.

However, high-voltage AC transmission links have disadvantages, which may compel a change to DC technology: (i) inductive and capacitive elements of overhead lines and cables put limits to the transmission capacity and the transmission distance of AC transmission links; (ii) this limitation is of particular significance for cables. Depending on the required transmission capacity, the system frequency and the loss evaluation, the achievable transmission distance for an AC cable will be in the range of 40–100 km. It will mainly be limited by the charging current; (iii) direct connection between two AC systems with different frequencies is not possible; and (iv) direct connection between two AC systems with the same frequency or a new connection within a meshed grid may be impossible because of system instability, too high short-circuit levels or undesirable power flow scenarios.

Nowadays, High-Voltage Direct Current (HVDC) transmission is widely recognized for long-distance, bulk power delivery, asynchronous interconnections with superior controllability and long submarine cable crossings. HVDC lines and cables are less expensive and have lower losses than those for three-phase AC transmission [12]. HVDC allows delivery of more power over fewer lines with narrower Right of Way (ROW), which is a strip of land where the transmission line is constructed, erected, operated and maintained. The maximum width of ROW is calculated by considering transmission line voltage, wind speed, sag, tower design, swing and other safety considerations. Usually for 11 kV transmission line, the ROW is 7 m. For 33 kV, it is 15 m. For 765 kV, the ROW is 67 m.

The advantages of a DC link over an AC link are as follows: (i) a DC link allows power transmission between AC networks with different frequencies or networks, which cannot be synchronized, for other reasons; (ii) no stability problems due to the transmission line length because no reactive power is needed to be transmitted; (iii) inductive and capacitive parameters do not limit the transmission capacity or the maximum length of a DC overhead line or cable. The conductor cross section is fully utilized because there is no skin effect; (iv) a digital control system provides accurate and fast control of the active power flow; (v) fast modulation of DC transmission power can be used to damp power oscillations in an AC grid and thus improve the system stability.

In DC transmission, only two conductors are needed for a single line. Only one conductor is enough using earth return, and by using two conductors and earth return, the capacity of the line is doubled. But, in case of AC transmission, at least three conductors are needed, and six conductors would be needed for double circuit line. The break-even distance is in the range of 500–800 km depending on a number of other factors, like country-specific cost elements, interest rates for

DOI: 10.1201/9781003231240-14

project financing, loss evaluation, cost of ROW, etc. An HVDC transmission system is basically environment-friendly because improved energy transmission possibilities contribute to a more efficient utilization of existing power plants. The land coverage and the associated ROW cost for an HVDC overhead transmission line is not as high as that of an AC line. This reduces the visual impact and saves land compensation for new projects. It is also possible to increase the power transmission capacity for existing ROW.

14.2   PRINCIPLE OF HVDC TRANSMISSION SYSTEM

Principles of HVDC transmission system is shown in Figure 14.1. In the HVDC station, the converter transformer steps up the generated AC voltages to the required level. The converter station rectifies AC to DC, which is then transmitted through overhead lines (or cables). At the receiving end of the converter station, an inverter converts the DC voltage back to AC, which is stepped down to the distribution voltage levels at various consumer ends.

14.3   COMPONENTS OF HVDC TRANSMISSION SYSTEM

Converters – HVDC transmission requires a converter at each end of the line. The sending end converter acts as a rectifier which converts AC power to DC power, and the receiving end converter acts as an inverter which converts DC power to AC power. This unit usually consists of two three-phase converters which are connected in series to form a 12-pulse converter. The converter consists of 12 thyristor valves, and these valves can be packaged as single valve or double valve or quadrivalve arrangements. Due to the evaluation of power electronic devices, the thyristor valves have been replaced by high power handling devices such as gate turn-off thyristors, Insulated Gate Bipolar Transistors (IGBTs) and light triggered thyristors. The valves are cooled by air, water or oil, and these are designed based on modular concept where each module consists of a series connected thyristor. Firing signals for the valves are generated in the converter controller and are transmitted to each thyristor in the valve through a fiber optic light guide system. The light signals are further converted into electrical signals using gate drive amplifiers with pulse transformers. The valves are protected using snubber circuits, gapless surge arrestors and protective firing circuits.

Converter Transformer with Suitable Ratio and Tap Changing – The transformers used before the rectification of AC in HVDC system are called as converter transformers. The different configurations of the converter transformer include three-phase–two winding, single-phase–three winding and single-phase–two winding transformers. The valve side windings of transformers are connected in star and delta with ungrounded neutral, and the AC supply side windings are connected in parallel with grounded neutral. The design of the control transformer is somewhat different from the one used in AC systems. These are designed to withstand DC voltage stresses and increased eddy current losses due to harmonic currents. The content of harmonics in a converter transformer is much higher compared to conventional transformer which causes additional leakage flux, and it results

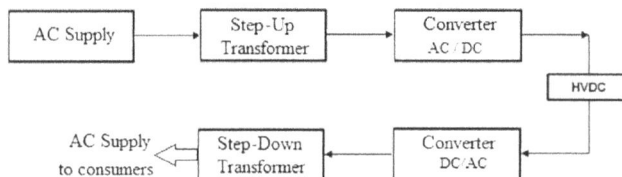

FIGURE 14.1   Principles of HVDC transmission system.

in the formation of local hotspots in windings. To avoid these hotspots, suitable magnetic shunts and effective cooling arrangements are required.

Filters at Both DC and AC Side – Due to the repetitive firing of thyristors, harmonics are generated in the HVDC system. These harmonics are transmitted to the AC network and led to the overheating of the equipment and also interference with the communication system. To reduce the harmonics, filters and filtering techniques are used. Types of filters include (i) AC filters – These are made with passive components, and they provide low impedance and shunt paths for AC harmonic currents. Tuned as well as damped filter arrangements are generally used in HVDC system; (ii) DC filters – Similar to AC filters, these are also used for filtering the harmonics. Filters used at DC end, usually smaller and less expensive than filters used in AC side. The modern DC filters are of active type in which passive part is reduced to a minimum. Specially designed DC filters are used in HVDC transmission lines in order to reduce the disturbances caused in telecommunication systems due to harmonics; (iii) High-frequency filters – These are provided to suppress the high frequency currents and are connected between converter transformer and the station AC bus. Sometimes these are connected between DC filter and DC line and also on the neutral side.

Smoothening Reactor in DC Side – Due to the delay in the firing angle of the converter station, reactive volt-amperes are generated in the process of conversion. Since the DC system does not require or generate any reactive power, this must be suitably compensated by using shunt capacitors connecting at both ends of the system.

Shunt Capacitors – It is a large series reactor, which is used on DC side to smooth the DC current as well as for protection purpose. It regulates the DC current to a fixed value by opposing sudden change of the input current from the converter. It can be connected on the line side, neutral side or at an intermediate location.

DC Transmission Lines – Two conductors with different polarity are used in HVDC systems to transfer the power from sending end to receiving end. The size of the conductors required in DC transmission is small for the same power handling capacity to that of AC transmission. Due to the absence of frequency, there is no skin effect in the conductors.

Line-Commutated Converters (LCC) – The line-commutated indicates that the conversion process relies on the line voltage of the AC system to which the converter is connected in order to effect the commutation from one switching device to its neighbor. LCC uses switching devices that are either uncontrolled (such as diodes) or that can only be turned on (not off) by control action, such as thyristors. Although HVDC converters can, in principle, be constructed from diodes, such converters can only be used in rectification mode, and the lack of controllability of the DC voltage is a serious disadvantage as each diode automatically turns ON when it is forward biased and turns OFF when it is reverse biased. Consequently, in practice present day all LCC HVDC systems use thyristor. A thyristor does not automatically turn ON at the instant in the AC cycles at which it becomes forward biased. After it has become forward biased, it waits till a gate pulse is impressed in its gate terminal. The flow of current in a valve ends when the voltage between anode and cathode becomes negative and at that time a firing pulse has no effect. This gives the controlled feature and has a control circuit block, to generate and supply "Gate Trigger Pulse" to the thyristor. The turn-on of the thyristors is controlled by a gate signal, while the turnoff occurs at the zero crossing of the AC current which is determined by the AC network voltage ("line commutation"). The LCC usually has a 12-pulse arrangement, in which two 6-pulse bridges are connected in series on the DC side.

Voltage Source Converters – Because thyristors can only be turned on (not off) by control action and rely on the external AC system to effect the turn-off process, the control system only has one degree of freedom – when to turn on the thyristor. In IGBT, both turn-on and turn-off can be controlled, giving a second degree of freedom. As a result, IGBTs can be used to make self-commutated converter.

14.4   TYPES OF HVDC TRANSMISSION SYSTEM

Monopolar – has only one conductor with negative polarity with respect to ground so as to reduce corona loss and radio interference. Ground provides the return path.

Bipolar – This link has two conductors, one operating with positive polarity and the other with negative polarity with respect to the earth. Here, the two converters of equal voltage rating are connected in series at each end of the DC line. The neutral points, i.e. the junction between converters, may be grounded at one end or at both ends. If it is grounded at both ends each pole can operate independently. If any of the links stop operating, the system operates as Monopolar mode with ground as return path.

Homopolar – This link has two conductors, both operating with same polarity usually with negative polarity with respect to the earth. The ground operates as a return path. In this system, the links are operated in parallel.

14.5   LEARNING OUTCOME

HVDC transmission offers several advantages compared to traditional alternating current transmission systems as the same allows more efficient bulk power transfer over long distances. HVDC transmission systems are becoming an integral part of the electrical power system, improving stability, reliability and transmission capacity. HVDC allows asynchronous power systems to be coupled with ease and gives the operator an ease to control power flow between the two. HVDC provides economical solution, when considering reactive power compensation requirements and overall system stability with an AC-based alternative.

In this chapter, readers have been able to learn all the above aspects.

15 Electrical Power Distribution Substation

15.1 REQUIREMENT OF DISTRIBUTION SUBSTATION

A substation is an assemblage of electrical equipments which may be grouped as: (i) Switching equipments – Switching is the operation of connecting and disconnecting of transmission lines or other components to and from the system. Switching events may be "planned" or "unplanned" (ii) Protection equipments, (iii) Measuring equipments and (iv) Control equipments and transformers. A distribution substation is a subsidiary station of an electricity generation, transmission and distribution system where voltage is transformed from high to low or the reverse using transformers. The substation may be installed at generating station itself; in that the substation will not have a separate entity but is part of generating station termed as generating substation, transmission substation, sub-transmission substation, distribution substation and switching substation. The substation may be extra-high-voltage (EHV) or high voltage (HV) or medium voltage (MV) or High-Voltage Direct Current (HVDC). Transmission substations are EHV substations.

The location of distribution substation is selected considering the factors like (i) as close to the load center as possible; (ii) that all the prospective loads may be conveniently reached without under voltage regulation; (iii) allowing access to the incoming transmission lines and outgoing distribution lines; (iv) space is available for the reasonable amount of expansion of the substation; and (v) site should be selected where municipal restrictions or property laws would permit the type of structure necessary for the substation.

Functions of substation are as follows: (i) change voltage from one level to another, (ii) regulate voltage to compensate for system voltage changes, (iii) switch transmission and distribution circuits into and out of the grid system, (iv) measure electric power qualities flowing in the circuits, (v) connect communication signals to the circuits, (vi) eliminate lightning and other electrical surges from the system, (vii) connect electric generation plants to the system, (viii) make interconnections between the electric systems of more than one utility, (ix) control reactive kilovolt-amperes supplied to and the flow of reactive kilovolt-amperes in the circuits.

15.2 TYPES OF SUBSTATIONS

a. **Based on Location**

Outdoor – All switchgear equipment, busbars and other switch yard equipment are installed outside, open to atmosphere, and are best suited where there is ample amount of space available for commissioning the equipment of the substation. The construction work required is comparatively less. In future the extension of the substation installation is easier. Time required for the erection of air insulated substation is less compared to indoor substation. All the equipment in switch yard is within view, and therefore the fault location is easier and related repairing work is also easy. There is practically no danger of the fault which appears at one point being propagated to another point for the substation installation because the equipment of the adjoining connections can be spaced liberally without any appreciable increase in the cost. Outdoor switch yards are more vulnerable to faults as they are located in outside atmosphere which has some influence from pollution, saline environment and other environmental factors like lightning.

DOI: 10.1201/9781003231240-15

Indoor – in which the apparatus is equipped inside the substation building.

Underground – can only be considered in large crowded cities.

Kiosk or Pole Mounted – They are small size distribution substations for rural electrification transformer stations.

b. **Based on Design Configuration**

Air Insulated Electrical Power Substation – All the substation equipments are installed in the outdoor. Clearances are the primary criteria for these substations and occupy a large area for installation.

Gas Insulated Electrical Power Substation – all the substation equipments are in the form of metal enclosed SF_6 gas modules. The modules are assembled in accordance with the required configuration. The various live parts are enclosed in the metal enclosures (modules) containing SF_6 gas at high pressure. Thus, the size of Power Substation reduces to 8%–10% of the Air Insulated Power Substation.

15.3   SUBSTATION COMPONENTS

Busbars – Various incoming and outgoing circuits are connected to busbars. Busbars receive power from incoming circuits and deliver power to outgoing circuits. The size of the busbar is important in determining the maximum amount of current that can be safely carried. Busbars are typically either flat strips or hollow tubes as these shapes allow heat to dissipate more efficiently due to their high surface area to cross-sectional area ratio.

Surge Arresters and Lightning Arresters – discharge the over voltage surges to earth and protect the equipment insulation from switching surges and lightning surges. Surge arresters are generally having low impulse ratio (ratio of breakdown voltage at surge frequency and breakdown voltage at normal power frequency). Surge arresters may be Silicon Carbide non-linear resistor type with spark-over gap or metal oxide, which is much more non-linear without spark-over gap type that discharges much more energy per unit volume. Lightning arresters are rod gap / sphere gap / horn gap / multiple gap / impulse protective gap / thyrite / metal oxide type.

Overhead Earth Wire Shielding – To protect the outdoor substation equipment from lightning strokes.

Isolators or Disconnecting Switches – Isolators are provided for isolation from live parts for the purpose of maintenance. Isolators are located at either side of the circuit breaker. Isolators are operated under no load. Isolator does not have any rating for current breaking or current making. Isolators are interlocked with circuit breakers. Types of Isolators are Central rotating, horizontal swing / Center-Break / Vertical swing / Pantograph type.

Earth Switch – is used to discharge the voltage on the circuit to the earth for safety. Earth switch is mounted on the frame of the isolators. Earth switch is located for each incomer transmission line and each side of the busbar section.

Current Transformer – Metering and protection used for stepping down current for measurement, protection and control.

Voltage Transformer – Metering and protection used for stepping down voltage for measurement, protection and control.

Circuit Breaker – is used for switching during normal and abnormal operating conditions.

Power Transformers – used to step up or step down AC voltages and to transfer electrical power from one voltage level to another. Tap changers are used for voltage control.

Shunt Reactors – are used for long EHV transmission lines to control voltage during low-load period or to compensate shunt capacitance of transmission line during low-load periods. Usually shunt reactors are un-switched.

Shunt Capacitors – are located at the receiving stations and distribution substations, used for compensating reactive power of lagging power factor and for improving the power factor.

They are also used for voltage control during heavy lagging power factor loads. Shunt capacitors are switched on during heavy loads and switched off during low loads.

Series Capacitor – are used for some long EHV AC lines to improve power transferability.

Series Reactor – Series reactors are used to limit short circuit current and to limit current surges associated with fluctuating loads.

DC Power Supply – All but the smallest substations include auxiliary power supplies. AC power is required for substations building small power, lighting, heating and ventilation, some communications equipment, switchgear operating mechanisms, anti-condensation heaters and motors. DC power is used to feed essential services such as circuit breaker trip coils and associated relays, supervisory control and data acquisition (SCADA) and communications equipment.

Neutral Grounding Equipment – is used to limit the short circuit current during ground fault. They are connected between neutral point and ground.

Line Trap – traps the high frequency component signals in the power lines.

Power Cable and Control Cable – Power cables are in compliance of rating of the power station equipment that are used to transmit power from one equipment to other. Control cable are low voltage cable used for connecting control equipments, relays and meters.

Relay and Control Panel – The control and relay panel is of cubical construction suitable for floor mounting. All protective, indicating and control elements are mounted on the front panel for ease of operation and control.

Illumination System – provides proper illumination to substation yard.

The function of transmission and distribution systems would be to take up power to the consumers from the source of power generation as economically and conveniently as possible. The distribution systems are designed on reliability assessment, substation evaluation on the basis of physical system description, performance criteria, reliability indices, failure mode and effect evaluation, accumulation of failure effects, etc. The distribution systems can be either primary or secondary or either radial or ring main.

Radial distribution system is used in less populated areas, where the primary feeder branches out to reach out the total area. It is (i) simple as fed at only one point; (ii) initial cost is low; (iii) reliability is less as in case of fault the whole system gets switched off affecting more number of consumers. In contrast, Ring main system that forms a loop over the area to be served is more reliable but is expensive. In this system voltage fluctuation is less.

15.4   SUBSTATION EARTHING

The objectives of a earthing system are: (i) to provide safety to personnel during normal and fault conditions by limiting step and touch potential, (ii) to provide means to carry electric currents into the earth under normal and fault conditions without exceeding any operating and equipment limits or adversely affecting continuity of service, (iii) to prevent damage to electrical/electronic apparatus, (iv) to dissipate lightning strokes and (v) to stabilize voltage during transient conditions and to minimize the probability of flashover during transients. There are two main types of earthing:

System Earthing/Neutral Earthing – Earthing of system is designed primarily to preserve the security of the system by ensuring that the potential on each conductor is restricted to such a value as is consistent with the level of insulation applied. From the point of view of safety, it is equally important that earthing should ensure efficient and fast operation of protective gear in the case of earth faults. Public supply systems are earth to facilitate detection of earth faults in the system and to control the fault current, since large fault currents can cause the potential rise of exposed parts of the power system to reach dangerous levels. The limitation of earthing to one point on each system is designed to prevent the passage of current through the earth under normal conditions and thus to avoid the accompanying risks of electrolysis and interference with communication circuits. With a suitable designed system, properly operated and maintained, earthing at several points are done.

Equipment Earthing (Safety) – The basic objectives are to ensure freedom from dangerous electric shock voltages exposure to persons in the area; to provide current carrying capability, both in magnitude and duration, adequate to accept the ground fault current permitted by the overcurrent protective system without creating a fire or explosive hazard to building or contents; and to contribute to better performance of the electrical system.

Earthing system is designed to achieve low earth resistance and also to achieve safe "Step Potential" and "Touch Potential". Step Potential is the potential difference between two points on the earth's surface, separated by distance of one pace, that will be assumed to be 1 m in the direction of maximum potential gradient. Touch Potential is the potential difference between a grounded metallic structure and a point on the earth's surface separated by a distance equal to the normal maximum horizontal reach, approximately 1 m. System of earthing may be:

Solid Earthing – The neutral is directly connected to the earth without having any intentional resistance between neutral and ground. The single-phase earth fault current in a solidly earthed system may exceed the three-phase fault current. The main advantage of solidly earthed systems is low over voltages, which makes the earthing design common at low voltage levels.

Resistance Earthing – Resistance is connected between the neutral and the ground to improve the earth fault detection. Resistance Grounding Systems limit the phase-to-ground fault currents, thereby reducing burning and melting effects in faulted electrical equipment like switchgear, transformers, cables, etc. and also reduce the mechanical stresses in circuits/equipments carrying fault currents.

Reactance Earthing – Reactance is connected between the neutral and ground.

Resonant Earthing – An adjustable reactor of correctly selected value to compensate the capacitive earth current is connected between the neutral and the earth. The coil is called Arc Suppression Coil or Earth Fault Neutralizer.

Effectively Earthed System – A system in which the value of the phase to earth voltage of the healthy phases during an earth fault, never exceed 1.39 times the pre-fault phase to ground voltage is effectively earthed system.

Pipe Earthing – In this method of earthing, a galvanized steel pipe of suitable length and diameter is buried vertically in the permanent wet soil under the ground. The length and diameter of the pipe are determined by the conditions of soil and the current to be carried. Normally minimum diameter and length of the pipe are maintained at 40 mm and 2.5 m respectively for ordinary condition of soil, and greater length is used for rocky and dry soil conditions. The depth under ground level at which the pipe is buried depends upon the moisture condition of soil, but it should not be less than 3.75 m under the ground. The earthing pipe is surrounded by alternative layers of charcoal and salt to keep moisture and thereby reduces the earth resistance.

Plate Earthing – In this method a metallic plate of sufficient size is buried in wet soil vertically under the ground. If copper plate is used for this purpose, the minimum dimension of the plates should be 60 cm×60 cm×3 mm, and if it is GI plate, then minimum dimension should be 60 cm×60 cm×6 mm.

The earthing resistance of an electrode is made up of: (i) resistance of the (metal) electrode, (ii) contact resistance between the electrode and the soil and (iii) resistance of the soil from the electrode surface outward in the geometry setup for the flow of current outward from the electrode to infinite earth. The type of soil largely determines its resistivity. Earth conductivity is, however, essentially electrolytic in nature and is affected by the moisture content of the soil and by the chemical composition and concentration of salts dissolved in the contained water.

15.5 TYPES OF BUS SYSTEMS OF SUBSTATION

Single Busbar System – This substation configuration consists of *all circuits* connected to a *single bus as shown in Figure 15.1*. It is also called radial system. A fault on the bus or between the bus and circuit breaker will result in an outage of the *entire bus* or substation.

The merits of such bus system are simple and low cost, simple to operate, simple protection scheme, small land area requirement and easily expandable. This type of configuration is used for distribution substations up to 33 kV and not used for large substations. The demerits are as follows: (i) fault of bus or any circuit breaker results in shut down of entire substation, (ii) difficult to do any maintenance, (iii) bus cannot be extended without completely de-energizing substations and (iv) less reliable.

Main and Transfer Busbar System – This substation configuration consists of a main bus and a transfer bus, which is normally de-energized. A tie CB is there to energize transfer bus, when required. All circuits are primarily connected to the main bus. The circuit breaker maintenance is achieved by closing the tie breaker. Figure 15.2 shows the Main and Transfer Busbar system.

The merit of such bus system is low initial and ultimate cost; any breaker can be taken out of service for maintenance (i) potential devices may be used on the main bus. This type of configuration is used for distribution substations for 110 kV where cost of duplicate bus bar system is not justified. The demerits are the following: (i) requires one extra breaker coupler, (ii) switching is somewhat complex when maintaining a breaker, (iii) protective relay scheme is quite complex and (iv) failure of a breaker or fault on the bus results in an outage of the whole substation.

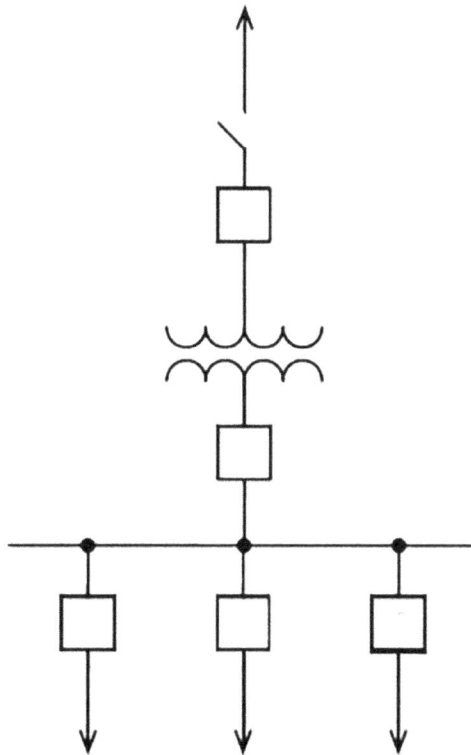

FIGURE 15.1   Single busbar system.

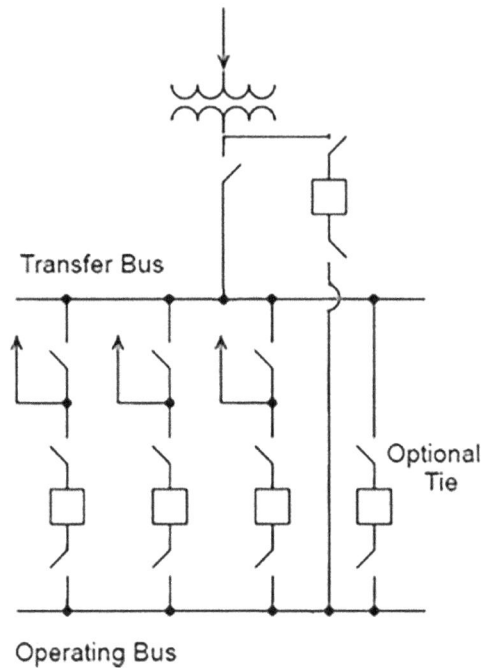

FIGURE 15.2 Main and transfer busbar system.

Double Bus and Single Breaker System – This substation configuration consists of two buses. Each circuit is equipped with a single breaker and is connected to both buses using isolators. A coupler breaker connects both the buses and can be closed for coupling the buses, allowing for more flexibility in operation. Maintenance of a bus or a circuit breaker in this arrangement can be accomplished without interrupting either of the circuits. This arrangement allows various operating options as additional lines are added to the arrangement; loading on the system can be shifted by connecting lines to only one bus. Figure 15.3 shows the Double Bus and Single Breaker system.

  The merits of such bus system are high flexibility and very high reliability, no interruption of service to any circuit from a bus fault and loss of one circuit per breaker failure. This type of configuration is widely used for high voltage. The demerits are as follows: (i) extra bus-coupler circuit breaker necessary; (ii) bus protection scheme is complicated.

Double Bus and Double Breaker System – This substation configuration consists of two buses. Each circuit is equipped with two breakers and is connected to both buses using isolators. A coupler breaker connects both the buses and can be closed for coupling the buses, allowing for more flexibility in operation. Usually both buses are energized. Substations with the double bus double breaker arrangement require twice the equipment as the single bus scheme. Figure 15.4 shows the Double Bus and Double Breaker system.

  The merits of such bus system are each has two associated breakers and has flexibility in permitting feeder circuits to be connected to any bus; any breaker can be taken out of service for maintenance; and high reliability. This type of configuration is not used for usual EHV substations due to high cost. The demerit is most expensive.

Ring Bus System – In this scheme, as indicated by the name, all breakers are arranged in a ring with circuits tapped between breakers. For a failure on a circuit, the two adjacent breakers will trip without affecting the rest of the system. Similarly, a single bus failure will only affect the adjacent breakers and allow the rest of the system to remain energized. Figure 15.5 shows the Ring Bus system.

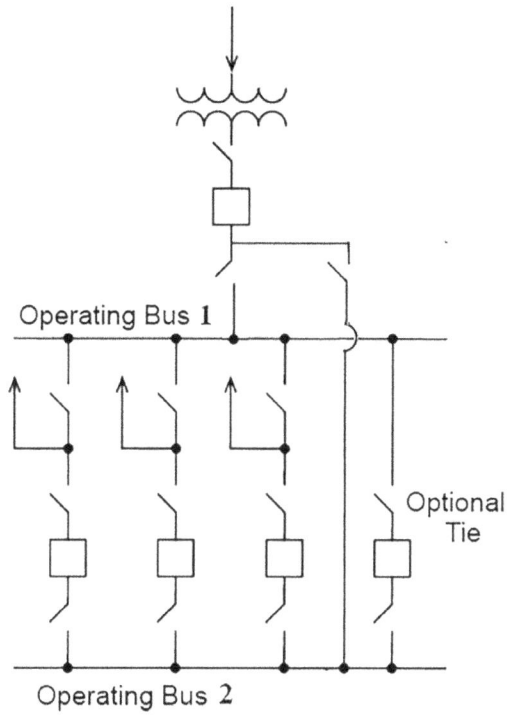

FIGURE 15.3 Double bus and single breaker system.

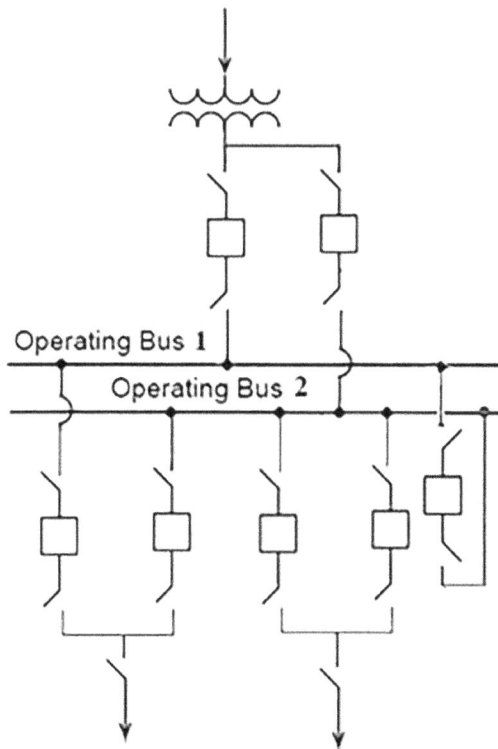

FIGURE 15.4 Double bus and double breaker system.

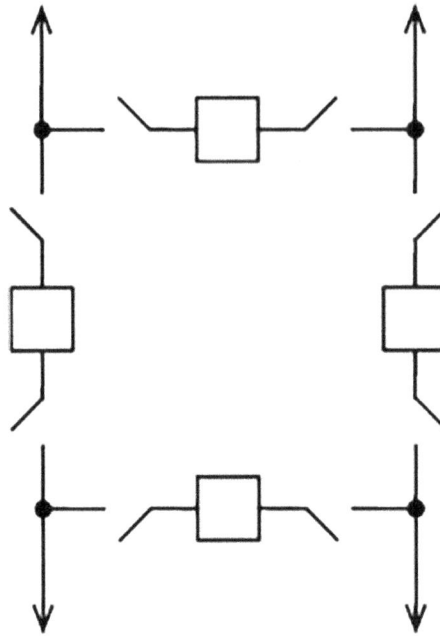

FIGURE 15.5   Ring bus system.

The merits of such bus system are flexible in operation, double feed to each circuit; no main buses; maintenance on a circuit breaker in this scheme can be accomplished without interrupting any circuit, including the two circuits adjacent to the breaker being maintained. The demerit is during fault, splitting of the ring may leave undesirable circuit combinations.

Breaker and Half System – This substation configuration consists of two main buses typically used at EHV stations. Both the buses are normally energized. Three breakers are connected between the buses. The circuits are terminated between the breakers as shown. In this bus configuration for two circuits, three breakers are required. Hence, it is called one and half scheme. The middle breaker is shared by both the circuits. Like the ring bus scheme, here also each circuit is fed from both the buses. Any of the breakers can be opened and removed for maintenance purposes without interrupting supply to any of the circuits. Figure 15.6 shows the Breaker and Half system.

The merits of such bus system are flexible operation and high reliability, isolation of either bus without service disruption, and bus fault does not interrupt service to any circuits. The demerits are as follows: (i) one-and-a-half breakers are needed for each circuit; (ii) more complicated relaying as the center breaker has to act on faults for either of the two circuits it is associated with.

15.6   DC DISTRIBUTORS

The part of system by which electric power is distributed among various consumers for local use is known as distribution system. The main parts are: (i) Feeders – Those electric lines which connect generating station (power station) or substation to distributors are called feeders. These are conductors which are never tapped. The feeder current always remains constant. (ii) Distributors – Those electric lines which connect distribution substations or feeding points are called distributors. These are conductors which are tapped to supply various loads of the consumers by service mains. (iii) Service Mains – a line which connects the consumer to the distributor.

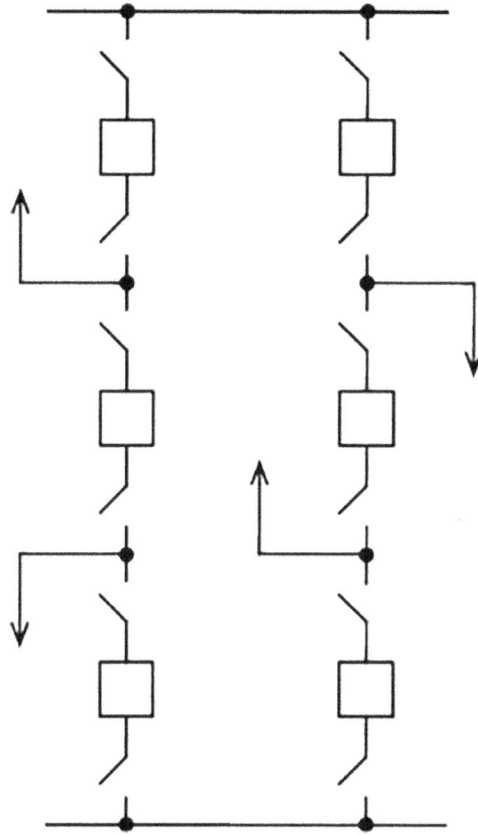

FIGURE 15.6 Breaker and half system.

The most general method of classifying DC distributor is the way they are fed by the feeders. On this basis, DC distributors are classified as:

Distributor Fed at One End – The distributor is connected to the supply at one end and loads are taken at different points along the length of the distributor. The current in the various sections of the distributor away from feeding point goes on decreasing and is maximum in section nearest to the feeding point. The voltage across the loads away from the feeding point goes on decreasing. In case a fault occurs on any section of the distributor, the whole distributor will have to be disconnected from the supply mains. Therefore, continuity of supply is interrupted. Figure 15.7 shows the distributor fed at one end.

Distributor Fed at Both Ends – The distributor is connected to the supply mains at both ends and loads are tapped off at different points along the length of the distributor. Here, the load voltage goes on decreasing as we move away from one feeding point say A reaches minimum value and then again starts rising and reaches maximum value when we reach

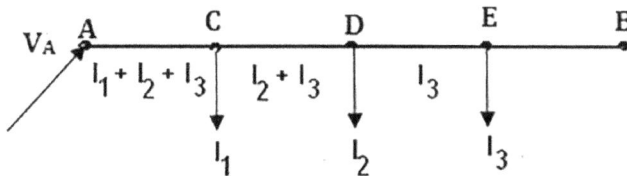

FIGURE 15.7 Distributor fed at one end.

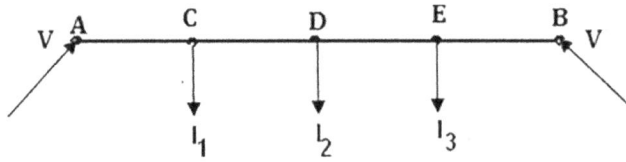

FIGURE 15.8 Distributor fed at both ends.

the other feeding point B. The minimum voltage occurs at some load point and is never fixed. It is shifted with the variation of load on different sections of the distributor. If a fault occurs on any feeding point of the distributor, the continuity of supply is maintained from the other feeding point. The area of cross section required for a doubly fed distributor is much less than that of a singly fed distributor. Figure 15.8 shows the distributor fed at both end.

Distributor Fed at the Center – The center of the distributor is connected to the supply mains. It is equivalent to two singly fed distributors, each distributor having a common feeding point and length equal to half of the total length.

Ring Distributor – The distributor is in the form of a closed ring. It is equivalent to a straight distributor fed at both ends with equal voltages, the two ends being brought together to form a closed ring. The distributor ring may be fed at one or more than one point.

Uniformly Loaded Distributor Fed at One End – A DC distributor may have (i) concentrated loading, (ii) uniform loading and (iii) both concentrated and uniform loading. Figure 15.9a shows the single line diagram of a two-wire DC distributor AB fed at one end A and loaded uniformly with i amperes per meter length. It means that at every 1 m length of the distributor, the load tapped is i amperes. Let l meters be the length of the distributor and r ohm be the resistance per meter run.

Consider a point C on the distributor at a distance of x meters from the feeding point A as shown in Figure 15.9b. Then current at point C is $il - ix = i(l - x)$ amps. Now, consider a small length dx near point C. Its resistance is $(r\,dx)$ and the voltage drop over length dx is $= i(l - x)rdx = ir(l - x)dx$.

Total voltage drop in the distributor up to point C is $V = \int_0^x ir(l-x)dx = ir\left[lx - x^2/2\right]$.

The voltage drop up to point B (i.e. over the whole distributor) can be obtained by putting x = l in the above expression. Voltage drops over the distributor AB. Hence, $V = ir\left[l^2 - l^2/2\right] = irl^2/2 = (i \times l)$ $(r \times l)/2 = IR/2$ where $i\,l = I$, the total current entering at point A, and $r\,l = R$, the total resistance of the distributor. Thus, in a uniformly loaded distributor fed at one end, the total voltage drop is equal to that produced by the whole of the load assumed to be concentrated at the middle point.

Uniformly Loaded Distributor fed at both end – Figure 15.10 shows the single line diagram of a two-wire DC distributor AB fed at both ends A and B. The distributor is loaded uniformly with i amperes per meter length. The distributor be fed at the feeding points A and B at equal voltages, say v volts. The total current supplied to the distributor is il. As the two end voltages are equal, current supplied from each feeding point is $il/2$.e. Consider a point C at a distance x meters from the feeding point A. Then current at point C is $il/2 - i(l - x) = i(l/2 - x)$. Consider a small length dx near point C. Its resistance is $r\,dx$ and the voltage drop over length dx is $dv = i(l/2 - x)rdx$.

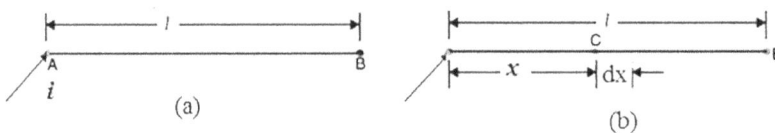

FIGURE 15.9 Uniformly loaded distributor fed at one end.

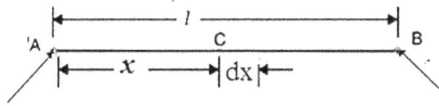

FIGURE 15.10   Uniformly loaded distributor fed at both ends.

Voltage drop at point C is $\int_0^x i(l/2-x)r\,dx = ir(lx/2 - x^2/2) = ir(lx - x^2)/2$. The point of minimum potential will be the mid-point. Therefore, maximum voltage drop will occur at mid-point,
i.e. where $x = l/2$. Maximum voltage drop will be $irl^2/8 = (il)(rl)/8 = IR/8$, where $i\,l = I$, the total
current entering at point A and $r\,l = R$, the total resistance of the distributor.

15.7   LEARNING OUTCOME

Electrical substation is an integral part of electricity transmission or sub-transmission and distribution or sub-distribution system as it plays the role of an interface between the two. Different topolo-
gies are being adopted for a substation in order to meet the reliability besides network security
which is the ability of a power system to supply electricity to customers uninterruptedly.

In this chapter, readers have been able to learn all the above aspects.

16 Power System Structure

16.1 OBJECTIVE OF POWER SYSTEM STRUCTURING

Electricity began as a monopoly utility as the same is capital intensive. The creation of infrastructure from generation to transmission and then distribution at the consumer end were very expensive, which provided extremely high barriers to entry. This had made one utility service provider to handle all functions of generation, transmission and distribution of electricity within a certain geographical area. Traditionally the name "vertically integrated utilities" was provided for the utility in the standard operating paradigm. The operational objectives were to provide quality power (voltage and frequency nearly constant) to a consumer, while ensuring reliability and overall economy (low cost). The price of power was "regulated" and based on actual costs. Power consumption increased with the flourishing of economy. This necessitated the further investment for the electricity.

Characteristics of a traditional Vertically Integrated electric industry are as follows: (i) During early days of the electric power industry with three links, generation, transmission and distribution in a single line, governments favored a regulated monopoly – vertically integrated utility structure. The price of power was "regulated" and based on actual costs; (ii) offered a risk-free way to finance the creation of electric industry; and (iii) establishment of electric industry required large capital for infrastructure building. Thus, for the purpose of risk minimization, a local monopoly where only the local electric utility can produce, move or sell commercial electric power within its service territory and stable market was assured; (iv) to prevent exploitation of consumers due to monopoly, the government brought in regulation. On the contrary, present days' concept competitive markets provide good incentives for efficiency and innovation. Additionally, monopoly attracts entry-level barrier: (v) the utility leaders could focus on building up their systems without having to worry about the competitors undercutting the prices to gain market share, etc.; (vi) offered electric utilities recognition and support from the government, which was necessary to solve problems like "Right of Way" (i.e. the "right" to an exclusive corridor to build a transmission line); (vii) obligation to serve – The utility must provide service to all electric consumers in its service territory, not just those that would be profitable; (viii) assured rate of return; and (ix) the central station concept because of significant economy of scale in power generation. Very large generators produced power at less than half the cost per kilowatt of small generator units, and the bigger the generator, the more economical the power it produced. Disadvantages of vertically integrated electric power industry are: (i) difficult to segregate the costs involved in generation, transmission or distribution; (ii) utilities often charged their customers an average tariff rate depending on their aggregated cost during a period.

Electric power sector has long been viewed as having economies of scale which implies average and marginal costs of production decline as the output of firm increases and of scope. However, during years, technological innovations led to the emergence of economically viable small-scale or "distributed" generation with improved efficiency. This has not only revolutionized traditional concept regarding economies of scale in power generation but also the extent to which distribution of electricity could be a competitive business. The need for induction of renewable sources of power into the generation network to mitigate the dependency of power generation using fossil fuels has become inevitable that has triggered the concept for initiating the idea of deregulation/restructuring [13–15]. Furthermore, computerized control systems have been developed for efficient monitoring power distribution, and generation system coupled with improved data communication system ensured the ability to monitor at various level in a cost-effective way and can control the units from remote operation centers. Another purpose of restructuring is to let the economic forces drive the price of electric supply and maximize social welfare via competition. Restructuring creates an

DOI: 10.1201/9781003231240-16

open environment by allowing electric supply authorities to compete and consumers to choose the supplier of electric energy. Competition is expected to bring innovation, efficiency and lower costs.

Restructuring requires the unbundling of three components – generation termed as Gencos, transmission termed as Transco and distribution termed as Discos – with a view to enhance function-specific efficiencies and ensuring better returns to generation and transmission businesses. It also requires the separation of transmission ownership from transmission control to ensure fair and non-discriminatory access to the transmission and ancillary services. The entity which controls the transmission system maintains real-time operation of the system and its grid stability. The market structure should allow long-term wholesale bilateral trading and a voluntary short-term spot market with transparent and justifiable prices of energy and ancillary services. The spot market would include both a day-ahead function to coordinate resource commitment and a real-time balancing function.

16.2   CONCEPT OF REGULATION AND DEREGULATION

Regulation means that the government has set down laws and rules that put limits on and define how a particular industry or company can operate. Deregulation in power industry is restructuring rules and economic incentives that government set up to control and drive the electric power industry [16,17].

The power industry across the globe is experiencing a radical change in its business as well as in an operational model, where the vertically integrated utilities are being unbundled and opened up for competition with private players. This enables an end to the era of monopoly. The arrangement of the earlier setup of the power sector was characterized by operation of a single utility generating, transmitting and distributing electrical energy in its area of operation. Thus, these utilities enjoyed monopoly in their area of operation. They were often termed as monopoly utilities. In this paradigm, it is difficult to segregate the cost involved in generation, transmission and distribution.

Deregulation is about removing control over the prices with introduction of market players in the sector. An overnight change in the power business framework with provision of entry to competing suppliers and subjecting prices to market interaction would not work successfully. There are certain conditions that create a conducive environment for the competition to work. These conditions need to be satisfied while deregulating or restructuring a system. "Deregulation" does not mean that the rules won't exist. The rules will still be there; however, a new framework would be created to operate the power industry. That is why the word "deregulation" finds its substitutes like "re-regulation", "reforms", "restructuring", etc.

In government-linked public utilities, factors other than the economics, for example, treatment of all public utilities at par, overstaffing, etc. resulted in a sluggish performance of these utilities. The economists started promoting introduction of a competitive market for electrical energy as a means of benefit for the overall power factor.

Impetus for deregulation of power industry was provided by the change in power generation technology. In the earlier days, cost-effective power generation was possible only with the help of mammoth thermal (coal/nuclear) plants. However, during the mid-1980s, the gas turbines started generating cost-effective power with smaller plant size. It was then possible to build the power plants near the load centers, and also, an opportunity was created for private players to generate power and sell the same to the existing utility. This technology change, supposed to have provided acceleration to the concept of independent power producers, supported the concept of deregulation further. This technology change is supposed to have provided acceleration to the concept of independent power producers. The deregulation of the industry has provided electrical energy with a new dimension where it is being considered as a commodity. The "commodity" status given to electrical power has attracted entry of private players in the sector. The private players make the whole business challenging from the system operator's point of view, as it now starts dealing with many players which are not under its direct control. This calls for introduction of fair and transparent set of rules for running the power business. The market design structure plays an important role in successful deregulation of power industry.

16.3 POWER SYSTEMS IN RESTRUCTURED ENVIRONMENT

Regardless of the market structures that may emerge in various parts of the world, one fact that seems always to be true is that transmission and generation services should be unbundled from one another giving rise to Gencos; in charge of power generation only including setting up new power generating stations and Transcos; in charge of power transmission only managing the transmission system. The generation market will become fully competitive, with many market participants who will be able to sell their bulk energy services (or demand side management) to Transcos. On the other hand, the operation of a transmission system, i.e. Transcos, is to deliver the bulk power from Gencos allowing open, non-discriminatory and comparable access to all suppliers and consumers of electrical energy. This function can be implemented by an entity called the Independent System Operator (ISO). ISO has responsibility for the reliability functions in its region of operation and for assuring that all participants have open and non-discriminatory access to transmission services through its planning and operation of the power transmission system. Additionally, distribution of electricity is taken away from transmission giving a separate entity Discos. Thus, generation, transmission and distribution of electricity have become three mutually exclusive operation with distinct barrier from one another.

16.4 POOLCO MODEL

Coordination through a well-designed pool-based electricity market can be a large part of the solution to the problems of promoting open access and competition. Analysis of the underlying conditions of electricity supply highlights the role of effective pooling arrangements in cutting through the complexity of the electricity system and exploiting the benefits of coordination for competition. Any efficient system for organizing the electricity market should include least-cost dispatch as a centerpiece. To be sure, the least-cost dispatch concentrates only on the short-run, and the greater part of the value of a competitive system is to be found in the long-run decisions that will control contracting and investment. A PoolCo is defined as a centralized market place that clears the market for buyers and sellers where electric power sellers/buyers submit bids and prices into the pool for the amount of energy that they are willing to sell/buy. Thus PoolCo model is comprised of competitive power providers as obligatory members of an independently owned regional power pool, vertically integrated transmission companies, vertically integrated distribution companies and a single and a separate entity for establishing bidding procedure, scheduling and dispatching generation resources, acquiring necessary services to assure system stability, administering settlement process and ensuring non-discriminatory access to the transmission grid [18,19].

The Pool provides (i) a source of firm back-up and top-up power to support either generators or suppliers offering long-term contracts to final customers; without access to a Pool, firm power could only be offered by generators owning a portfolio of plant and to the extent that firm power is a necessary requirement of consumers, and the competitiveness of both the generation market and the final supply would be limited; (ii) a ready market for generators unable to sell their power under contract or wanting a market for spill or excess production; (iii) a reference price for long- or short-term contracts struck outside the Pool which provide participants with price stability not immediately available inside the Pool; (iv) a reference price to be used in signaling the optimal development of generation and transmission capacity of the system.

16.4.1 ELECTRICITY MARKET

Prior to the restructuring of the electricity sector, generation, transmission and distribution were part of a single entity. Price for selling of electricity to the consumers was decided by the single utility service provider with the regulations which are generally imposed by the government or the government authority without any undue advantage to any particular entity at the cost of end users.

Post restructuring created generation, transmission and distribution as different entities. These entities sell and buy electricity with the intention of making profits. Competition in the electric industry generally means competition only in the production or generation of electricity as the end user or the consumer has the choice of open access to different producers through transmission network. The transportation functions which include transmission and distribution are natural monopolies and thus cannot be competitive as it doesn't make economic sense to build multiple sets of competing transmission systems. Generally, generation accounts for 35%–50% of the final cost of delivered electricity, whereas transmission accounts for about 5%–15% and distribution accounts for about 30%–50% of final cost of electricity.

In the post-restructuring era, electricity is purchased, sold and traded in wholesale and retail markets with the market participants Gencos (generating companies), Transco (transmission companies), Discos (distribution companies), customers and ISO. Power can be transferred directly between a Genco and Disco or Genco to Disco via Transco or from one Genco to Transco to another Transco then to a Disco competing in the free market to sell the generated electricity [20–23].

Central auction (PoolCo or Power Exchange) and bilateral trading are two basic market trading mechanisms of wholesale electricity market. This is shown in Figure 16.1. In central auction or PoolCo, Gencos and Discos submit their bids in price and volume in the forward market to inject their power into and out of the Pool. Sellers compete for the right to inject power into the grid, not for specific customers. On the other hand, buyers compete for buying power, and if their bids are too low, they may not get any power. Market is cleared when equilibrium is set between incremental (marginal) supply and demand curves. This pool serves as a centralized place for trading electricity, and there is no direct transaction between Gencos and end users. At times, PoolCo provides a market for Gencos unable to sell their output via contract to a specific customer or who are looking for market for excess production.

The price for wholesale electricity can be predetermined by a buyer and seller through a bilateral contract or it can be set by organized wholesale markets. In the bilateral trading model, sellers (Gencos) and buyers (Transco/Discos) enter into contract for electrical energy. This model permits direct contracts between customers and generators without entering into pooling arrangements. Gencos can also be buyers, when they do not generate enough power on their own. Both parties request the ISO or Power System Operator (PSO) to provide transmission. If security of the system is not threatened, the ISO/PSO dispatches the entire requested amount. In the bilateral trading ISO/PSO is outside the ambit of price fixation, and the contract price information is, therefore, limited to the parties involved and according to their own financial terms. Therefore, this methodology is often referred to as direct access method.

Besides the Central auction and bilateral trading, another method called Power Exchange Market method often called spot price pool, where Gencos put their bids and pricing in the power market along with the Discos who put their buying bids along with price. Based on Market Clearing Price, ISO will select the selling bids given by Gencos and buying bids proposed by Discos, without discriminating the Power Exchange Market.

Prior to the post restructuring era, the transmission charges were pooled for the region and paid for by beneficiaries of the regions based on regional postage stamp method. The transmission charges

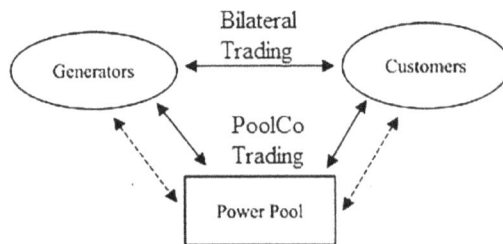

FIGURE 16.1 Central auction and bilateral trading.

were shared on the basis of energy drawls by all users of a system in a region and were paid at the same price per MW of allocated transmission capacity irrespective of the transmission distance and network configuration. If R_t is the transmission price for transaction t, then $R_t = T_C \left(P_t / P_{\text{peak}} \right)$, where T_C is the total transmission charges, P_t is the power transmitted and P_{peak} the entire system condition.

Contract path method of transmission pricing is based upon the assumption that the power transacted is confined to flow along a specific transmission system path between the point of supply of the transmission service provider and to the point of delivery of the beneficiary. The contract path interconnects the points of injection and receipt. A portion or all charges associated with the transmission facility in the contract path are then allocated to the beneficiary which is defined virtually without having power flow studies.

Distance-based MW-mile method of transmission pricing is based upon the magnitude of the transacted power and the airline distance in mile between the transacted power delivery in MW and receipt point without considering the actual network condition.

The power flow-based MW-mile methodology allocates the charges for each transmission facility to transmission transactions based on the extent of use of that facility by these transactions and considering the actual network condition. The allocated charges are then added up over all transmission facilities to evaluate the total price for use of transmission system. The cost assigned to a customer for the transaction is on the basis of the extent of use. For this reason, this methodology is also called facility-by-facility method.

Short Run Incremental Cost pricing methodology entails evaluating and assigning all the incremental (new) operating costs associated with a new transmission transaction. The transmission transaction operating costs can be estimated using an optimal power flow model that accounts for all operating constraints including transmission system (static or dynamic stability) constraints and generation scheduling constraints.

Long Run Incremental Cost pricing methodology entails evaluating all long-run costs (operating and reinforcement costs) necessary to accommodate a transmission transaction and assigning such costs to that transaction. The reinforcement cost is determined as change in costs between long-term transaction plans and the present transaction cost.

In Short Run Marginal Cost (SRMC) pricing methodology, the marginal operating cost of the power system due to a transmission transaction is calculated first. Marginal operating cost is the cost of accommodating a marginal increase in the transacted power. The marginal operating cost per MW of transacted power can be estimated as the difference in the optimal cost of power at all points of delivery and receipt of that transaction. The marginal operating cost is then multiplied by the magnitude of the transacted power to yield the SRMC for the transmission transaction.

In Long Run Marginal Cost (LRMC) pricing methodology, the marginal operating cost and reinforcement costs of the power system are used to determine the prices for a transaction. In SRMC, the transmission capacity is assumed to be fixed, whereas in LRMC, the transmission capacity is assumed to be variable. In long run consideration, all costs are variable including all production factors, and there is no fixed cost.

Point of Connection transmission charge pricing methodology was introduced for sharing of Inter State Transmission Systems (ISTS) charges and losses among the Designated ISTS Customers depending on their location and sensitivity to their distances from load centers (generators) and generation (customers) and the direction of the node in the grid.

16.5 CONCEPT OF DISTRIBUTED AND DISPERSED GENERATION

The term "distributed generation", or DG, refers to the small-scale generation of electric power by a unit site close to the load point of consumption being served. DG entails using many small generators, of 2–50 MW output, situated at numerous strategic points throughout cities and towns, so that each provides power to a small number of consumers nearby. While these small generators might be solar or wind turbine units, generating units in this category are most often highly efficient gas turbines in small combined cycle plants, because these are the most economical choices. Although small compared to traditional central station generators, such 2–500 MW generating units are large, both physically and electrically compared to the needs of individual energy consumers, producing power for between 50 and 400 homes.

Dispersed generation refers to use of smaller generating units of less than 500 kW output and often sized to serve individual homes or businesses, which is decentralized and feeding into the distribution level power-grid. These units are small enough to fit into garages or, like central air-conditioners, on a pad behind a house and are highly efficient and used for reliability enforcement. Micro gas turbines, fuel cells, diesel, and small wind and solar PV generators make up this category. Many of the benefits from these decentralized system from the fact that the generating units are inherently modular, which makes distributed power highly flexible. It can provide power where it is needed and when it is needed. The generators can be quieter and less polluting than large power plants, which make them suitable for on-site installation at some customer locations. They also have low maintenance cost and has quick startup and as a result can be used to meet up peak power demand.

The types of DG are: (i) Type 1: This type of DG is capable of delivering only active power such as photovoltaic, micro turbines and fuel cells, which are integrated to the main grid with the help of converters/inverters. (ii) Type 2: DG capable of delivering both active and reactive power. DG units based on synchronous machines (cogeneration, gas turbine, etc.) come under this type. (iii) Type 3: DG capable of delivering only reactive power. Synchronous compensators such as gas turbines are examples of this type and operate at zero power factors. (iv) Type 4: DG capable of delivering active power but consuming reactive power. Mainly induction generators, which are used in wind farms, come under this category. However, doubly fed induction generator systems may consume or produce reactive power, i.e. operate similar to synchronous generator.

16.6   ENVIRONMENTAL ASPECT OF ELECTRIC GENERATION

The varieties of fuels used to generate electricity have some impact on the environment. Fossil fuel power plants release air pollution, require large amounts of cooling water and can mar large tracts of land during the mining process. Nuclear power plants are generating and accumulating copious quantities of radioactive wastes that currently lack any repository.

The generation of electricity is the single largest source of CO_2 emissions. Burning coal produces far more CO_2 than oil or natural gas. Some methods of electricity production produce no or few CO_2 emissions – solar, wind, geothermal, hydropower and nuclear systems particularly. Power plants fueled by wood, agricultural crop wastes and livestock wastes, and methane collected from municipal landfills release CO_2 emissions but may contribute little to global climate change since they also can prevent even greater releases of both CO_2 and methane. Biomass fuels that depend on forest resources must be evaluated carefully since the stock of forests worldwide represents a storehouse for CO_2. If forests are harvested for fuel to generate electricity, and are not replaced, global climate change could be accelerated.

Competition in the electricity industry offers consumers for the first time the opportunity to directly influence the environmental footprint of electric power production. Suppliers are assembling electricity resource portfolios that are significantly cleaner than the status quo. By selecting one of these resource portfolios, which boost the amount of renewable energy sources in the fuel mix, consumers can help ensure that the emissions of pollutants that cause acid rain are reduced.

Electric power production accounts for roughly a quarter of all particulate matter emitted. Particulate emissions correlate strongly with emissions of the pollutants that contribute to acid rain and smog, namely, sulfur dioxide and nitrogen oxides. Among the renewable resources, solar, wind and hydropower technologies emit zero air emissions. Biomass and geothermal fuels, as well as state-of-the-art natural gas facilities, may emit tiny amounts of air toxics that include heavy metals like mercury and chromium, organic chemicals like benzene, and dioxins.

Withdrawal of large volumes of surface water for either power plant cooling or hydropower generation can kill fish, larvae and other organisms trapped against intake structures (impinged), or swept up (entrained) in the flow through the different sections of a power plant. The use of water to generate power at hydropower facilities imposes unique, and by no means insignificant, ecological impacts. The diversion of water out of the river removes water for healthy in-stream ecosystems. Stretches below dams are often completely de-watered. Fluctuations in water flow from peaking operations create a "tidal effect", disrupting the downstream riparian community that supports its unique ecosystem. A dam's impoundment slows water flow, which hinders natural downstream migration of many fish species. By slowing river flow, dams also allow silt to collect on river and reservoir bottoms and bury fish spawning habitat. Silt trapped above dams accumulates heavy metals and other pollutants. Disrupting the natural flow of sediments in rivers also leads to erosion of riverbeds downstream of the dam and increases risks of floods.

Power plant sites may become sacrifice zones, sealed off from any future land use due to contamination linked to the operation of a power plant. The land impacts of hydropower facilities depend on individual dam design, location and operation. Land use and ecosystem impacts of facilities that use large impoundments can be severe. The dam and reservoir may transform the landscape, obliterate sensitive land resources and permanently alter regional land use patterns.

16.7 LEARNING OUTCOME

Power sector reforms which are multiyear initiatives are happening to invite more and more private investments in power generation to bridge the demand-supply gap. Further a paradigm shift involving the re-design of power system structures is inevitable to provide access to renewable-based energy system that offers flexibility in services. The reform is being done keeping in view bringing more competition so that power prices become competitive ensuring transparency alongside there should have a provision of ensuring recovery of cost of service from consumers to make the power sector sustainable. The reform will also provide physical security and reliability of power as every country aspires to provide reliable, affordable and sustainable electricity to its citizens. Most developing countries continue to operate with vertically integrated national power utilities that operate as monopolies. Restructuring is intended separating out generation from transmission and transmission from distribution and creating multiple generation and distribution utilities.

In this chapter, readers have been able to learn all the above aspects.

17 Economic Operation of Energy Generating Systems

17.1 INTRODUCTION TO ECONOMIC OPERATION OF ENERGY GENERATIONS SYSTEMS

The commercial success of any electrical undertaking depends upon the cost and reliability. The primary consideration is to provide a station and electrical system at minimum capital cost consistent with sound engineering which results in minimum operating costs throughout its useful life. Economic operation of power generating systems calls for the selection of the best operating configuration that gives maximum operating economy or minimum operating cost as certain load demand existing at any point of time in a power system may be supplied in an infinite operating configurations. The steady-state operating condition of power system is governed by load flow analysis. In the load flow problem, if the specified variables P_{Gi} and $|V_i|$ of generator buses are allowed to vary in a region constrained by practical considerations (upper and lower limit active and reactive power generations, bus voltage limits), then for a certain $P - Q$ values at the load buses, there result an infinite number of load flow solutions each pertaining to one set of values specified P, V (control variables). The total operating cost includes fuel, salary and wages engaged in production, input raw material, depreciation and maintenance cost which are directly related to the value of the output power. The conventional method of power generation excluding hydro plant installation is divided into two parts, namely the boiler plant, which converts the heat energy of the fuel into steam, and the turbine plant, which converts the energy in the steam into electricity. Considerations are also to be given to the factors influencing the cost of generation are the efficiency of the generators and associated equipment including transmission/distribution losses. Present day practice implies that the efficiency attainable be balanced against capital and operating costs of the plant required to attain this efficiency.

Economical operation and performance together with convenience of access to ail essential apparatus and auxiliaries and ease of control are important features. Reactive power generation has no appreciable influence on the fuel consumption, and the fuel cost is critically dependent on real power generation. Fuel cost characteristics (fuel cost against net active power output) of different units may be different giving different economic efficiency. So the problem of selecting the optimum operating configuration reduces to the problem of finding optimal combination of generating units to run and allocate their real power generations.

Ramp rate that influences how quickly a plant can increase or decrease power output, in MW/h or in [% of capacity per unit time]; Ramp time that provides the amount of time a generator takes from the moment it is turned on to the moment it can start providing energy to the grid at its lower operating limit (LOL) in time scale generally minutes or hour; Capacity that indicates the maximum output of a plant, in MW; LOL that indicates the minimum amount of power a plant can generate once it is turned on, in MW; Minimum Run Time that indicates the shortest amount of time a plant can operate once it is turned on, in time scale generally minutes or hour; Start-up and Shut-down Costs that indicate the costs involved in turning the plant on and off are the factors that dictate the successful operation of a plant economically.

The problem of selection of combination of units solved well before its actual implementation is a problem of operation planning generally known as Unit Commitment (UC). The problem related to the allocation of the power outputs of the generators connected to the system at a particular time in

DOI: 10.1201/9781003231240-17

a manner which minimizes the operating (fuel) cost of the system in a real time problem is known as Economic Dispatch. The fuel cost is meaningful in case of thermal (or, nuclear) units, but for hydro units fuel cost as such is not meaningful. The same is also applicable to power from renewable. Obviously, power generation by hydro units is much cheaper from fuel consumption point of view and can give much better operating economy. But the operation of such plants is dependent on the availability of water which is however restricted and subject to seasonal variations. In those systems, where thermal and hydro sources are available, economy can be achieved by properly coordinating the two types of generations and is called Hydrothermal Scheduling. All the generating units in a system do not participate in economic dispatch. Nuclear units and very large steam units are run at constant MW setting as it is desirable to maintain the output of such units at as constant a level as possible. These units are called Base Load Units. Rest of the units that participate in economic dispatch are called Controllable Units. Fuel costs in base load units then appear as a fixed cost and do not appear in the economic dispatch problem. Economic dispatch aims at allocating the electricity load demand to the committed generating units in the most economic or profitable way, while continuously respecting the physical constraints of the power system. To obtain the most economic schedule of generation taking into account a number of system limitations, such as the heat rate curves, generation limits or ramping limitations of the generating units, limitations of the transmission lines or reliability preventive parameters of the system (e.g. the power reserve). It is the short-term determination of the optimal output of a number of generating units.

17.2   ECONOMIC OPERATION OF THERMAL SYSTEM, ITS BASIC OPERATION AND STRUCTURING

A power system is having N controllable units (already running) and supplying a total P_{Load}. The fuel cost of each unit is F_i and the active power generated by of each unit is P_i; [24]. The total fuel cost $F_T = F_1 + F_2 + \cdots + F_N = \sum_1^N F_i(P_i)$. Consideration is to minimize F_T subject to constraint that the sum of the powers generated must be equal to the received load. Assumption for this is that transmission losses are neglected. Hence the constraint function is $\varphi = 0 = P_{\text{Load}} - \sum_1^N P_i$.

The necessary condition for an extreme value of the objective function is to add the constraint function φ to the objective function F_T after multiplying it by an undetermined multiplier λ. This is known as Lagrange function $\mathcal{L} = F_T + \lambda\varphi$. The necessary condition for an extreme value of the objective function result when we make the first derivative of the $\mathcal{L}$ to each of the independent variable is equal to zero. In this case there are $N+1$ variable, the N values of output power P_i and Lagrange multiplier. The derivative of $\mathcal{L}$ w.r.t λ gives back constraint function.

$$\frac{\partial \lambda}{\partial P_i} = \frac{dF_i(P_i)}{dP_i} - \lambda = 0, \text{ or } \frac{dF_i}{dP_i} = \lambda \tag{17.1}$$

That is the necessary condition for the existence of the minimum cost operating condition for the thermal power system is that the incremental cost rates of all the units must be equal to some undetermined value λ. Additionally, the power output of each unit must be greater than or equal to minimum power permitted and must be equal or less than maximum power permitted to the particular unit.

The fuel cost is generally obtained by best fitting a polynomial to the experimental data. With a second-order polynomial the fuel cost characteristics becomes $C_i = (a_i/2)P_{Gi}^2 + b_i P_{Gi} + d_i = C_i(P_{Gi})$.

$$\text{Using } \frac{dF_i}{dP_i} = \lambda \text{ indicates } a_i P_{Gi} + b_i = \lambda, \text{ or } P_{Gi} = \lambda - \frac{b_i}{a_i} \tag{17.2}$$

The above equation is called the coordination equation. Simply stated, for economic generation scheduling to meet a particular load demand, when transmission losses are neglected and generation limits are not imposed, all plants must operate at equal incremental production costs, subject to the constraint that the total generation be equal to the demand.

The power output of any generator has a maximum value dependent on the rating of the generator. The upper limit of P_{Gi} is set by thermal limits on the turbine generator unit, while the lower limit is set by boiler and/or other thermodynamic considerations. The output of a generating unit must not exceed a specified maximum value; also the power output must be at or above its Minimum Stable Generation value.

17.3   UNIT COMMITMENT

It is neither economical nor technically feasible to run all the available units all the time. In the problem of economic load dispatch, the objective is to distribute the load in real time between the units already running in such a manner that at every instant (in the face of varying demand, i.e. operating points) the operating cost becomes minimum. Basically UC problem is to find the least-cost dispatch of available generation resources to meet an estimated electric power demand over a given time horizon satisfying the operational constraints on transmission system and generation resources.

UC is primarily used to determine the minimum production cost that includes fuel costs, maintenance costs and startup costs, and schedule for thermal generating units. It is an optimization problem, concerned with slow responding thermal generating units as such units take a longer time to start. The starting procedure for a unit that should be committed at a particular time must start long before it actually comes on-line. Therefore, a commitment schedule must be operated ahead of its actual on-line operation. Therefore, UC problem is a problem of operation planning. The purpose of the planning is to determine a schedule called UC schedule which will tell us beforehand when and which units to start and shut down during the operation over a pre-specified time horizon (time range of planning) ahead of current time such that total operating cost for that period (cost functional, i.e. time integral instantaneous cost) becomes minimum. Additionally, as electricity generation from renewable resources increases, UC faces challenges due to the high level of uncertainty in variable renewable resources.

Which units would be on-line at a particular time (to meet the demand at that time) were assumed to be fixed prior. Selection of units for running (committing units to operation) at various time of operation is the problem known as UC. UC implies that there are a number of subset of the complete demand which of the subset be used in order to provide the minimum operating cost. The objective function of UC is to minimize overall cost, while the constraints shared by all units running at a time are the total demand at that time and the reserve capacity available at that time from the units already running. The constraints of individual units are minimum/maximum power capacity and ramping constraints.

The operating cost of thermal plant is high, though their capital cost is low compared to the operating cost of hydroelectric plant which is low, though their capital cost is high. So it has become economical as well as convenient to have both thermal and hydro plants in the same grid. Systems without any thermal generation are fairly rare. The hydroelectric plant can be started quickly, and it has higher reliability and greater speed of response. Hence hydroelectric plant can take up fluctuating loads. The optimal scheduling of large hydrothermal systems is an important aspect of power scheduling and economic operation. Techniques developed for scheduling hydrothermal systems may be used in some systems by assigning a pseudo-fuel cost to some hydroelectric plant. Then the schedule is developed by minimizing the production "cost" as in a conventional hydrothermal system. Hydrothermal UC coordination problem relates utilizing hydro potential satisfying hydro constraints in such a way that the cost of produced electricity from thermal resources during a scheduling period of time is lowermost and with lowest environmental impact.

The pumped hydro plant can be used as a spinning reserve unit to safeguard the system against forced generation outages if the reserve availability is very marginal. The pumped storage hydro plants are designed to save fuel cost, i.e. during peak load periods, it discharges water for generation at hours of high demand and high costs, displacing high-cost fossil generation. During light load periods, i.e. period of low demand and low cost, water is pumped from the lower to the upper reservoir utilizing the economical energy generated by base load units. A pumped storage unit can therefore smooth peak loads, provide reserve and play an important role in reducing total generation cost.

17.4   RESERVES IN POWER GENERATION

Improvements in system reliability can be achieved by using better components or incorporating redundancy. The utilities generally have several reserves. Reserves are a necessity. Generation redundancy is attained by providing generating capacity above that needed for maximum load demand and transfers. This spare capacity represents the reserve of generation necessary to keep the risk of power shortages below an acceptable level. The installed capacity reserve relates to the long-term ability of the system to meet the expected demand requirements, while the operating reserve relates to the short-term ability to meet a given load.

Power system reliability is dependent on the reserves available in the system and relates to the issues of delivery of quality power to all points of utilization within accepted standards, availability of power with adequacy in supply and supply outage of power having security, i.e. the ability to respond to disturbances. System Average Interruption Frequency Index (SAIFI) which is the average frequency of sustained interruptions per customer over a predefined area is a measure of reliability. It is the total number of customer interruptions divided by the total number of customers served. System Average Interruption Duration Index (SAIDI) which is the length of time of duration of interruptions per consumers is another method of measuring power system reliability. SAIDI is equal to total duration of sustained interruptions in a time period say in a month or year/ total number of consumers. The third method is Customer Average Interruption Duration Index (CAIDI), which is the average time needed to restore service to the average customer per sustained interruption. CAIDI is equal to total duration of sustained interruptions in a time period, say in a month or year/total number of interruptions. Alternatively CAIDI is the ratio of SAIDI and SAIFI.

A variety of indices have been developed to measure reliability and its cost in power systems area such as (i) Loss of Load Probability (LOLP) that evaluates the probability that the system load might exceed the available generating capacity. Loss of load occurs only when the system load exceeds the generating capacity in service. The probabilities of generating capacity levels are combined with the probabilities of load magnitudes to get the LOLP [25–27]. LOLP does not give indications about the frequency of occurrence or likely duration of a generation deficit. (ii) Loss of Load Expectation (LOLE) evaluates the expected number of hours per month or year that an electricity production park cannot meet its demand. LOLE values are generally deduced from a much longer-term average; (iii) Expected Frequency of Load Curtailment and Expected Duration of Load Curtailment.

To have reliability, operating reserves are kept which is the installed capacity committed to provide for unexpected plant outages. The reserves are of various types as follows:

Spinning Reserve – Generation capacity that is on-line but unloaded and that can respond within 10 minutes to compensate for generation or transmission outages. The cost of spinning reserves is more than the non-spinning reserves. Non-spinning reserves consist of fast-start units. "Frequency-responsive" spinning reserve responds within 10 seconds to maintain system frequency. Spinning reserves are the first type used when shortfalls occur. Spinning reserve means part loaded generating capacity with some reserve margin that is synchronized to the system and is ready to provide increased generation at short notice pursuant to dispatch instructions or instantaneously in response to a frequency drop.

Supplemental Reserve – Generation capacity that may be off-line or that is comprised of a block of "curtailable" and/or "interruptible" load and that can be available within 10 minutes. Unlike spinning reserve capacity, supplemental reserve capacity is not "synchronized" with the grid (frequency). Supplemental reserves are used after all spinning reserves are on-line.

Backup Supply – Generation that can pick up load within an hour. Its role is, essentially, a backup for reserves. Backup supply may also be used as back up for commercial energy sales.

17.5 CONTINGENCY ANALYSIS

For practical system operation, apart from ensuring the satisfactory operation of the system at a particular operating condition, it is also equally important to make sure that the system operates with adequate level of security. "Security" implies the ability of the system to operate within system constraints (on bus voltage magnitudes, current and power flow over the lines) in the event of outage (contingency) of any component (generator or transmission line). If the system is operating at high loading (light loading) conditions, then the post-contingency system condition would be highly stressed (lightly stressed). Therefore, the post-contingency values of different quantities (voltages, current/power flow) depend on the present operating condition. In case the post-outage (post contingency) does not involve any violation of any operating constraints, the system is said to be operating securely. Otherwise, the system is said to enter an emergency operating condition. Therefore, for detecting the possibility of appearances of emergency operating condition, analysis of the post-contingency scenario or contingency analysis (CA) of the system needs to be carried out which provides information to the operator about the security. Therefore, CA is one of the "security analyses".

Most power systems are operated in such a way that any single contingency will not leave other components heavily overloaded, so that cascading failure is avoided. The contingencies may be grouped into two broad categories: (i) Power outage – that includes loss of generating unit, sudden change in load that may create stability problems or sudden change of power flow in an intertie; (ii) Network outage – that includes outages of a transmission line or equipment like transformer.

For carrying out contingency analysis, outages of all the elements (preferably) connected to the power system need to be carried out one-by-one corresponding to any particular operating condition to keep the system operating when components fail. However, in any dynamic system like the power system, the operating point changes frequently with change in loading/generating conditions. With the change in system operating conditions, the CA exercise needs to be carried out again at the new operating point in real time and in a faster way. Thus, for proper monitoring of system security, a large number of outage cases need to be simulated repeatedly over a short span of time. Ideally, these outage cases should be studied with the help of full AC load flow solutions. However, analysis of thousands of outage cases with full AC power flow technique will involve a significant amount of computation time, and as a result, it might not be possible to complete this entire exercise before the new operating condition emerges. Therefore, instead of using full non-linear AC power flow analysis, approximate, but much faster, techniques using either DC load flow analysis or based on linear sensitivity factors are used to estimate the post-contingency values of different quantities of interest.

Following the analysis, the system control function traditionally used in power system operation consists (i) the energy management system (EMS) that use real-time data for the automated monitoring, operation and control of power system; (ii) the supervisory control and data acquisition (SCADA) that is distributed among various remote sites; and (iii) the communications interconnecting the EMS and the SCADA.

SCADA is a system that has been around since the 1960s which consists of different hardware and software elements that function in tandem to enable operators to supervise and control.

SCADA starts with data acquisition (DA) that indicates the information is gathered from the remote equipment and sent to the operator for monitoring/supervision purposes (S), which also enables the operators to control (C). SCADA consists of remote terminal units located in the power stations and substations that act as a field data interface devices, a communications system to transfer data between field data interface devices and the Energy Control Center, a central host computer server(s) and a set of standard and/or customized software system.

17.6 LEARNING OUTCOME

Present day power system is invariably supplied by a number of power plants. There is a growing complexity as the more flexible generation technologies like renewable sources are feeding into the system. To achieve good quality of service with minimum cost per unit of energy supplied, it is necessary to have a judicial mix of different sources. The purpose of economic operation of power system is to reduce the operating cost of generation to the minimum. The system control engineer has to consider a number of factors while interchanging energy from one power station to another station. Hydrothermal scheduling plays an important role in the operation planning of power system satisfying constraints related to the two systems. Power systems must always have some amount of operating reserve that can instantly respond to a sudden increase in the electric load or a sudden decrease in the renewable power output. Operating reserve provides a safety margin that helps ensure reliable electricity supply despite variability in the electric load and the renewable power supply.

In this chapter, readers have been able to learn all the above aspects.

18 Automatic Generation and Control

18.1 INTRODUCTION TO AUTOMATIC GENERATION CONTROL

Automatic generation and control (AGC) is a control function in power system especially in power generation, whose purpose is the tracking of load variations while maintaining system frequency to the nominal value at all times, net tie-line interchanges and optimal generation levels close to scheduled (or specified) values.

In conventional mode, electric power is generated by converting mechanical energy into electrical energy. The rotor mass, which contains turbine and generator units, stores kinetic energy due to its rotation. This stored kinetic energy accounts for sudden increase in the load. Let the mechanical torque input be denoted by T_m and the output electrical torque by T_e. Neglecting the rotational losses, a generator unit is said to be operating in the steady state at a constant speed when the difference between these two elements of torque is zero. In this case the accelerating torque $T_a = T_m - T_e$ is zero. When the electric power demand increases suddenly, the electric torque increases. However, without any feedback mechanism to alter the mechanical torque, T_m remains constant. Therefore, the accelerating torque T_a becomes negative causing a deceleration of the rotor mass. As the rotor decelerates, kinetic energy is released to supply the increase in the load. Also note that during this time, the system frequency, which is proportional to the rotor speed, also decreases.

It can be thus inferred that any deviation in the frequency for its nominal value of 50 or 60 Hz is indicative of the imbalance between T_m and T_e. The frequency drops when $T_m < T_e$ and rises when $T_m > T_e$. The primary objective of automatic generation control is to regulate the frequency to the specified nominal value and to maintain the interchange of power between control areas at the scheduled values by adjusting the output of selected generators. This function is commonly referred to as load frequency control. A secondary objective of the AGC is to distribute the required change in generation among units to minimize operating costs.

18.2 LOAD FREQUENCY CONTROL (SINGLE AREA CASE)

The load-frequency control (LFC) is used to restore the balance between load and generation in each control area by means of speed control maintaining the frequency at its nominal value. The power exchanges between different control areas are controlled by LFC [27–29]. The swing equation of the synchronous generator is $M\left(d^2\delta / dt^2\right) = P_m - P_e = P_a$; where P_m, P_e and P_a, respectively, are the mechanical, electrical and accelerating power in MW.

Expressing the speed deviation in pu, the equation can be written as

$$\frac{d^2\Delta\omega}{dt^2} = \frac{1}{2H}\left(\Delta P_m - \Delta P_e\right) \tag{18.1}$$

In general power system loads are a composite of a variety of electrical devices. For resistive loads, the electrical power is independent of frequency. In the case of motor loads, the electrical power is changes with frequency due to changes in motor speed. The overall frequency-dependent characteristic of a composite load may be expressed as $\Delta P_e = \Delta P_L + D\Delta\omega_r$, where $\Delta P_L =$ non-frequency sensitive load change, $D\Delta\omega_r =$ frequency sensitive load change and $D =$ load damping constant and is expressed as a percent change in load for 1% change in frequency. Typical values of D are 1%–2%.

DOI: 10.1201/9781003231240-18

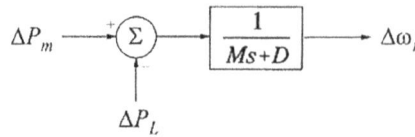

FIGURE 18.1 The block diagram representation of the generator and load.

The system block diagram including the effect of load damping and generator load model is shown in Figure 18.1.

Load changes can cause the system frequency to drift. Whenever there is a mismatch in power, speed changes. The purpose of the frequency governing mechanism is to sense the machine speed and adjust the input valve in order to change the mechanical power output. This action stops once the power mismatch is made zero. Ideally, this adjustment will compensate for the load changes and restore the frequency back to its nominal value. The turbine governor is the primary frequency controller for a synchronous machine. The adjective isochronous is used which means constant speed. An isochronous governor adjusts turbine valve/gate to bring the frequency back to the nominal or scheduled value. An isochronous governor works satisfactorily when a generator is supplying an isolated load or when only one generator in a multi-generator system is required to respond to changes in load. For power load sharing between generators connected to the system, speed regulation or droop characteristics must be provided. Droop can be defined as the percentage change in speed for a change in load. In steady-state conditions, synchronous machine's primary frequency control transfer function $G(s)$ depends on the type of turbine and on the control, and in steady-state conditions, it is equal to $1/R$, where R is called the regulating constant. R determines the per unit change in rated power output, ΔP, for a given change in frequency, i.e. $R = \Delta\omega / \Delta P$.

At 100% load, the generation is also 100% and frequency (or speed) is also 100%. When load reduces, frequency increases, as generation remains the same. When load reduces by 50%, frequency increases by 2%, in the characteristic shown. When load reduces by 100%, frequency increases by 4%. In other words, 4% rise in frequency should reduce power generation by 100%. This 4% is called "droop" of 4%. Figure 18.2 shows the ideal steady-state characteristic of a governor with speed droop characteristic.

In an isolated power system, the function of AGC is to restore frequency to the specified nominal value by adding a reset signal or integral control signal which acts on the load reference settings of the governors of units of AGC. The integral control action ensures zero frequency error in the steady state. This is shown in Figure 18.3.

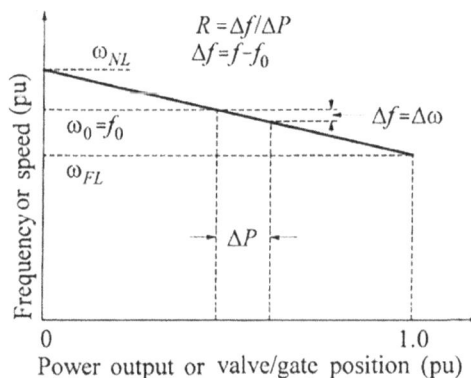

FIGURE 18.2 Ideal steady-state characteristic of a governor with speed droop characteristic.

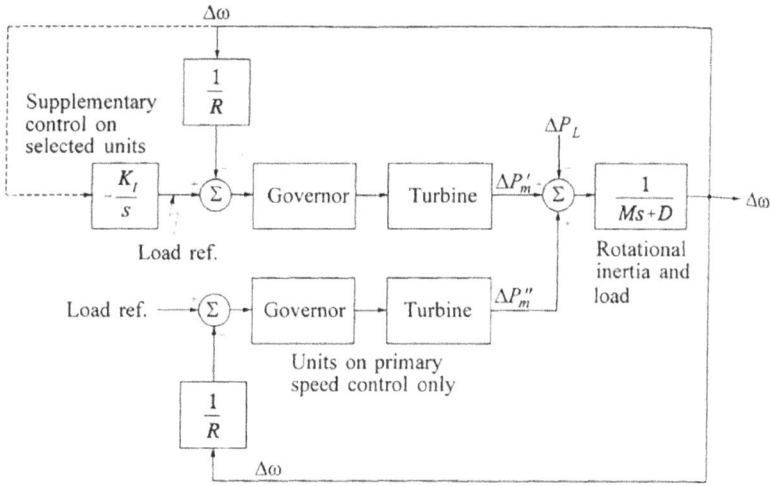

FIGURE 18.3   Integral control with AGC.

Since the function of AGC is only to bring the frequency to the nominal value, it is achieved using the supplementary loop which uses the integral controller to change the reference power setting so as to change the speed set point. The integral controller gain K_1 needs to be adjusted for satisfactory response (in terms of overshoot, settling time) of the system.

18.3   LOAD FREQUENCY CONTROL OF TWO AREA SYSTEMS

Let us consider an interconnected system, consisting of two areas connected by a tie line of reactance X_{tie} as shown in Figure 18.4a. For load frequency studies, each area may be represented by an equivalent circuit consisting of a voltage source behind an equivalent reactance as viewed from the tie bus. The electrical equivalent of the system is shown in Figure 18.4b [27,30–32].

The power flow on the tie line from area 1 to area 2 is $P_{12} = (E_1 E_2 / X_T) \sin(\delta_1 - \delta_2) = (E_1 E_2 / X_T) \sin \delta_{12}$; where P_{12} power exchanged power from area 1 towards area 2 via the tie line, δ_1 and δ_2 are the power angles of end voltages E_1 and E_2 and $\delta_{12} = \delta_1 - \delta_2$. For a small deviation in the tie-line flow from the nominal value, linearizing about an initial point represented by $\delta_1 = \delta_{10}$ and $\delta_2 = \delta_{20}$,

$$P_S = \frac{dP_{12}}{d\delta_{12}}\bigg|_{\delta_{120}} = \frac{E_1 E_2}{X_T} \cos \Delta\delta_{120} \tag{18.2}$$

Tie-line power deviation then will be $\Delta P_{12} = P_S(\Delta\delta_{12}) = P_S(\Delta\delta_1 - \Delta\delta_2) = T\Delta\delta_{12}$, where T is the synchronizing torque coefficient given by $T = (E_1 E_2 / X_T) \cos(\delta_{10} - \delta_{20})$. The steady-state frequency deviation $(f - f_0)$ is the same for two areas. For a total load change ΔP_L, $\Delta f = \Delta\omega = \Delta\omega_1 = \Delta\omega_2$. Consider the steady-state values following an increase in area 1 load by ΔP_{L1}, then

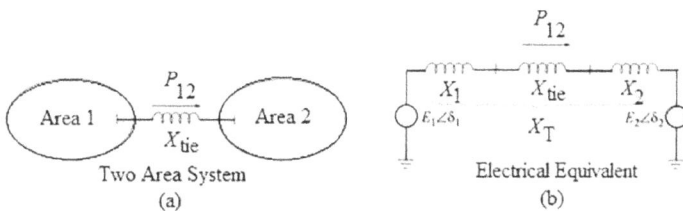

FIGURE 18.4   Two area system and electrical equivalent.

$\Delta P_{m1} - \Delta P_{12} - \Delta P_{L1} = \Delta \omega D_1$ and $\Delta P_{m1} + \Delta P_{12} = \Delta \omega D_2$. Change of the mechanical power is given by $\Delta P_{m1} = -\Delta \omega / R_1 = -\Delta f / R_1$ and $\Delta P_{m2} = -\Delta \omega / R_2 = -\Delta f / R_2$.

A positive ΔP_{12} represents an increase in power transfer from area 1 to area 2. This in effect is equivalent to increasing the load of area 1 and decreasing the load of area 2; therefore feedback ΔP_{12} has a negative sign for area 1 and a positive sign of area 2. The steady-state frequency deviation $(f - f_0)$ is the same for the two areas. For a total load change of ΔP_L,

$$\Delta f = \Delta \omega_1 = \Delta \omega_2 = -\Delta P_L \bigg/ \left[\left(\frac{1}{R_1} + \frac{1}{R_2} \right) + (D_1 + D_2) \right] \tag{18.3}$$

So, $\Delta f \left[(1/R_1) + D_1 \right] = -\Delta P_{12} - \Delta P_{L1}$ and $\Delta f \left[(1/R_2) + D_2 \right] = \Delta P_{12}$ (18.4)

$$\Delta f = \frac{-\Delta P_{L1}}{\left(\dfrac{1}{R_1} + \dfrac{1}{R_2} \right) + (D_1 + D_2)} = \frac{-\Delta P_{L1}}{(\beta_1 + \beta_2)} \quad \text{and} \quad \Delta P_{12} = \frac{-\Delta P_{L1} \left(\dfrac{1}{R_2} + D_2 \right)}{(\beta_1 + \beta_2)} = \frac{-\Delta P_{L1} \beta_2}{(\beta_1 + \beta_2)} \tag{18.5}$$

where β_1 and β_2 are the composite frequency response characteristics of areas 1 and 2, respectively, and are known as frequency bias factors. An increase in area 1 load by ΔP_{L1} results in a frequency reduction in both areas and a tie-line flow of ΔP_{12}. A negative ΔP_{12} is indicative of flow from area 2 to area 1. The tie-line flow deviation reflects the contribution of the regulation characteristics $(1/R + D)$ of one area to another. Similarly, for a change in area 2 load by ΔP_{L2}, we have

$$\Delta f = \frac{-\Delta P_{L2}}{(\beta_1 + \beta_2)} \quad \text{and} \quad \Delta P_{12} = -\Delta P_{21} = \frac{\Delta P_{L2} \beta_1}{(\beta_1 + \beta_2)} \tag{18.6}$$

This relationship establishes the basis of LFC of interconnected system.

The block diagram representation is indicated in Figure 18.5, where each area is represented by an equivalent inertia M, load-damping constant D, turbine, and governing system with an effective speed droop R. The tie line is represented by a synchronizing torque coefficient T.

The LFC of an interconnected power system has two prime considerations: (i) the maintenance of frequency and power exchange over inter area tie lines on scheduled values, with the objective of maintaining the area generation and demand balance by adjusting the outputs of regulating units in response to deviation of frequency tie-line exchange. Area control error is a simple control strategy based upon three conditions in the normal state; (i) the frequency should be kept approximately at the nominal value, (ii) the tie-line flow should be maintained at about schedule and (iii) each area should absorb its own load change, and a change of power in area 1 should be met by increase in generation in both areas and a reduction in frequency. The area control error is the linear combination of the tie-line flow deviation and the frequency deviation. The control strategy uses a signal which is a function of frequency and tie power deviation. When this signal is used for control, frequency and tie power deviations are corrected to zero.

18.4   LEARNING OUTCOME

The transmission system grew from local and regional grids to a large interconnected network that was managed by coordinated operating and planning procedures. Peak demand and energy consumption grew at predictable rates, and technology evolved in a relatively well-defined operational and regulatory environment. With the growing demand the need for generation flexibility comes from the need to control the system frequency, so it is helpful to briefly review frequency control in

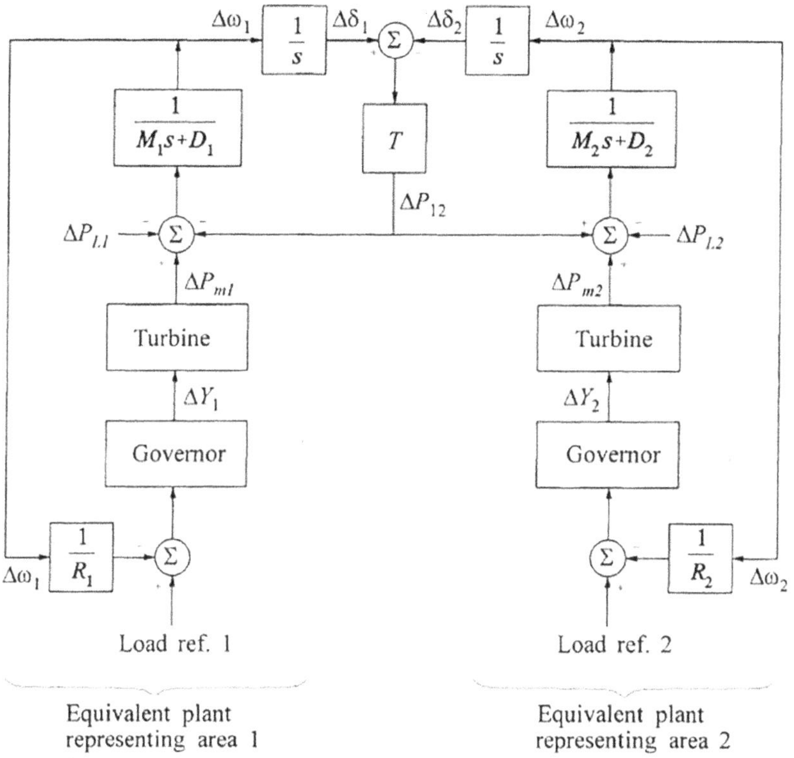

FIGURE 18.5 Two area system block diagram.

the power system. In normal operation, the control area follows a predetermined schedule of power exchange with its neighboring areas. Schedules are prepared based on the forecasted load and are negotiated well in advance; they are an input to system operation in real time. The residual generation/load imbalance is handled by AGC.

In this chapter, readers have been able to learn all the above aspects.

19 Compensation in Power System

19.1 CONCEPT OF REACTIVE POWER

Reactive power has its origin in the phase shift between applied voltage and current waveforms. When a device consumes real power such that the voltage and current waveforms are in phase with each other, the device consumes zero reactive power. When the current defined "into" a device lags the voltage, it consumes reactive power. The amount of reactive power consumed by the device depends on the phase shift between the voltage and current.

19.2 EFFECTS OF REACTIVE POWER FLOW THROUGH TRANSMISSION LINE

Effects of reactive power can be summarized as: (i) loads consume reactive power, so this must be provided by some source; (ii) the generation of reactive power can limit the generation of real power and (iii) the delivery system (transmission lines and transformers) consumes reactive power, so this must be provided by some source (even if the loads do not consume reactive power). However, all transmission lines do provide some reactive power from their shunt line charging which offsets their consumption of reactive power in their series line losses; (iv) the flow of reactive power from the supplies to the sinks causes additional heating of the lines and voltage drops in the network; (v) losses in all power system elements from the power station generator to the utilization devices increase due to reactive power drawn by the loads, thereby reducing transmission efficiency; (vi) due to the reactive power flow in the lines, the voltage drop in the lines increases due to which low voltage exists at the bus near the load and makes voltage regulation poor; (vii) the operating power factor reduces due to reactive power flow in transmission lines; (viii) low power factor due to reactive power flow in line conductors necessitates large-sized conductor to transmit same power when compared to the conductor operating at high power factor; (ix) reactive power in the lines directly affects Kilovolt Ampere (KVA) rating of the system equipment carrying the reactive power and hence the size and cost of the equipment directly; and (x) reactive component of the current prevents the full utilization of the installed capacity of all system elements and hence reduces their power transfer capability.

Still we need reactive power because reactive power (VARs) is required to maintain the voltage to deliver active power (watts) through transmission lines. When there is not enough reactive power, the voltage sags down and it is not possible to deliver the required power to load through the lines. Adequate reactive power control solves power quality problems like flat voltage profile maintenance at all power transmission levels, and improvement of power factor, transmission efficiency and system stability.

19.2.1 Importance of Reactive Power

Importance of reactive power: (i) voltage control in an electrical power system is important for proper operation for electrical power equipment to prevent damage such as overheating of generators and motors, to reduce transmission losses and to maintain the ability of the system to withstand and prevent voltage collapse; (ii) decreasing reactive power causes voltage to fall while increasing it causes voltage to rise. A voltage collapse may occur when the system tries to serve much more load than the voltage can support; (iii) when reactive power supply lowers voltage, as voltage drops, current must increase to maintain power supplied, causing system to consume more reactive power and the voltage drops further. If the current increases too much, transmission lines go offline,

DOI: 10.1201/9781003231240-19

overloading other lines and potentially causing cascading failures; (iv) if the voltage drops too low, some generators will disconnect automatically to protect themselves. Voltage collapse occurs when an increase in load or less generation or transmission facilities causes dropping voltage, which causes further reduction in reactive power from capacitor and line charging, and still further voltage reductions. If voltage reduction continues, this will cause additional elements to trip, leading further reduction in voltage and loss of the load. The result in these entire progressive and uncontrollable declines in voltage is that the system is unable to provide the reactive power required supplying the reactive power demands.

19.3   COMPENSATION

A power network is mostly reactive. A synchronous generator usually generates active power that is specified by the mechanical power input. The reactive power supplied by the generator is dictated by the network and load requirements. A generator usually does not have any control over it. However, the lack of reactive power can cause voltage collapse in a system. It is therefore important to supply/absorb excess reactive power to/from the network. Voltage and reactive power control involves proper coordination among the voltage and reactive power control equipment in the distribution system to obtain an optimum voltage profile and optimum reactive power flows in the system. The way the flow of reactive power is controlled is called compensation.

> **Effects of Compensation** – (i) Reduction in reactive component of circuit current, (ii) Maintenance of voltage profile within limits, (iii) Reduction of copper losses in the system due to reduction of current, (iv) Reduction in investment in the system per KW of load supplied, (v) Decrease in KVA loading of generators, (vi) Improvement of power factor and (vii) Reduction in KVE demand charges for large consumers.
>
> **Load Compensation** – It is the management of reactive power to improve power quality, i.e. voltage profile and power factor. Here the reactive power flow is controlled by installing shunt capacitors and reactors (compensating devices) at the load end bringing about proper balance between generated and consumed reactive power and is the most effective means in improving the power transfer capability of the system and voltage stability.
>
> **Line Compensation** – Shunt capacitors raise the load power factor which greatly increases the power transmitted over the line as it is not required to carry the reactive power. There is a limit to which transmitted power can be increased by shunt compensations as it would require very large capacitor bank, which would be impractical. For increasing power transmitted over the line, other and better means can be adopted through use of series capacitors. The series capacitor compensates the inductive reactance of transmission line and thereby virtually compensates the line length.
>
> **Shunt Compensation** – The ideal shunt compensator is an ideal current source. There are two types of shunt compensation: active and passive. Shunt capacitive compensation is used to improve the power factor. Whenever an inductive load is connected to the transmission line, power factor lags because of lagging load current. To compensate, a shunt capacitor is connected which draws current leading the source voltage. The net result is improvement in power factor. For passive compensation, shunt capacitors are either permanently connected to the system, or switched, and they contribute to voltage control by modifying characteristics of the network. Shunt inductive compensation is used either when charging the transmission line or when there is very low load at the receiving end. Due to very low or no load, very low current flows through the transmission line. Shunt capacitance in the transmission line causes voltage amplification (Ferranti Effect). The receiving end voltage may become double the sending end voltage (generally in case of very long transmission lines). To compensate, shunt inductors are connected across the transmission line. The power transfer capability is thereby increased depending upon the

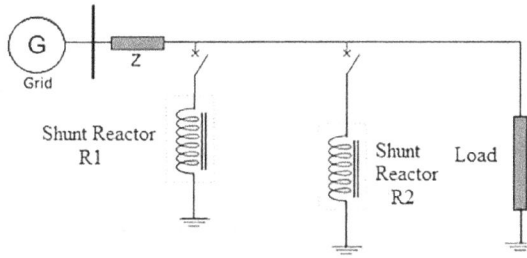

FIGURE 19.1 Connection diagram of shunt reactor.

power equation. Connection diagram of shunt reactor is shown in Figure 19.1. A reactor of sufficient size is permanently connected to the line (Reactor – R1) to limit the overvoltages that also eliminate the switching transient.

Static Shunt Compensation Devices: Static VAR Compensator (SVC) – is a parallel combination of controlled shunt reactor and capacitor to regulate the voltage at a bus quickly and reliably. Typical SVCs include Thyristor-Controlled Reactor (TCR), Thyristor-Switched Reactor (TSR), Thyristor-switched capacitors (TSCs) and Mechanically switched capacitor (MSC). SVC helps in steady-state voltage control, dynamic voltage control during disturbance, reduction of temporary and dynamic overvoltage, improving transient stability and damping of power oscillations.

In TCR, the reactor is connected in series with a bidirectional thyristor switch. The impedance of the device can be continuously changed by varying the conduction angle of thyristors. Figure 19.2 shows a TCR single-phase equivalent circuit in which the shunt reactor is dynamically controlled from a minimum value (practically zero) to a maximum value by means of conduction control of the bidirectional thyristor valves. By this controlled action, the SVC can be seen as a variable shunt reactance and the effective inductive reactance X_L controlled by the thyristor switching. The thyristor conducts on alternate half cycle of the supply frequency depending on the firing angle α_{svc}, which is measured from a zero crossing of voltage.

The instantaneous current supplied by SVCs is given by:

$$i = \begin{cases} \left(\sqrt{2}V \,/\, X_L \right)\left(\cos\alpha_{svc} - \cos\omega t \right) \alpha_{svc} \le \omega t \le \alpha_{svc} + \epsilon \\ 0 \;\; \alpha_{svc} + \epsilon \le \omega t \le \alpha_{svc} + \pi \end{cases}$$

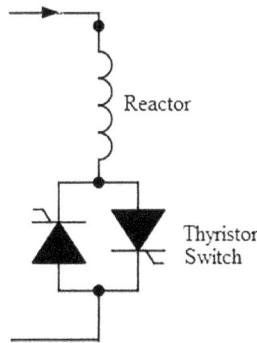

FIGURE 19.2 Thyristor controlled reactor.

V is rms value of SVC voltage at the point of common coupling, X_L is rms value of SVC total inductance, α_{svc} is firing delay angle and ε is SVC conduction angle given by $2(\pi - \alpha_{svc})$. It can be seen that as the delay angle α_{svc} increases, the conduction angle ϵ of the valve decreases. The SVC's control of the output current is based on the control of the firing delay of thyristors. Hence, the maximum injected current is obtained by a firing delay of $90°$ (full conduction). Meanwhile, the firing angle delays between $90°$ and $180°$ electrical degrees only indicate a partial current contribution. This fact contributes to enhancing the device's inductance and makes it possible, at the same time, to decrease its contribution of reactive power and current. With the reactor turned "off" through thyristor switching, the parallel combination will act as a shunt capacitor that will supply reactive power to the system.

The TSR is same as TCR, but thyristor is either in zero- or full-conduction, where equivalent reactance is varied in stepwise manner.

A TSC scheme consists of a capacitor bank split up into approximately sized units, each of which is switched on and off by using thyristor switches. Each single-phase unit consists of a capacitor (C) in series with a bidirectional thyristor switch and a small inductor (L) as shown in Figure 19.3. The purpose of the inductor is to limit switching transients, to damp inrush currents and to prevent resonance with the network. In three-phase applications, the basic units are connected in delta. The switching of capacitors excites transients which may be large or small depending on the resonant frequency of the capacitors with the external system. The thyristor firing controls are designed to minimize the switching transients. This is achieved by choosing the switching instant when the voltage across the thyristor switch is at minimum, ideally zero.

When speed is not the main focus, MSCs are a simple and low-speed solution that provide grid stabilization and voltage control under heavy load conditions, while mechanically switched reactors provide stabilization under low load conditions.

Static Synchronous Compensator (STATCOM) – A STATCOM generally consists of a voltage source converter that is used to convert the DC input voltage to an AC output voltage, a DC capacitor for supply of constant DC voltage of the voltage source converter, an inductive reactance which is a transformer connected between the output of voltage source converter and power system and a harmonic filter, which has operating principle, similar to those of rotating synchronous compensators (i.e. generators), but with relatively faster operation. Since in the STATCOM, the voltage source is created from a DC capacitor, STATCOM has very little active power capability. For dynamic compensation that facilitates enhanced voltage stability by providing reactive power

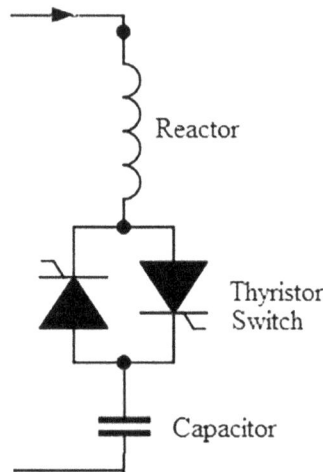

FIGURE 19.3   Thyristor switched capacitor.

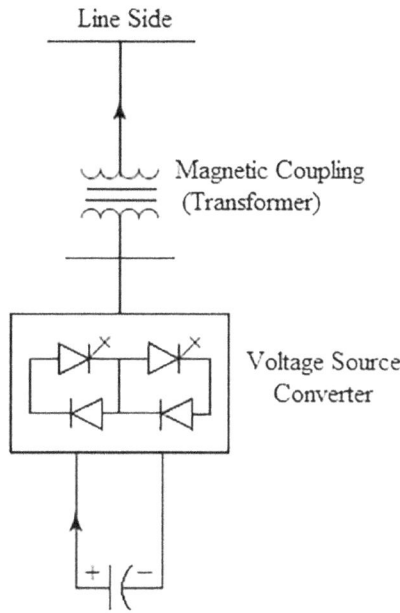

FIGURE 19.4   STATCOM schematic diagram.

support to the power system, STATCOM is preferred over SVCs in view of faster response and requirement of less space. The reactive power compensation provided by STATCOM is more than SVCs because at a low voltage limit, the reactive power drops off as the square of the voltage for the SVCs, but drops off linearly with the STATCOM. STATCOM at reduced voltage can still maximum current, whereas SVC current capability reduces in proportion to voltage. As a result, STATCOM has superior dynamic response, and for comparison, STATCOM may be rated for 75% of SVC rating for same performance in response to line fault. Active voltage control is possible with a STATCOM. This could further help with system stability control. The schematic diagram of STATCOM is shown in Figure 19.4 using voltage source convertor.

The Demerits of Shunt Compensation – (i) The usage of shunt capacitor banks suffers from the following drawbacks that shunt capacitors do not affect current/power factor beyond their point of application; (ii) the reactive power supplied by the shunt capacitor banks is directly proportional to the bus voltage and (iii) when the reactive power required is less on light loads, capacitor bank output will be high. The disadvantage can eliminate a number of capacitors in parallel and then capacitance can be varied by switching ON/OFF depending upon load current.

Series Compensation – In series compensation, capacitors are connected in series with the transmission and distribution lines. This reduces the transfer reactance between buses to which the line is connected, increases the maximum power that can be transmitted and reduces the effective reactive power losses. Although series capacitors are not usually implemented for voltage control, they do contribute to improving the system voltage and reactive power balance. The shunt capacitors, shunt reactors and series capacitors provide passive compensation to the power system, whereas the active compensation is provided by the synchronous condensers, SVC and STATCOM. Series compensation increases transmission capacity. The power transfer capacity of a line is given by $P = \sin\delta$, where V_S is the sending end voltage, V_R is the receiving end voltage, X is the total reactance of the line and δ is the phase angle between V_S and V_R. Power transfer without and with compensation by

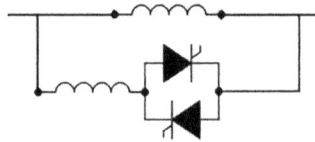

FIGURE 19.5 TCSR schematic diagram.

installing a capacitor of value X_C is $P_{WC} = (V_S V_R / X_L) \sin\delta$ and $P_C = \left[V_S V_R / (X_L - X_C)\right]\sin\delta$. Thus, $P_C / P_{WC} = X_L / (X_L - X_C) = 1/(1 - K)$ where $K = X_C / X_L$. K is the degree of compensation that lies between 40% and 70%. Thus $P_C > P_{WC}$.

Series Capacitor – connected in series with the line.

Thyristor-controlled series reactor (TCSR) is an inductive reactance compensator that consists of a series reactor in parallel with thyristor switched reactor as shown in Figure 19.5.

Thyristor-controlled series capacitor (TCSC) is composed of a series compensating capacitor in parallel with a thyristor controlled reactor as shown in Figure 19.6, which offers continuous control of the transmission-line series compensation level.

Static synchronous series compensator (SSSC) is serially connected instead of shunt. It is able to transfer both active and reactive power to the system, permitting it to compensate for the resistive and reactive voltage drops – maintaining high effective X/R that is independent of the degree of series compensation. The source connected across the DC capacitor provides voltage and compensates for device losses. SSSC can be used to reduce the equivalent line impedance and enhance the active power transfer capability of the line. SSSC can also inject a voltage component, which is of the same magnitude but opposite in phase angle with the voltage developed across the line through the controlled source voltage. Thus, the effect of the voltage drop on power transmission is offset. The schematic diagram of SSSC is shown in Figure 19.7 using voltage source convertor.

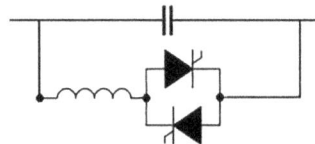

FIGURE 19.6 TCSC schematic diagram.

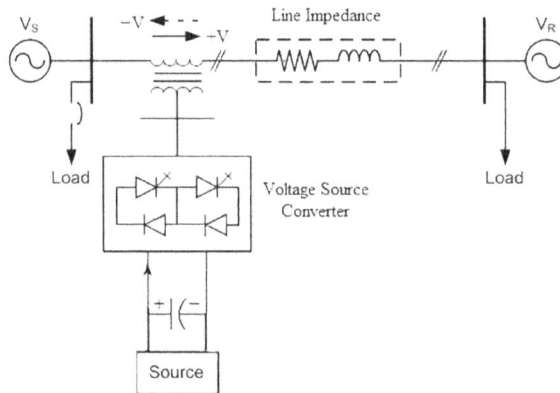

FIGURE 19.7 SSSC schematic diagram.

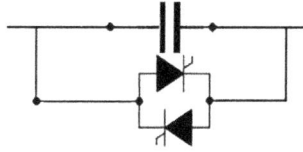

FIGURE 19.8 TSSC schematic diagram.

Thyristor-switched series capacitor (TSSC) is composed of a number of modules of series compensating capacitor in parallel with a thyristor controlled circuit as shown in Figure 19.8. The variable series compensation is highly effective in both controlling power flow in the line and in improving transient stability. The capacitor once inserted into the line will be charged by the line current from zero to maximum during the first half cycle and discharged from maximum to zero during successive half cycle until it is bypassed again.

19.4 POWER FACTOR

Power quality is essential for efficient equipment operation, and power factor contributes to this. Power factor also indicates how efficiently incoming power is used in an electrical installation. The value to which power factor to be improved so as have net annual saving is known as the most economical power factor. Power factor can be improved by installation of capacitors

Consider a consumer is taking a peak load of P Kw at a power factor $\cos\varphi_1$ and charged at a rate of Rs. x per KVA of maximum demand per annum. Suppose the consumer improves power factor $\cos\varphi_2$ by installing power factor correction equipment. Let the expenditure incurred Rs. y per KVAr. KVA maximum demand at $\cos\varphi_1 = P / \cos\varphi_1 = P\sec\varphi_1$. KVA maximum demand at $\cos\varphi_2 = P / \cos\varphi_2 = P\sec\varphi_2$. The power triangle in Figure 19.9 shows apparent power demands on a system before and after adding capacitors.

Actual saving by installing power factor correcting equipment is $xP(\sec\varphi_1 - \sec\varphi_2)$. Reactive power at $\cos\varphi_1 = P\tan\varphi_1$ and at $\cos\varphi_2 = P\tan\varphi_2$. Cost of power factor correction equipment $= yP(\tan\varphi_1 - \tan\varphi_2)$. Then net annual saving $S = xP(\sec\varphi_1 - \sec\varphi_2) - yP(\tan\varphi_1 - \tan\varphi_2)$. If φ_2 is variable, then $dP / d\varphi_2 = 0$; gives $\sin\varphi_2 = y / x$, or $\cos\varphi_2 = \sqrt{1 - (y / x)^2}$.

Effects of Poor Power Factor – (i) poor power factor causes an enhanced value of reactive power; (ii) the generation of reactive power can limit the generation of real power; (iii) the flow of reactive power from the supplies to the sinks causes additional heating of the lines and voltage drops in the network, so enhanced reactive power implies enhanced heating; (iv) losses in all power system elements from the power station generator to the utilization devices increase due to reactive power drawn by the loads, thereby reducing transmission efficiency; (v) due to reactive power flow in the lines, the voltage drop increases due to

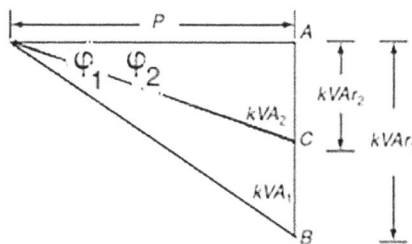

FIGURE 19.9 Power triangle.

which voltage regulation becomes poor; (vi) the low power factor due to reactive power flow in line conductors necessitates large-sized conductor to transmit same power when compared to the conductor operating at high power factor; (vii) the reactive power in the lines directly affects KVA rating of the system equipment carrying the reactive power and hence the size and cost of the equipment directly; (viii) reactive component of the current prevents the full utilization of the installed capacity of all system elements and hence reduces their power transfer capability.

19.5   LEARNING OUTCOME

The voltage instability is reactive power imbalance in the power system network and occurs when there is a sudden and unpredicted increase or decrease in reactive power demand in the system. To stabilize the line voltage, reactive power (VAR) compensation is required, which is control of reactive power to enhance power system network performance.

In this chapter, readers have been able to learn all the above aspects.

Questions & Answers

1. The pulverized fuel system are nowadays universally used in thermal power plants than conventional fuel firing method (stoke firing) due to
   a. high thermal efficiency
   b. better control as per the load demand as the rate of combustion can be controlled easily and immediately
   c. Both a) and b) (*)
   d. None of the above
2. The amount of air which is used to carry the coal and to dry it before entering into combustion chamber
   a. is known as "primary air" (*)
   b. is known as "secondary air"
   c. is known as "forced air"
   d. None of the above
3. The amount of air which is supplied separately for the complete combustion of the coal
   a. is known as "secondary air" (*)
   b. is known as "primary air"
   c. is known as "forced air"
   d. None of the above
4. The fineness of the coal should be such as
   a. 70% of it should pass through 200 mesh sieve and 98% through 50 mesh sieve (*)
   b. 98% of it should pass through 200 mesh sieve and 70% through 50 mesh sieve
   c. 50% of it should pass through 200 mesh sieve and 50% through 50 mesh sieve
   d. None of the above
5. Advantages of Pulverized Coal Firing
   a. Breaking a given mass of coal into smaller pieces exposes more surface area for combustion
   b. Greater surface area of coal per unit mass of the coal allows faster combustion as more coal surface is exposed to heat and oxygen
   c. Wide variety and low grade coal can be burned more easily
   d. All of the above
6. Disadvantages of Pulverized Coal Firing
   a. requires many additional auxiliary equipment and its operation cost is also high compared to stoke firing system
   b. The system produces fly ash
   c. The possibility of the explosion is more as coal burns like gas
   d. All of the above
7. The steam turbines are mainly
   a. impulse turbine
   b. impulse-reaction turbine
   c. Either a) or b)
   d. None of the above
8. The functions of a condenser are
   a. to provide lowest economic heat rejection temperature for steam
   b. to convert exhaust steam to water for reserve thus saving on feed water requirement
   c. to introduce make up water
   d. All of the above

9. The function of the Economizer is
    a. to improve the efficiency of boiler by extracting heat from flue gases to heat water and send it to boiler drum (*)
    b. to improve the efficiency of turbine by extracting heat from exhaust steam to heat water and send it to turbine
    c. Either a) or b)
    d. None of the above
10. Advantages of Economizer
    a. used to save fuel and increase overall efficiency of boiler plant
    b. reducing size of boiler – as the feed water is preheated in the economizer
    c. Either a) or b)
    d. None of the above
11. The function of air preheater
    a. The heat carried out with the flue gases coming out of economizer is utilized for preheating the air before supplying to the combustion chamber
    b. supply of hot air for drying the coal in pulverized fuel systems to facilitate satisfactory combustion of fuel in the furnace
    c. Either a) or b)
    d. None of the above
12. Coal-based thermal power plant works on the principal of
    a. Modified Rankine Cycle
    b. Carnot Cycle
    c. Either a) or b)
    d. None of the above
13. In fire tube boilers
    a. hot gases are passed through the tubes and water surrounds these tubes (*)
    b. hot gases are passed outside the tubes and water inside the tubes
    c. Either a) or b)
    d. None of the above
14. In water tube boilers
    a. hot gases are passed through the tubes and water surrounds these tubes
    b. hot gases are passed outside the tubes and water inside the tubes (*)
    c. Either a) or b)
    d. None of the above
15. Advantage of Water Tube Boiler
    a. Larger heating surface can be achieved by using more number of water tubes
    b. Due to convectional flow, movement of water is much faster than that of fire tube boiler; hence, rate of heat transfer is high which results in higher efficiency
    c. Very high pressure in order of 140 kg/cm^2 can be obtained smoothly
    d. All of the above
16. The difference between the heat of steam per unit weight at the inlet to turbine and the heat of steam per unit weight at the outlet to turbine represents the heat given out in the steam turbine which is converted to mechanical power. The heat drop per unit weight of steam is called
    a. enthalpy drop (*)
    b. entropy drop
    c. heat drop
    d. None of the above

17. Electrostatic precipitator removes dust or other finely divided particles
    a. from flue gases (*)
    b. from steam coming from turbine
    c. from air entering into boiler
    d. None of the above

18. Advantages of Fire Tube Boiler
    a. Compact in construction
    b. Fluctuation of steam demand can be met easily
    c. Cheaper than water tube boiler
    d. All of the above

19. Disadvantages of Fire Tube Boiler
    a. Due to large water the required steam pressure rising time is quite high
    b. Output steam pressure cannot be very high since the water and steam are kept in same vessel
    c. The steam received from fire tube boiler is not very dry
    d. All of the above

20. The heat addition (in the boiler) and rejection (in the condenser) in Rankine Cycle
    a. are isobaric (*)
    b. are isothermal
    c. Either a) or b)
    d. None of the above

21. Proximate Analysis indicates the behavior of coal when it is heated. This is done to ascertain
    a. moisture content (*)
    b. volatile matter and ash content
    c. All the above
    d. None of the above

22. One gram sample of coal is subjected to a temperature of about 1050°C for a period of 1 hour, the loss in weight of the sample gives the
    a. moisture content (*)
    b. volatile matter content
    c. ash content
    d. None of the above

23. One gram sample of coal is subjected to a temperature of about 9500°C in a covered platinum crucible for a period of 7 minutes, the loss in weight of the sample gives the
    a. moisture content
    b. volatile matter content (*)
    c. ash content
    d. None of the above

24. One gram sample of coal is subjected to a temperature of about 7200°C in an uncovered crucible until the coal is completely burned and a constant weight is reached, which is the
    a. moisture content
    b. volatile matter content
    c. ash content (*)
    d. None of the above

25. The ratio of fixed carbon to volatile matter indicates the
    a. rank of coal (*)
    b. moisture content
    c. ash content
    d. None of the above

26. The basic fuel for nuclear reaction is derived from
    a. natural uranium
    b. natural gas
    c. natural coal
    d. None of the above

27. Nuclear fission is initiated when the nucleus of a uranium atom is hit by
    a. a neutron (*)
    b. a proton
    c. an electron
    d. None of the above
    [Nuclear fission is initiated when the nucleus of a uranium atom is hit by a neutron and split into two halves. In addition to splitting the nucleus into two halves, the process releases two or more neutrons and a large amount of energy in the form of heat. The released neutrons in turn hit more uranium nuclei and release more neutrons to produce a chain reaction. The fission process produces a large amount of energy for power generation.]

28. A moderator in the nuclear power plant
    a. slows down the high energy neutrons to low energy neutrons (*)
    b. accelerates the low energy neutrons to high energy neutrons
    c. accelerates the chemical reaction
    d. None of the above

29. Moderators are
    a. ordinary water
    b. graphite
    c. heavy water
    d. All of the above

30. A fissile nuclide is
    a. U235
    b. U234
    c. U238
    d. None of the above

31. U-238 can induce fission after absorbing
    a. a high energy neutron (fast neutron) (*)
    b. a low energy neutron (thermal neutron)
    c. a high/low energy neutron
    d. None of the above

32. In hydroelectric power plant
    a. kinetic energy due to flow of water is used for generating electricity
    b. potential energy due to the height of water is used for generating electricity (*)
    c. Both a) and b) are used
    d. None of the above

33. The type of hydropower turbine selected for a project is based on the
    a. height of standing water – referred to as "head"
    b. the flow, or volume of water, at the site
    c. Both a) and b) (*)
    d. None of the above

34. Reaction turbines are generally used for sites
    a. with lower head and higher flows than compared with the impulse turbines (*)
    b. with lower head and lower flows than compared with the impulse turbines
    c. with higher head and higher flows than compared with the impulse turbines
    d. with higher head and lower flows than compared with the impulse turbines

35. Hydropower stations are found to be
    a. economical choice to meet peak load in the grid due to its unique capabilities of quick starting and stopping (*)
    b. uneconomical choice to meet peak load in the grid due to its unique capabilities of quick starting and stopping
    c. economical choice to meet peak load in the grid due to its incapability of quick starting and stopping
    d. uneconomical choice to meet peak load in the grid due to its incapability of quick starting and stopping

36. In Impulse Turbine
    a. the pressure of liquid does not change while flowing through the rotor of the machine (*)
    b. the pressure of liquid increases while flowing through the rotor of the machine
    c. the pressure of liquid decreases while flowing through the rotor of the machine
    d. None of the above

37. In Reaction Turbine
    a. the pressure of liquid changes while it flows through the rotor of the machine (*)
    b. the pressure of liquid does not change while it flows through the rotor of the machine
    c. the pressure of liquid changes while it flows through the nozzle of the machine
    d. None of the above

38. What is demand factor?
    a. Ratio of connected load to maximum demand
    b. Ratio of average demand to connected load
    c. Ratio of maximum demand to the connected load (*)
    d. None of the above

39. The load factor is
    a. always less than unity (*)
    b. less than or equal to unity
    c. always greater than unity
    d. None of the above

40. Coincidence factor is
    a. the ratio of total maximum demand to the sum of individual maximum demands (*)
    b. the ratio of the sum of individual maximum demands to total maximum demand
    c. ratio of maximum demand to the connected load
    d. None of the above

41. Coincidence factor is reciprocal of
    a. diversity factor (*)
    b. demand factor
    c. capacity factor
    d. None of the above

42. Flat rate tariff is charged on what basis?
    a. Connected load
    b. Units consumed (*)
    c. Maximum demand
    d. None of the above

43. Two part tariff is charged on what basis?
    a. Units consumed
    b. Maximum demand
    c. Both (a) and (b) (*)
    d. None of the above

44. A 25 MVA, 33 KV transformer has a PU impedance of 0.9. The PU impedance at a new base 50 MVA at 11 KV will be
    a.   15.2
    b.   16.2 (*)
    c.   17.2
    d.   None of the above

45. A transmission line has a pu reactance of 30%. If the working voltage is now increased to 110% of its original voltage keeping the MVA rating of the line remaining the same, the pu reactance of the line will now be
    a.   33%
    b.   36.3%
    c.   24.8% (*)
    d.   None of the above

46. If base MVA for a three-phase system is 100 and line-to-line base voltage is 11 kV, then base impedance is
    a.   1.21
    b.   1.30
    c.   1.11
    d.   None of the above (*)

47. Skin effect results in
    a.   reduced effective resistance but increased effective internal reactance of the conductor
    b.   increased effective resistance but reduced effective internal reactance of the conductor (*)
    c.   reduced effective resistance as well as effective internal reactance
    d.   increased effective resistance as well as effective internal reactance

48. Skin depth is given by where f is the frequency, μ, the permeability and σ (= $1/\rho$), the conductivity of the conductor
    a. $\delta = \sqrt{(1/\pi f\mu\sigma)}$ (*)
    b. $\delta = \sqrt{(f\mu\sigma/\pi)}$
    c. $\delta = \sqrt{(0.5/\pi f\mu\sigma)}$
    d.   None of the above

49. Skin depth
    a.   varies as the inverse square root of the conductivity (*)
    b.   varies as the square root of the conductivity
    c.   proportional to conductivity
    d.   None of the above

50. Skin effect
    a.   reduces effective cross section area of the conductor (*)
    b.   increases effective cross section area of the conductor
    c.   does not have any effect on the effective cross section area of the conductor
    d.   None of the above

51. Skin effect
    a.   attenuates the higher frequency components of a signal more than the lower frequency components (*)
    b.   attenuates the lower frequency components of a signal more than the higher frequency components
    c.   attenuates both higher and lower frequency components of a signal
    d.   None of the above

52. The conductors of a 10 km long, single-phase, two-wire transmission line are separated by a distance of 1.5 m. The diameter of each conductor is 1 cm. If the conductors are copper, the inductance of the circuit is
    a. 50.0 mH
    b. 45.3 mH
    c. 23.8 mH (*)
    d. 19.6 mH [L=2×2×10–7 ln (D/r') since two wire]

53. A single-phase two conductor line operates at 50 Hz. The diameter of each conductor is 20 mm and spacing between conductors is 3 m. What is the inductance of each conductor per km?

$$D = 3 \text{ m.}, r = 10 \text{ mm} = 0.01 \text{ m}, r' = e^{-1/4}r = 0.7788 \ r = 0.7788 \times 0.01 = 0.7788 \times 10^{-2} \text{m}$$

$$L = 2 \times 10^{-7} \left(\ln[D/r'] \right) \text{H/m} = 2 \times 10^{-7} \left(\ln\left[3/0.7788 \times 10^{-2} \right] \right) \text{H/m} = 1.19 \text{ mH/km}$$

54. A single-phase two conductor line operates at 50 Hz. The diameter of each conductor is 20 mm and spacing between conductors is 3 m. When conductor material is steel of relative permeability 50, then what will be the loop inductance per km of the line?
    Loop inductance with steel conductors

$$= 2(L_{in} + L_{ext}) = 2\left[0.5 \times 10^{-7} \mu_{r(int)} + 2 \times 10^{-7} \ln D / r \right]$$

$$= 10^{-7} \left[50 + 4\ln(3 / 0.01) \right] = 7.281 \text{ mH/km}$$

55. Conclusion for voltage regulation for short transmission line
    a. When the load power factor is lagging or unity or leading such that $IR\cos\varphi_R > IX\sin\varphi_R$, then the voltage regulation is positive, i.e. $V_R < V_S$
    b. When the load power factor is leading such that $IR\cos\varphi_R < IX\sin\varphi_R$, then the voltage regulation is negative, i.e. $V_R > V_S$
    c. For a given V_R and I, the voltage regulation increases with decrease in p.f. for lagging loads
    d. For a given V_R and I, the voltage regulation decreases with decrease in p.f. for leading loads

56. Constant power locus of a transmission line at a particular sending end and receiving end voltage is
    a. circle (*)
    b. straight line
    c. ellipse
    d. None of the above

57. Condition for maximum regulation is
    a. $\tan\varphi = X/R$
    b. $\tan\varphi = R/X$
    c. $\cos\varphi = X/R$
    d. None of the above

58. A medium line with parameters A, B, C, D is extended by connecting a short line of impedance Z in series. Find the overall A, B, C, D parameters.

For the medium transmission line:

$$V_s = AV_{R1} + BI_{R1} \text{ and } I_s = CV_{R1} + DI_{R1}$$

For the short transmission line: as A = 1, B = Z, C = 0, D = 1; $V_{R1} = V_{R2} + ZI_{R2}$ and $I_{R1} = I_{R2}$

This gives:

$$V_s = AV_{R1} + BI_{R1} = AV_{R2} + AZI_{R2} + BI_{R2} = AV_{R2} + (AZ + B)I_{R2}$$

$$I_s = CV_{R1} + DI_{R1} = CV_{R2} + CZI_{R2} + DI_{R2} = CV_{R2} + (CZ + D)I_{R2}$$

Thus, ABCD parameters are given by $\begin{bmatrix} A & (AZ+B) \\ C & (CZ+D) \end{bmatrix}$

59. What is VSWR of transmission line?

VSWR or Voltage Standing Wave Ratio is defined as the ratio of the maximum voltage to the minimum voltage in standing wave pattern along the length of a transmission line structure.

When a transmission line is terminated with impedance Z_L, that is not equal to the characteristic impedance of the transmission line, Z_0, not all of the incident power is absorbed by the termination. Part of the power is reflected back so that phase addition and subtraction of the incident and reflected waves creates a voltage standing wave pattern on the transmission line.

$$\text{VSWR} = \frac{V_{max}}{V_{min}} = \frac{V_i + V_r}{V_i - V_r}; V_i = \text{incident voltage wave magnitude and}$$

$$V_i = \text{reflected voltage wave magnitude}$$

The reflection coefficient is defined as V_r / V_i. VSWR can have any value from 1 to infinity.

60. What is the need for slack bus?

The slack bus is needed to account for transmission line losses as the total power generated must be equal to sum of power consumed by loads and losses. In a power system, only the generated power and load power are specified for the buses. The slack bus is assumed to generate the power required for losses. Since the losses are unknown, the real and reactive powers are not specified for slack bus. They are estimated through the solution of line flow equations.

61. What is swing bus (slack bus/reference bus)?

A bus is called swing bus when the magnitude and phase of bus voltage are specified for it. The swing bus is the reference bus for load flow solution, and it is required for accounting for the line losses. Usually one of the generator buses is selected as the swing bus.

62. What is the effect of acceleration factor in the load flow solution algorithm?

In load flow solution by iterative methods, the number of iterations can be reduced if the correction voltage at each bus is multiplied by some constant. The multiplication of the constant will increase the amount of correction to bring the voltage closer to the value it is approaching. The multiplier that accomplishes this improved convergence is called acceleration factors.

63. Why do we go for iterative methods to solve load flow problems?

The load (or power) flow equations are nonlinear algebraic equations and so explicit solution is not possible. The solution of nonlinear equations can be obtained only by iterative numerical techniques.

64. What do you mean by a flat voltage start?

In iterative methods of load flow solution, the initial voltage of all buses except slack bus is assumed as $1 + j0$ p.u. This is referred to as flat voltage start.

65. What are the disadvantages of Gauss-Seidel method?

The disadvantages of Gauss-Seidel method are as follows: (i) requires large number of iterations to reach convergence, (ii) not suitable for large systems and (iii) convergence time increases with size of the system.

66. What is Jacobian matrix? How the elements of Jacobian matrix are computed?

The matrix formed from the derivatives of load flow equations is called Jacobian matrix, and it is denoted by J, which is used for solving power flow problem using Newton-Raphson method. The elements of Jacobian matrix will change in each iteration. In each iteration, the elements of the Jacobian matrix are obtained by partially differentiating the load flow equations with respect to unknown variable and then evaluating the first derivatives using the solution of previous iteration.

67. What are the advantages of Newton-Raphson method?

The advantages of Newton-Raphson method are as follows: (i) faster, more reliable and the results are accurate, (ii) requires less number of iterations for convergence, (iii) numbers of iterations are independent of the size of the system and (iv) suitable for large system.

68. What are the disadvantages of Newton-Raphson method?

The disadvantages of Newton-Raphson method are: (i) programming is more complex, (ii) memory requirement is more and (iii) computational time per iteration is higher due to larger number of calculations per iteration.

69. What is the advantage of Fast Decoupled Load Flow (FDLF) analysis in terms of convergence?

In FDLF method, the convergence is geometric, two to five iterations are normally required for practical accuracies and speed for iterations of the FDLF is nearly five times that of NR method or about two-thirds that of the GS method.

70. What is Fast Decoupled Load Flow analysis?

An important characteristic of any practical electric power transmission system operating in steady state is the strong interdependence between real powers and bus voltage angles and between reactive powers and voltage magnitudes. This interesting property of weak coupling between $P - \delta$ and $Q - V$ variables gave the necessary motivation in developing the FDLF, in which $P - \delta$ and $Q - V$ problems are solved separately.

71. The bus admittance (Y_{bus}) is given below and find the buses having shunt elements.

$$j \begin{bmatrix} -5 & 2 & 2.5 & 0 \\ 2 & -10 & 2.5 & 4 \\ 2.5 & 2.5 & -9 & 4 \\ 0 & 4 & 4 & -8 \end{bmatrix}$$

In first row, since $Y_{11} + Y_{12} + Y_{13} + Y_{14} \neq 0$, bus 1 is having shunt element. In second row, since $Y_{21} + Y_{22} + Y_{23} + Y_{24} \neq 0$, bus 2 is having shunt element. In third row, since $Y_{31} + Y_{32} + Y_{33} + Y_{34} = 0$, bus 3 is not having shunt element. In fourth row, since $Y_{41} + Y_{42} + Y_{43} + Y_{44} = 0$, bus 4 is not having shunt element.

72. The elements of Y_{bus} matrix contain
    a. short circuit driving point admittance
    b. short circuit transfer admittance
    c. Both a) and b) (*)
    d. None of the above

73. What are the elements of Z bus matrix?
    a. Open circuit driving point impedance
    b. Open circuit transfer impedances
    c. Both a) and b) (*)
    d. None of the above

74. What are the consideration points of Fast Decoupled Load Flow?
    The consideration points are
    a. Under normal steady state operation, the voltage magnitudes are all nearly equal to 1.0
    b. As the transmission lines are mostly reactive, the conductance is quite small compared to the susceptance $G_{ij} \ll B_{ij}$
    c. Under normal steady-state operation the angular differences among the bus voltages are very small, i.e. $\theta_i - \theta_j \approx 0$

75. What is the difference between isolators and electrical circuit breakers?
    Isolators are mainly for switching purpose under normal conditions, but they cannot operate in fault conditions. Actually they are used for isolating the CBs for maintenance, whereas CB gets activated under fault conditions according to the fault detected.

76. What is meant by trip free feature in the circuit breaker?
    The operating mechanism will start operating for closing operation upon receipt of signal for closing. Meanwhile a fault has taken place and a relay closes the trip circuit of the breaker. The trip free mechanism permits the circuit breaker to be tripped by the protective relay even if it is under the process of closing. In trip free feature, the contacts of the circuit breaker must return to the open position and remain there when an opening operation follows a closing operation, regardless of whether the closing signal, force or action is maintained in the event of a fault in the system.

77. What is meant by first pole to clear factor in the circuit breaker?
    First pole to clear factor of a circuit breaker is the ratio of power frequency recovery voltage across the first pole to clear the arc to the normal phase to ground voltage when all the three poles of breaker are open. It is denoted as k_{PP}. If V_{CB} be the recovery voltage across the first pole of CB and V_{PH} be the phase to ground voltage, then $k_{PP} = V_{CB} / V_{PH}$. The value of k_{pp} depends on the system grounding. For system with non-effectively neutral earthing, $k_{PP} = 1.5$ while for system with effectively neutral earthing, $k_{PP} = 1.3$.

78. What is meant by current chopping in the circuit breaker?
    Current Chopping in circuit breaker is defined as a phenomenon in which current is forcibly interrupted before the natural current zero. Current Chopping is mainly observed in vacuum circuit breaker and Air Blast Circuit Breaker. When a vacuum circuit breaker interrupts current at power frequency, the current rarely comes smoothly to zero. Before current zero is reached, when the instantaneous current is a few amperes, the arc becomes unstable, causing the current to be interrupted abruptly and prematurely. Such a premature interruption of the alternating current before its natural zero is referred to as current chopping.

79. The chopping current in OCB is given by, where $\lambda =$ chopping number and $C_L =$ system capacitance $= 10-50$ nF
    a. $I_0 = \lambda$ (*)
    b. $I_0 = \lambda /$
    c. $I_0 = \sqrt{C_L} / \lambda$
    d. None of the above

80. What is meant by re-ignition and re-strikes?

When the circuit breaker is interrupting a fault, it results in arcing in the interrupting medium. During the process of interruption, the arcing medium is trying to regain its insulation property. For the interruption to be successful, the interrupting medium should withstand this fast rising recovery voltage. Thus, there is a race in the interrupting medium to go from conducting state to insulating state, with the TRV, if the rate of rise of TRV is more than speed with which the medium returns to insulating state the arcing medium breakdown causing current to continue to flow in the circuit breaker, if speed of medium is higher the interruption is successful. This process of establishment of current is called re-ignition and refers to re-ignition of arc in the circuit breaker. Re-ignition generally occurs almost immediately after the current zero and is generally because the arc plasma containing conducting ions reestablishes current.

81. A three-phase alternator has the line voltage 11 kV. The alternator is connected to a circuit breaker. The inductive reactance up to the circuit breaker is 5 Ω / perphase. The distributed capacitance up to circuit breaker between phase and neutral is 0.01 µF. Determine a) peak restriking voltage, b) time for peak restriking voltage, c) frequency of oscillation and d) Max R. R. R. V.

$L = 5 / \omega = 5 / 314 = 0.0159$ mH; $V_L = 11$ kV; $V_P = 11 / \sqrt{3} = 6.35$ kV rms. Hence, $E_m =$. $\sqrt{2}V_{rms} = 9$ kV Peak restriking voltage $= 2 \times E_m = 18$ kV.

Time for peak restriking voltage: $t / \sqrt{LC} = \pi$, or, $t = \pi\sqrt{LC} = 12.6\pi \times 10^{-6} = 39.5$ µs. $f_n = 1 / 2\pi\sqrt{LC} = 1 / 2\pi \times 12.6 \times 10^{-6} = 12{,}637$ c/s. Max R. R. R. V $= t / \sqrt{LC} = E_m / 12.6 \times 10^{-6} = 714$ V/µs

82. What is meant by rated short circuit making current in the circuit breaker?

The rated short circuit making current of a circuit breaker is the peak value of first current loop of short circuit current which the circuit breaker is capable of making at its rated voltage. Rated short circuit making current is 2.55 times rated short circuit breaking current.

83. What is resistance switching?

A deliberate connection of a resistance in parallel with the contact space or arc is called resistance switching for dampening the over voltage transients due to current chopping, capacitive current breaking, etc. The value of resistance r across the contact gap at which the frequency of TRV becomes zero is called Critical Damping Resistance.

The magnitude resistance r for resistance switching is given by $r = 0.5\sqrt{L/C}$.

84. What are the advantages of resistance switching?

(i) to reduce the rate of rise of re striking voltage and the peak value of re-striking voltage, (ii) to reduce the voltage surges due to current chopping and capacitive current breaking and (iii) to ensure even sharing of re-striking voltage transient across the various breaks in multi-break circuit breakers.

85. What is meant high resistance arc quenching in the circuit breaker?

In high resistance method of arc quenching, arc resistance is made to increase with time by lengthening of arc or cooling of arc or reducing cross section of arc or splitting of arc (by letting the arc pass through a narrow opening or by having smaller area of contacts) so that current is reduced to a value insufficient to maintain the arc. Consequently, the current is interrupted or the arc is extinguished. It is employed only in d.c. circuit breakers and low-capacity a.c. circuit breakers. The arc resistance can be increased with the help of arc runners, which are horn-like blades of conducting material or by splitting the arc by arc splitters, which are plates made of resin bonded fiber glass.

86. **What is meant by low resistance arc quenching in the circuit breaker?**

    In low resistance method or zero current method, arc resistance is kept low until current is zero where the arc extinguishes naturally and is prevented from restriking in spite of the rising voltage across the contacts. All modern high power a.c. circuit breakers employ this method for arc extinction. In an a.c. system, current drops to zero after every half-cycle. At every current zero, the arc extinguishes for a brief moment. However, the arc appears again with the rising current wave. At current zero, the space between the contacts is deionized quickly by introducing fresh un-ionized medium. This increases the dielectric strength of the contact space to such an extent that the arc does not continue after current zero.

87. **What is fusing factor?**

    It is the ratio of minimum fusing current to the current rating and is greater than 1.0. The minimum fusing current is the current at which the fuse will melt. The current rating is the rms value of current which the fuse can carry continuously.

88. **What is Preece Equation?**

    W. H. Preece investigated the fusing (melting) current of a wire. The fusing current: $I = a*d^{3/2}$ where I is the fusing current, d is the diameter of the wire in inches, and a Preece's Coefficient for the particular metal in use and is a constant that depends on the material. He determined that $a = 10,244$ for copper.

    Near the fusing threshold, the heat loss (I^2R) and heat generated (πhdl) are approximately equal, where h is the heat loss per unit area from radiation or convection, d is the wire diameter and l is the wire length. So we can set the heat generated equal to the heat dissipation as follows:

    $$I^2R = I^2\left(\rho l/A\right) = I^2 4\rho l/\pi d^2 = \pi hdl \Rightarrow I^2 = \pi^2 d^3 hl/4\rho l = \pi^2 d^3 h/4\rho \Rightarrow I = a*d^{3/2},$$

    where $a = (\pi/2)\sqrt{(h/\rho)}$.

89. **What is Arc-Suppression Coil or Peterson Coil?**

    Arc-Suppression Coil or Peterson Coil is an earthing reactor so designed that its reactance is such that the reactive current to earth under fault conditions balances the capacitance current to earth flowing from the lines so that the earth current at the fault is limited to practically zero.

90. **What is touch potential?**

    The touch potential is the potential difference between a grounded metallic structure and a point on the earth's surface separated by a distance equal to the normal maximum horizontal reach, approximately one meter.

91. **What is step potential?**

    The step potential is the potential difference between two points on the earth's surface, separated by distance of one pace, that will be assumed to be one meter in the direction of maximum potential gradient.

92. **What are effective earthing, non-effective earthing and coefficient of earthing?**

    Effective earthing is also called solid earthing that is without inserting any intentional resistance or reactance with the neutral point and earth. In this case co-efficient of earthing is more than 80%.

    When neutral point to earth connection is made through resistance or reactance, then the system is said to be non-effectively earthed. In this case coefficient of earthing is greater than 80%. High resistance earthing uses a resistance that limits the earth fault current to 10 A. Coefficient of earthing is the ratio of highest rms voltage of healthy phase to earth and line-to-line normal rms voltage.

93. What are unit protection system and non-unit protection system?

A unit protection system is one in which only faults occurring within its protected zone are isolated. Faults occurring elsewhere in the system have no influence on the operation of a unit system. A non-unit system is a protection system which is activated even when the faults are external to its protected zone.

Unit system has absolute discrimination and its zone of protection is absolutely defined. Examples are differential protection, pilot wire and carrier current protection. A non-unit system does not possess absolute discrimination. The discrimination is obtained by time grading, current grading or a combination of time and current grading. Example – distance protection.

94. What is meant by pickup level and reset level of relay?

Pickup level of actuating signal means the value of actuating quantity (voltage or current) which is on threshold above which the relay initiates to be operated. If the value of actuating quantity is increased, the electromagnetic effect of the relay coil is increased, and above a certain level of actuating quantity, the moving mechanism of the relay just starts to move.

Reset level means the value of current or voltage below which a relay opens its contacts and comes in original position.

95. What is meant by plug-setting multiplier of relay?

The plug-setting multiplier, PSM, is defined as: I_{relay}/PS, where I_{relay} is the current through the relay operating coil and PS is the plug-setting of the relay.

96. A 1.0 A relay (i.e. a relay with current coil designed to carry 1.0 A on a continuous basis) whose plug has been set at 0.5 A, i.e. at 50%. If, for a certain fault, the relay current is 5.0 A. What is PSM of the relay?

PSM of the relay is $(5.0/0.5) = 10$, since $I_{relay} = 5.0 = 5.0\,A$ and $PS = 50\% = 0.5$.

97. What is meant by 1.3 seconds normal inverse IDMT relay?

The relay operation time is 1.3 seconds at TMS = 1.0 and PSM = 10.

98. What is meant by Annual System Load Factor?

The Annual System Load Factor is the ratio of the energy availability in the system to the energy that would have been required during the year if the annual peak load met was incident on the system throughout the year. This factor depends on the pattern of utilization of different categories of load.

99. What is meant by Reserve Margin of a system?

Reserve margin of a system is defined as the difference between the Installed Capacity and the peak load met as a percentage of the peak load met. This factor depends on a number of parameters, major ones being the mode of power generation, i.e. hydro, thermal, renewable and the availability of the generating stations which primarily is a function of forced and planned shutdown of the generating units, capacity of the Discoms to procure power.

100. What is ToD (Time of Day) Metering/ToU (Time of Use) Pricing?

ToD/ToU metering is a billing method in which depending on the expected load on the grid, a billing day is divided into several time zones. The duration of each time zone is programmable and the user can define the time zones as per his requirements. The meter records the energy consumed in different time zones in separate registers and exhibits accordingly. Consumption in each of the time zone is charged at different rates. The tariff rates for different time zones are fixed in such a way that a consumer pays more for energy used during peak hours than for off-peak hours. It becomes the responsibility of the consumer to either restrict his energy usage or pay accordingly. This encourages consumers to shift load during cheaper time periods of the day. TOD metering helps consumers to manage their consumption which in turn helps the utility in managing the peak demand.

TOD metering helps in shifting the loads mostly to off-peak hours, resulting in reduction in the peak demand requirement of the utilities by flattening their load curve. It would also improve the financial health of the utilities as the utilities would not have to buy costly power during peak hours.

Hence, TOD metering system is very useful for utilizing the available electrical energy in an optimum way. This also helps the utilities to plan their distribution infrastructure appropriately. This can be implemented in all consumer categories be it domestic, commercial, industrial or even BPL consumers.

Presently, most of the States have implemented this type of metering for industrial and commercial consumers.

101. What is ramp rate of power generation?

Power plant flexibility is recognized as a vital tool to manage variability in electric loads and provide grid support services. One measure of this flexibility is ramp rate – the rate at which a power plant can increase or decrease output. Flexible generating units help provide stability to the electric grid by ramping output up or down as demand and system loads fluctuate. Because solar and wind generation can change within minutes, electric grid operators rely on power plants that can provide additional load (or curtail load) on the same timescale as variations in renewable output. The increase or reduction in output per minute is called the ramp rate and is usually expressed as megawatts per minute (MW/min) or in the percentage of rated load per minute (% P/min). In general, ramp rates greatly depend on the generation technology.

Higher ramp rates allow operator to adjust the net power more rapidly to meet changing demand. However, higher ramping rate normally implies rapid changes in firing temperatures and results in higher thermal stresses of the components. A typical large fossil-fired thermal generator may be able to ramp 1% of its capacity in 1 minute. Smaller units and combustion turbines can typically ramp faster. Hydro units typically have very fast and accurate ramping capability.

102. What is meant by flexibility in power system?

Flexibility expresses the extent to which a power system can modify electricity production or consumption in response to variability, expected or otherwise.

103. What is spinning reserve?

Spinning reserve means part loaded generating capacity with some reserve margin that is synchronized to the system and is ready to provide increased generation at short notice pursuant to dispatch instructions or instantaneously in response to a frequency drop. Alternatively, spinning reserve means the capacity which can be activated on decision of the system operator and which is provided by devices which are synchronized to the network and are able to effect a change in the active power.

104. What is Surge Impedance Loading?

It is the unit power factor load over a resistance line such that series reactive loss (I^2X) along the line is equal to shunt capacitive gain (V^2Y). Under these conditions the sending end and receiving end voltages and current are equal in magnitude but different in phase position.

105. What is CPMC?

It is combined protection, monitoring and control system incorporated in the static system.

106. Belted/unscreened cables do have the advantage that their impulse withstand values
    a.   are generally higher than equivalent screened cables (*)
    b.   are generally lower than equivalent screened cables
    c.   are generally equal to equivalent screened cables
    d.   None of the above

107. XLPE cable has
    a. high dielectric strength
    b. good mechanical strength and non-hygroscopic
    c. greater current carrying capacity and overload and short circuit performance
    d. All the aforesaid properties (*)
108. Conductor screen used in cable is
    a. to maintain a uniform electric field
    b. to minimize electrostatic stresses
    c. Both a) and b)
    d. None of the above
109. A 36 kV, 3 core, 300 mm² Cu conductor cable is to be laid in ambient air temperature of 35°C. The rating is given in manufacturers' tables as 630 A at 25°C and a derating factor of 0.9 is applicable for 35°C operation. What is the cable rating for the 35°C application?
    a. 567 A (*)
    b. 700 A
    c. 630 A
    d. 315A

References

1. T. Basso, "IEEE 1547 and 2030 Standards for Distributed Energy Resources Interconnection and Interoperability with the Electricity Grid", Tech. Rep. NREL/TP-5D00-63157, December 2014.
2. Renewables 2016: Global Status Report, accessed on Mar. 6, 2017. [Online]. Available: http://www.ren21.net/wp-content/ uploads/2016/10/REN21_GSR2016_FullReport_en_11.pdf.
3. Surekha R. Desmukh, G. A. Vaidya, Sanjay Kulkarni – "Availability Based Tariff: A Reliable and Economical Experience of Western Grid", National Power Systems Conference, NPSC, 2004, pp. 522–527.
4. "Electricity Act 2003", Ministry of Power, Government of India, New Delhi, India, June 2003.
5. J. Mohan, D. Nadana Moorthy, B.V. Manikandan – "Reliability Analysis of Grid Using Availability Based Tariff (ABT) Method", 2011 International Conference on Recent Advancements in Electrical, Electronics and Control Engineering.
6. I. J. Nagrath, D. P. Kothari –*Modern Power System Analysis*, Tata McGraw Hill Education Private Limited, Third edition.
7. B. Bhushan, "International Interconnections Based on Experience in South Asia; Indian Perspective", Proceeding of IEEE Power Engineering Society Winter Meeting, 2001, pp. 7–8.
8. S. Mukhopadhyay, "Interconnection of Power Grids in South Asia", Proceeding of IEEE Power Engineering Society General Meeting, 2003, pp. 2184–2185.
9. M V Deshpande – "Elements of Electrical Power Transmission and Distribution Design", P. V. G Prakashan, Poona.
10. John J. Grainger, William D. Stevenson, Jr. –*Power System Analysis*, McGraw-Hill, Inc.
11. J. C. Chan, M. D. Havtley, L. J. Hiivala Alcatel Canada Wire Inc – "Performance Characteristics of XLPE versus EPR as Insulation for High Voltage Cables", *IEEE Electrical Insulation Magazine*, May/June 1993, Vol. 9, No. 3.
12. Michael P. Bahrman – "Overview of HVDC Transmission", 142440178X/06/$20.00 ©2006 IEEE.
13. A.R. Abhyankar, S. A. Khaparde, "Introduction to Deregulation in power industry".
14. Harleen Kaur, Puneet Kuma, Anuja Sharma, Nikhil Kamaiya – "Power System Restructuring & Competitive Wholesale Electricity Markets in Deregulated Environment", International Conference on Innovative Applications of Computational Intelligence on Power, Energy and Controls with their Impact on Humanity (CIPECH14); 28 & 29 November 2014.
15. Yong-Hua Song, Xi-Fan Wang – *Operation of Market-Oriented Power Systems*, Springer.
16. Sonaxi Bhagawan Raikar, Kushal Manoharrao Jagtap – "Role of Deregulation in Power Sector and Its Status in India", 2018 National Power Engineering Conference (NPEC), IEEE.
17. "Power sector reform in India" http://shodhganga.inflibnet.ac.in/bitstream/10603/111787/7/chapter4.pdf.
18. Peng Wang, Yi Ding, Lalit Goel – "Unreliability and Responsibility of Generation and Transmission Companies in Restructured Power Systems with Poolco Market".
19. William W. Hogan – "An Efficient Bilateral Market Needs a Pool", Hearings, August 4, 1994 San Francisco, CA.
20. Mathew J. Morey – "Power Market Auction Design: Rules and Lessons In Market-based Control for the New Electricity Industry", Envision Consulting Alexandria, VA 22314-4813.
21. Hung-po Chao, Robert Wilson – "Design of Wholesale Electricity Markets", Electric Power Research Institute.
22. "Report of Task Force to Review Framework of Point of Connection (PoC) Charges" – March, 2019, Central Electricity Regulatory Commission.
23. "Formulating Pricing Methodology for Inter-State Transmission in India" – February, 2010, Central Electricity Regulatory Commission.
24. Allen J. Wood, Bruce F. Wollenberg –*Power Generation, Operation and Control*, John Wiley & Sons, Inc.
25. Giorgio Gambirasio – "Computation of Loss-of-Load Probability", *IEEE Transactions on Reliability*, April 1976, Vol. R-25, No. 1.
26. Isa S. Qamber – "Loss of Load Probability Effect on Four Power Stations", 2019 International Conference on Fourth Industrial Revolution (ICFIR).

27. P Kundur – *Power System Stability and Control*, McGraw-Hill, Inc.
28. Anoop Kumar Mohanta, Ritesh Dash, Chinmaya Behera, Priya Brata Behera – "Load Frequency Control of a Single Area System: An Experimental Approach", 2015 International Conference on Circuit, Power and Computing Technologies [ICCPCT].
29. Serhat Duman, Nuran Yorukeren, Ismail Hakki Altas, – "Load Frequency Control of a Single Area Power System using Gravitational Search Algorithm", 978-1-4673-1448-0/12/$31.00 ©2012 IEEE.
30. Samah A. Rahim, Serien Ahmed, Mustafa Nawari – "A Study of Load Frequency Control for Two Area Power System Using Two Controllers", 2018 International Conference on Computer, Control, Electrical, and Electronics Engineering (ICCCEEE).
31. H. Saadat – *Power System Analysis*, McGraw-Hill, 1999.
32. A. Sharifi, K. Sabahi, M. Aliyari Shoorehdeli, M.A. Nekoui, M. Teshnehlab – "Load Frequency Control in Interconnected Power System using Multi-Objective PID Controller", 2008 IEEE Conference on Soft Computing in Industrial Applications (SMCia/08), June 25–27, 2008, Muroran, Japan.

Index

For Product Safety Concerns and Information please contact our EU
representative GPSR@taylorandfrancis.com
Taylor & Francis Verlag GmbH, Kaufingerstraße 24, 80331 München, Germany